111.50

ELEMENTS

OF NUMERICAL

ANALYSIS

ELEMENTS

OF NUMERICAL

ANALYSIS

PETER HENRICI

Professor of Mathematics
Eidgenössische Technische Hochschule
Zürich, Switzerland

John Wiley & Sons, Inc. New York · London · Sydney

WILEY INTERNATIONAL EDITION

Library of Congress Catalog Card Number: 64–23840

PRINTED IN THE UNITED STATES OF AMERICA

10 11 12 13 14

ISBN 0 471 37241 2

to George Elmer Forsythe

PREFACE

The present book originated in lecture notes for a course entitled "Numerical Mathematical Analysis" which I have taught repeatedly at the University of California, Los Angeles, both in regular session and in a Summer Institute for Numerical Analysis, sponsored regularly by the National Science Foundation, whose participants are selected college teachers expecting to teach similar courses at their own institutions. The prerequisites for the course are 12 units of analytic geometry and calculus plus 3 units of differential equations.

The teaching of numerical analysis in a mathematics department poses a peculiar problem. At a time when the prime objectives in the instruction of most mathematical disciplines are rigor and logical coherence, many otherwise excellent textbooks in numerical analysis still convey the impression that computation is an art rather than a science, and that every numerical problem requires its own trick for its successful solution. It is thus understandable that many analysts are reluctant to take much interest in the teaching of numerical mathematics. As a consequence these courses are frequently taught by instructors who are not primarily concerned with the mathematical aspects of the subject. Thus little is done to whet the computational appetite of those students who feel attracted to mathematics for the sake of its rich logical structure and clarity, and our schools fail to turn out computer-trained mathematicians in the large numbers demanded by modern science and technology.

Contrary to the view of computation as an art, I have always taken the attitude that numerical analysis is primarily a *mathematical* discipline. Thus I have tried to stress unifying principles rather than tricks, and to establish connections with other branches of mathematical analysis. Indeed, if looked at in this manner, numerical analysis provides the instructor with a wonderful opportunity to strengthen the student's grasp of some of the basic notions of analysis, such as the idea of a sequence, of a limit, of a recurrence relation, or of the concept of a definite integral. To achieve a balance between practical and theoretical content, I have made — for the first time in a numerical analysis text — a clearcut distinction between algorithms and theorems. An algorithm is a computational procedure; a theorem is a statement about what an algorithm does.

In addition to standard material the book features a number of modern algorithms (and corresponding theorems) that are not yet found in most textbooks of numerical analysis, such as: the quotient-difference algorithm, Muller's method, Romberg integration, extrapolation to the limit, sign wave analysis, computation of logarithms by differentiation, and Steffensen iteration for systems of equations. In the last chapter I have given an elementary theory of error propagation that is sufficiently general to cover many algorithms of practical interest.

Another novelty for a textbook in numerical analysis is my attempt to treat the theory of *difference* equations with the same amount of rigor and generality that is usually given to the theory of *differential* equations. Thus this important research tool becomes available for systematic use in a number of contexts. In fact, difference equations form one of the unifying themes of the book.

On the basis of my teaching experience I have found it necessary to include a rather extensive chapter on complex numbers and polynomials. The modern trend of replacing the course in classical theory of equations by a course in linear algebra has had the rather curious consequence that, at the level for which this book is intended, students now are less familiar with the basic properties of the complex field than they used to be.

The book contains about 300 problems of varying computational and analytical difficulty. (Some of the more demanding problems are marked by an asterisk.) In addition, a small number of *research problems* have been stated at the end of some of the chapters. Some of these problems are in the form of non-trivial theoretical assignments requiring library work. The others pose practical questions of some general interest and are intended to stimulate undergraduate research participation. Their solution usually presents a non-trivial challenge in experimental computation.

I have omitted all references to numerical methods in *algebra* and matrix theory, because I feel that this topic is best dealt with in a separate course. As its theoretical foundations are quite different, to treat it with the same attitude that I have tried to adopt towards numerical *analysis* would roughly have doubled the size of the book.

For a similar reason I have omitted material on programming and programming languages. It goes without saying, however, that no curriculum in numerical mathematics can be complete without permitting the student to acquire some experience in actual computation. At the Institute for Numerical Analysis, this experience is acquired in a simultaneous three unit programming course (one unit lectures, two units laboratory). The subject matter in this book is deliberately arranged so that the easy-to-program algorithms occur early, and a course based on it can easily be synchronized with a programming course.

A preliminary version of the book has been used on a trial basis at a number of schools, and I have been fortunate to receive a number of comments and constructive criticisms. In particular I wish to express my gratitude to Christian Andersen, P. J. Eberlein, Gene H. Golub, M. Melkanoff, Duane Pyle, T. N. Robertson, Sydney Spital, J. F. Traub, and Carroll Webber for suggestions which I have been able to incorporate in the final text. I also wish to thank Thomas Bray and Gordon Thomas of the Boeing Scientific Research Laboratories for assistance in planning some of the machine-computed examples. Finally I record my debt to my wife, Eleonore Henrici, for her unflinching help, far beyond the call of conjugal duty, in preparing both the preliminary and the final version of the manuscript.

I dedicate this book to my mentor and former collegue, George E. Forsythe, who had a decisive influence on my view of the whole area of mathematical computation.

Zurich, Switzerland   P. HENRICI
June 1964

CONTENTS

---

* Sections that may be omitted at first reading without essential loss of continuity.

INTRODUCTION

chapter 1 what is numerical analysis?

1.1 Attempt of a Definition

Unlike other terms denoting mathematical disciplines, such as calculus or linear algebra, the exact extent of the discipline called numerical analysis is not yet clearly delineated. Indeed, as recently as twenty years ago this term was still practically unknown. It did not become generally used until the Institute of Numerical Analysis was founded at the University of California in Los Angeles in 1947. But even today there are widely diverging views of the subject. On the one hand, numerical analysis is associated with all those activities which are more commonly known as *data processing*. These activities comprise—to quote only some of the more spectacular examples—such things as the automatic reservation of airline seats, the automatic printing of paychecks and telephone bills, the instantaneous computation of stock market averages, and the evaluation and interpretation of certain medical records such as electroencephalograms. On the other hand, the words "numerical analysis" have connotations of endless arithmetical drudgery performed by mathematical clerks armed with a pencil, a tremendous sheet of paper, and the indispensable eraser.

Between these extreme views we wish to steer a middle course. As far as this volume is concerned, we shall mean by numerical analysis the *theory of constructive methods in mathematical analysis*. The emphasis is on the word "constructive." By a constructive method we mean a procedure that permits us to obtain the solution of a mathematical problem with an arbitrary precision in a finite number of steps that can be performed rationally. (The number of steps may depend on the desired accuracy.) Thus, the mere proof that nonexistence of the solution would lead to a logical contradiction does not represent a constructive method.

A constructive method usually consists of a set of directions for the performance of certain arithmetical or logical operations in predetermined

3

order, much as a recipe directs the housewife to perform certain chemical operations. As with a good cookbook, this set of directions must be complete and unambiguous. A set of directions to perform mathematical operations designed to lead to the solution of a given problem is called an *algorithm*. The word algorithm was originally used primarily to denote procedures that terminate after a finite number of steps. Finite algorithms are suitable mainly for the solution of problems in algebra. The student is likely to be familiar with the following two examples:

1. The Euclidean algorithm for finding the greatest common divisor of two positive integers.
2. The Gaussian algorithm for solving a system of n linear equations with n unknowns.

The problems occurring in analysis, however, usually cannot be solved in a finite number of steps. Unlike the recipes of a cookbook, the algorithms designed for the solution of problems in mathematical analysis thus necessarily consist of an *infinite* (although denumerable) sequence of operations. Only a finite number of these can be performed, of course, in any practical application. The idea is, however, that the accuracy of the answer increases with the number of steps performed. If a sufficient number of steps are performed, the accuracy becomes arbitrarily high.

Problems

  1. Formulate a (finite) algorithm for deciding whether a given positive integer is a prime.
  2. Formulate an algorithm for computing $\sqrt{2}$ to an arbitrary number of decimal places. (The algorithm is necessarily infinite. Why?)

1.2 A Glance at Mathematical History

Confronted with the definition of a constructive method given above, an unspoiled student is likely to ask: Is not all mathematics constructive in the indicated sense?

There indeed was a time when most of the work done in mathematics was not only inspired by concrete questions and problems but also aimed directly at solving these problems in a constructive manner. This was the period of those classical triumphs of mathematics that fill the layman with awe to the present day: The prediction of celestial phenomena such as eclipses of the sun or the moon, or the accurate prophecy of the appearance of a comet. Those predictions were possible, because the solutions of the underlying mathematical problem were not merely shown to *exist*, but were actually *found* by constructive methods.

The high point of this classical algorithmic age was perhaps reached in the work of Leonhard Euler (born 1707 in Basel, died 1781 in St. Petersburg). Euler was possessed of a faith in the all-embracing power of mathematics which today appears almost naive. At the age of twenty, before he had ever seen the ocean, he won a competition of the Paris Academy of Sciences with an essay on the best way to distribute masts on sailing vessels. Innumerable numerical examples are dispersed in the (so far) seventy volumes of his collected works, showing that Euler always kept foremost in his mind the immediate numerical use of his formulas and algorithms. His infinite algorithms frequently appear in the form of series expansions.

After Euler's time, however, the faith in the numerical usefulness of an algorithm appears to have decreased slowly but steadily. While the problems subjected to mathematical investigation increased in scope and generality, mathematicians became interested in questions of the existence of their solutions rather than in their construction. It is true that up to 1900 most existence theorems were proved by methods which we would call constructive; however, the computational demands made by these methods were such as to render absurd the idea of actually carrying through the construction. It is hardly conceivable, for instance, that Emile Picard (1856–1941) ever thought of going through the motions required by his iteration method for solving, say, a nonlinear partial differential equation.

In view of the feeling of algorithmic impotence which must have pervaded the mathematical climate at that time, it is easily understandable that mathematicians were increasingly inclined to use purely logical rather than constructive methods in their proofs. An early significant instance of this is the Bolzano-Weierstrass theorem (see Buck [1959], p. 10) where we are required infinitely many times to decide whether a given set is finite or not. To make this decision even once is, in general, not possible by any constructive method. During the second half of the 19th century the nonconstructive, purely logical trend in mathematics was rapidly picking up momentum. Some main stations of this development are marked by the names of Dedekind (1831–1916), Cantor (1845–1918), Zermelo (1871–1953). In spite of some countertrends inspired by outstanding mathematicians such as Hermann Weyl (1885–1955), the logical point of view appeared to be steering towards an almost absolute victory near the end of the first half of the 20th century. Mathematics finally seemed at the threshold of making true the proud statement of Jacobi (1804–1851) that: "Mathematics serves but the honor of the human spirit."

By a strange coincidence, algorithmic mathematics has been liberated from the vincula of numerical drudgery at the very moment when pure mathematics finally seems to have liberated itself from the last ties of algorithmic thought. From the beginnings of the art of computation to the early 1940's the speed of computation had been increased by a factor of about ten, due to the invention of various computing devices which today seem primitive. Since the early 1940's the speed of computation has increased a millionfold due to the invention of the electronic digital computer. To put into effect even the most complicated algorithms presents no difficulty today. As a consequence, the demand for algorithms of all kinds has increased enormously.

Problem

3. Look up Dedekind's axiom concerning the so-called Dedekind sections (see Taylor [1959], p. 447) and give three examples of its application in elementary calculus.

1.3 Polynomial Equations: An Illustration

To illustrate the types of problems and concepts we wish to deal with, we will consider the problem of solving *polynomial equations*. We are all familiar with the quadratic equation

(1-1) $$x^2 + a_1 x + a_2 = 0,$$

where a_1 and a_2 denote arbitrary real numbers. It is well known that the solutions of this equation are given by the formula

(1-2) $$x = -\tfrac{1}{2}a_1 \pm [(\tfrac{1}{2}a_1)^2 - a_2]^{1/2};$$

in certain cases the computation of the square root leads to complex numbers (see chapter 2). Formula (1-2) states more than the mere existence of a solution of equation (1-1); in fact it indicates an algorithm which permits us to calculate the solution. More precisely, the problem of computing the solution is reduced to the simpler problem of computing a square root. If an instrument for computing roots is available (this could be a table of logarithms, a computing machine especially programmed for the purpose, or the reader of this book armed with the algorithm he developed when solving problem 2), then any quadratic equation can be solved. Is the same also true for equations of higher degree?

Let there be given an arbitrary polynomial of degree n,

(1-3) $$p(x) = x^n + a_1 x^{n-1} + a_2 x^{n-2} + \cdots + a_n;$$

the problem is to find the solutions of the equation $p(x) = 0$. In the golden age of algorithmic mathematics Scipione dal Ferro (1496–1525) and Rafaello Bombelli (*L'Algebra*, 1579) discovered that in the cases $n = 3$ and $n = 4$ there exist algorithms for finding all solutions of equation (1-3) that merely require an instrument for computing roots. Through several centuries attempts were made to solve equations of higher than the fourth degree in a similar manner, but all these efforts remained fruitless. Finally Galois (1811–1832) proved, in a paper written on the eve of his premature death in a duel, that it is not possible in the case $n > 4$ to compute the solutions of equation (1-3) with an instrument that merely calculates roots. This, then, is a typical instance of nonexistence of a certain type of algorithm. Modern numerical analysis, too, knows of such instances of nonexistence of algorithms.

If a problem cannot be solved with an algorithm of a certain type, this does not mean that the problem cannot be solved at all. The problem is merely that of discovering a new algorithm. Today there are no practical limits to mathematical inventiveness in the discovery of new algorithms. It must be admitted, though, that the talent for discovering algorithms is in no way confined to mathematicians. Some of the most effective algorithms of numerical analysis have been discovered by aerodynamicists such as L. Bairstow (1880–   ); by astronomers such as P. L. Seidel (1821–1896); by the meteorologist L. F. Richardson (1881–1953); and by the statistician A. C. Aitken (1895–   ).

The problem of finding the solutions of a polynomial equation of arbitrary degree has attracted mathematicians of many generations such as Newton (1643–1727), Bernoulli (1700–1782), Fourier (1768–1830), Laguerre (1834–1886), and even today significant contributions to the problem are made almost every year. One of the simplest algorithms for solving equation (1-3) is due to Daniel Bernoulli. If a sequence of numbers $z_1, z_2, z_3, \ldots$ is determined by setting $z_k = 0$ for $k < 0$, $z_0 = 1$, and calculating z_k for $k > 0$ by means of the recurrence relation

$$(1\text{-}4) \quad z_k = -a_1 z_{k-1} - a_2 z_{k-2} - \cdots - a_n z_{k-n} \quad (k = 1, 2, \ldots),$$

then the sequence of quotients

$$(1\text{-}5) \qquad\qquad q_k = \frac{z_{k+1}}{z_k}$$

—if certain hypotheses are fulfilled—tends to a solution of equation (1-3) (see chapter 7). Here we have a typical example of an infinite algorithm, for the sequence $\{z_k\}$ never terminates. The recursive nature of Bernoulli's process—the same formula (1-4) is evaluated over and over again—also is typical of many processes of modern numerical analysis.

1.4   How to Describe an Algorithm

An algorithm is not well defined unless there can be no ambiguity whatsoever about the operation to be performed next. Obvious causes for the breakdown of an algorithm are divisions by zero, or (in the real domain) square roots of negative numbers. More generally, an algorithm will always break down if a function is to be evaluated at a point where it is undefined. All such occurrences must be foreseen and avoided in the statement of the algorithm.

EXAMPLES

3.   Let an algorithm be defined as follows: Choose z_0 arbitrarily in the interval $(0, 2)$, and compute $z_1, z_2, \ldots$ by the recurrence relation

$$z_{k+1} = \frac{1}{2 - z_k}.$$

This algorithm is not well defined, since the formula does not make sense when $z_k = 2$, which happens to be the case for instance, if $z_0 = \frac{5}{4}$ and $k = 3$. The algorithm can be turned into a well-defined one by a statement such as the following: "Whenever $z_k = 2$, set $z_{k+1} = 13$."

4.   Bernoulli's algorithm described in §1.3 is well defined to the extent that the right-hand side of equation (1-4) always has a meaning and the sequence $\{z_k\}$ thus always exists. However, it still could be the case that infinitely many z_k's are zero. The corresponding elements q_k would then be undefined. (Consider, for example, the polynomial $p(z) = z^2 - 1$.) Under suitable hypotheses on the polynomial (1-3) it can be shown, however, that the quotients q_k are always defined for k sufficiently large (see chapter 7).

Many simple algorithms can be perfectly well described by means of the conventional symbolism of algebra, supplemented, if necessary, by English sentences. In this manner we shall for instance describe most of the algorithms given in this book. Ordinary algebraic language is not the only language, however, in which an algorithm can be expressed. As a matter of fact, recent experience has shown that for very involved algorithms, especially those occurring in numerical linear algebra, the traditional mathematical language is sometimes grossly inadequate.

Ordinary mathematical language has one further disadvantage: It cannot be understood directly by computing machines without first being "coded." In the interest of breaking down the communication barrier between man and machine as well as between man and man, it becomes very desirable to describe algorithms in a language that can be understood by both.

An algorithmic language with this property, called FORTRAN (= *for*mula *tran*slator), was created around 1957 by the International Business Machines Corporation (IBM). It is very widely used not only on IBM machines, but on other machines as well. FORTRAN is a completely adequate tool (especially in view of some recent refinements) for describing and communicating to the machine the vast majority of algorithms currently used in computational practice. It is strongly suggested that the student of this book familiarize himself with FORTRAN and use it to solve the computational problems posed in subsequent chapters. Excellent introductions to FORTRAN are available (McCracken [1961]).

FORTRAN was designed primarily as a practical research tool. Its great advantages are its simplicity and its wide circulation. From a theoretical point of view, though, the FORTRAN language has some serious shortcomings, and some of its rules appear highly artificial. Realizing this, an international team of computer scientists gathered in 1957 with the aim of creating a universal algorithmic language that would be satisfactory from a theoretical point of view and that would not suffer from any artificial limitations. The result of their efforts was the language known as ALGOL (= *algo*rithmic *l*anguage). A revised version called ALGOL 60 has found wide acceptance, especially in Europe. At the time this is being written, there are not yet many machines in the United States equipped to read ALGOL, although there are reasons to believe that this will change in the future. ALGOL is described in an official document of the Algol committee (Naur [1960]); there are also some less formal, but more readable, introductions to the language (McCracken [1962], Schwarz [1962]).

1.5 Convergence and Stability

Once an algorithm is properly formulated, we wish to know the exact conditions under which the algorithm yields the solution of the problem under consideration. If, as is most commonly the case, the algorithm results in the construction of a sequence of numbers, we wish to know the conditions for *convergence* of this sequence. The practitioner of the art of computation is frequently inclined to judge the performance of an algorithm in a purely pragmatic way: The algorithm has been tried out in a certain number of examples, and it has worked satisfactorily in 95 per cent of all cases.

Mathematicians tend to take a dim view of this type of scientific investigation (although it is basically the standard method of research in such vital disciplines as medicine and biology). It is indeed always desirable to base a statement about the performance of a given algorithm on logic rather than empirical evidence. Such logically provable statements are called *theorems* in mathematics. As we shall see, theorems about the convergence of algorithms can be stated in a good many cases.

EXAMPLE

5. The following is a necessary and sufficient condition for the convergence of Bernoulli's method mentioned in §1.3: Among all zeros of maximum modulus of the polynomial (1-3) there exists exactly one zero of maximum multiplicity (see chapter 2 for an explanation of these concepts). If this condition is fulfilled, the sequence $\{q_n\}$ converges to that zero of maximum modulus and multiplicity.

Once the question of convergence is settled, numerous other questions can be asked about the performance of an algorithm. One might want to know, for instance, *how fast* the algorithm converges. Or one may wish to know something about the size of the error, if the algorithm is artificially terminated after a finite number of steps. The latter question can be interpreted in two ways: Either one wishes to know how big the error is *at most*, or how big the error is *approximately*. The answer to the first question is given by an *error bound*; the answer to the second by an *asymptotic formula*. Mathematicians, who like to think in categories, usually give preference to error bounds. However, an error bound which exceeds the true error by a factor 10^6 is practically useless, whereas an approximate formula, while not representing a guaranteed bound, still can be very useful from a practical point of view. (Our scientists, if they depended on guaranteed error bounds, would never have dared to put a manned satellite in orbit.) Last but not least, the study of the asymptotic behavior frequently reveals information which enables one to *speed up* the convergence of the algorithm. Examples of this will occur in almost every chapter of this volume.

Even the theoretical convergence of an algorithm does not always guarantee that it is practically useful. One more requirement must be met: The algorithm must be *numerically stable*. In high school we learned how to express the result of multiplying two six-digit decimal fractions (which in general has twelve digits) approximately in terms of a six-digit fraction by a process known as *rounding*. Although the resulting fraction is not the exact value of the product, we readily accept the minor inaccuracy in view of the greater manageability of the result. If several multiplications are to be performed in a row, rounding becomes a practical necessity, as it is impossible to handle an ever-increasing number of decimal places. In view of the fact that the individual errors due to rounding are small, we usually assume that the accuracy of the final result is not seriously affected by the individual rounding errors.

Modern electronic digital computers, too, work with a limited number of decimal places. The number of arithmetic operations that can be performed per unit time, however, is about a million times as large as that performed in manual computation. Although the individual rounding errors are still small, their cumulative effect can, in view of the large number of arithmetic operations performed, grow very rapidly like a cancer, and completely invalidate the final result. In order to be sound, an algorithm must remain immune to the accumulation of rounding errors. This immunity is called *numerical stability*.

The concept of numerical stability can be very well illustrated by means of Bernoulli's algorithm discussed in §1.3. As mentioned above, this algorithm will, under certain conditions, yield the zero of maximum modulus of the polynomial (1-3). This algorithm is stable in the sense that it furnishes the zero to the same number of significant digits as are carried in the computation of elements of the sequence $\{z_k\}$. We shall have the opportunity to discuss the following extension of the Bernoulli method: If we form the quantities

$$q_k' = \frac{q_{k+1} - q_k}{q_k - q_{k-1}}$$

then under certain hypotheses (which can be specified) the sequence $\{q_k'\}$ tends to the zero of next smaller modulus of the polynomial p (see chapter 8). We have thus again an instance of a convergent algorithm or, to put it differently, of a logically impeccable mathematical theorem. In spite of all this the algorithm just formulated is practically useless. The quantities q_k' are affected by rounding errors to an extent which completely spoils convergence.

In spite of the above example, the reader should not form the impression that stability is an invariant property which an algorithm either has or does not have. In the last analysis, the stability of an algorithm depends on the computing machine with which it is performed. The modification of Bernoulli's algorithm described above is unstable if performed on an ordinary computing machine but it could be rendered stable on a machine equipped to compute with a variable number of decimal places. In a similar way it is possible to increase the stability of certain algorithms in numerical linear algebra by performing certain crucial operations with increased precision.

Rather than trying to set up absolute standards of stability, it should be the aim of a theory of numerical stability to predict in a quantitative manner the extent of the influence of the accumulation of rounding errors if certain hypotheses are made on the individual rounding errors.   Such a theory will be descriptive rather than categorical.   Ideally, it will predict the outcome of a numerical experiment, much as physical theories predict the outcome of physical experiments.   A particularly useful model of error propagation is obtained if the individual errors are treated as if they were statistically independent random variables.   In chapter 16 we will show how this model can be applied to a large variety of algorithms discussed elsewhere in this volume.

chapter 2 complex numbers and polynomials

One of the recurring topics in the following chapters will be the solution of polynomial equations. Complex numbers are an indispensable tool for any serious work on this problem.

2.1 Algebraic Definition

The reader is undoubtedly aware of the fact that within the realm of real numbers the operations of addition, subtraction, multiplication, and division (except division by zero) can be carried out without restriction. However, the same is not true for the extraction of square roots. The equation

$$(2\text{-}1) \qquad x^2 = a$$

is not always soluble. If $a < 0$, there is no real number x which satisfies equation (2-1), because the square of any real number is always nonnegative.

Historically, complex numbers originated out of the desire to make equation (2-1) always solvable. This was achieved by simply *postulating* the existence of a solution also for negative values of a. Actually, only the solution of one special equation, namely the equation

$$(2\text{-}2) \qquad x^2 = -1$$

has to be postulated. Following Euler, we denote a solution of this equation by i. The symbol i is called imaginary unit; it is a "number" satisfying

$$i^2 = -1.$$

Postulating further that i can be treated as an ordinary number, we can now solve any equation (2-1) with $a < 0$.

EXAMPLE

1. A solution of $x^2 = -25$ is given by $x = i5$, for

$$(i5)^2 = i^2 5^2 = (-1)25 = -25.$$

Another solution is $x = -i5$, for

$$(-i5)^2 = (-1)^2 i^2 5^2 = 1(-1)25 = -25.$$

Not only special equations of the form (2-1), but any quadratic equation of the form (2-3)

(2-3) $x^2 + 2bx + c = 0$

with real coefficients b and c can now be solved. As is well known, the method consists in writing the term on the left in the form of a complete square plus correction term:

$$x^2 + 2bx + c = (x + b)^2 + c - b^2.$$

Equation (2-3) demands that

$$(x + b)^2 = b^2 - c.$$

For $b^2 - c \geq 0$ we obtain the solutions familiar from elementary algebra. If $b^2 - c < 0$, then $c - b^2 > 0$, and our equation is equivalent to

$$(x + b)^2 = -(c - b^2).$$

According to the above, it has the "solutions"

$$x + b = \pm i\sqrt{c - b^2}$$

or

$$x = -b \pm i\sqrt{c - b^2}.$$

EXAMPLE

2. The equation $x^2 + 6x + 25 = 0$ is equivalent to

$$(x + 3)^2 + 25 - 9 = 0$$

or

$$(x + 3)^2 = -16.$$

It therefore has the solutions

$$x = -3 \pm i4.$$

Any expression of the form $a + ib$, where a and b are real, is called a *complex number* and may henceforth be denoted by a single symbol such

as z.   If $z = a + ib$, we call a the *real part* of z and b the *imaginary part*
of z.   In symbols this relationship is expressed as follows:

$$a = \operatorname{Re} z, \qquad b = \operatorname{Im} z.$$

Two complex numbers are considered equal if and only if both their real
and their imaginary parts are equal.

What has been gained by the introduction of complex numbers besides
the ability of solving all quadratic equations in what at the moment may
appear as a rather formalistic manner?   Certainly nothing has been lost,
because all real numbers are contained in the set of complex numbers.
They are simply the complex numbers with imaginary part zero.   Also,
nothing has been spoiled, because we shall see that the ordinary rules of
arithmetic still hold for complex numbers.

To justify this statement, let

$$z_1 = a + ib$$
$$z_2 = c + id$$

be any two complex numbers.   We *define*

(2-4) $$\begin{cases} z_1 + z_2 = a + ib + c + id = a + c + i(b + d) \\ z_1 - z_2 = a + ib - (c + id) = a - c + i(b - d). \end{cases}$$

Two complex numbers are added (subtracted) by adding (subtracting)
their real and imaginary parts.   The role of the "neutral element," i.e.,
of the number ω satisfying $z + \omega = z$ for all z, is played by the complex
number†

$$0 = 0 + i0.$$

Multiplying two complex numbers formally, we obtain

$$z_1 z_2 = (a + ib)(c + id) = ac + i(ad + bc) + i^2 bd.$$

But $i^2 = -1$.   We thus define:

(2-5) $$z_1 z_2 = ac - bd + i(ad + bc).$$

For the special case of real numbers ($b = d = 0$) the above definitions
reduce to the ordinary sum and product of real numbers.   Moreover, the
following rules, familiar from ordinary arithmetic, still hold for any
complex numbers z_1, z_2, z_3:

---

† Two different kinds of zeros are used in this equation.   On the right, we have twice
the real number zero.   On the left, we have the complex number zero, whose real
and imaginary parts are the real number zero.   As no misunderstandings can arise,
one does not try to make a graphical distinction between the two zeros.

The commutative laws

$$z_1 + z_2 = z_2 + z_1, \qquad z_1z_2 = z_2z_1,$$

the associative laws

$$(z_1 + z_2) + z_3 = z_1 + (z_2 + z_3), \qquad z_1(z_2z_3) = (z_1z_2)z_3,$$

and the distributive law

$$z_1(z_2 + z_3) = z_1z_2 + z_1z_3.$$

The proof of these relations is an immediate consequence of our definitions and of the corresponding rules for real numbers.

We have yet to define the *quotient* z_2/z_1 of two complex numbers. The real case shows that some restriction has to be imposed here: The case $z_1 = 0$ will have to be excluded. Is it possible, however, to define z_2/z_1 for all $z_1 \neq 0$? To answer this question, we recall the meaning of the quotient. In the real case we mean by $x = b/a$ the (unique) solution of the equation $ax = b$. Analogously, we understand by $z = z_2/z_1$ the solution of

$$z_1 z = z_2.$$

If $z = x + iy$, this amounts to

$$(a + ib)(x + iy) = c + id,$$

or, after separating real and imaginary parts,

(2-6)
$$\begin{cases} ax - by = c \\ bx + ay = d \end{cases}$$

The two relations (2-6) can be regarded as a system of two linear equations for the two unknowns x and y. This system has a unique solution for arbitrary values of c and d if and only if its determinant is different from zero. The determinant is

$$\begin{vmatrix} a & -b \\ b & a \end{vmatrix} = a^2 + b^2;$$

it will be different from zero if and only if at least one of the numbers a and b is different from zero, that is, if $z_1 \neq 0$. If $z_1 \neq 0$, the solution of equation (2-6) is easily found to be

$$x = \frac{ac + bd}{a^2 + b^2}, \qquad y = \frac{ad - bc}{a^2 + b^2}.$$

We thus define

(2-7)
$$\frac{z_2}{z_1} = \frac{ac + bd}{a^2 + b^2} + i\,\frac{ad - bc}{a^2 + b^2}.$$

The same result could have been obtained by formal manipulation, as follows: Let

$$z = \frac{z_2}{z_1} = \frac{c + id}{a + ib}.$$

In the fraction on the right we multiply both numerator and denominator by $a - ib$. In the denominator we obtain

$$(a + ib)(a - ib) = a^2 + b^2 + i(ba - ab) = a^2 + b^2,$$

and carrying out the multiplication in the numerator we again find equation (2-7). This procedure cannot replace the proof that $z_1 z = z_2$ has the solution (2-7), because it assumes the existence of the solution; once existence has been established, however, it is very useful for the computation of the numerical value of z_2/z_1.

EXAMPLES

3. $\dfrac{1 + i2}{3 + i4} = \dfrac{(1 + i2)(3 - i4)}{(3 + i4)(3 - i4)} = \dfrac{11 + i2}{9 + 16} = \dfrac{11}{25} + i\dfrac{2}{25}.$

4. The reciprocal $z = 1/(a + ib)$ of a complex number $a + ib \neq 0$ is calculated as follows:

$$z = \frac{1}{a + ib} = \frac{a - ib}{(a + ib)(a - ib)} = \frac{a - ib}{a^2 + b^2} = \frac{a}{a^2 + b^2} - i\frac{b}{a^2 + b^2}.$$

5. To express

$$z = \cfrac{1}{i + \cfrac{1}{i + \cfrac{1}{i + 1}}}$$

in the form $a + ib$. Successive applications of the above method of reduction yield

$$\cfrac{1}{i + \cfrac{1}{i + \cfrac{1}{i + 1}}} = \cfrac{1}{i + \cfrac{1}{i + \cfrac{1 - i}{2}}} = \cfrac{1}{i + \cfrac{2}{i + 1}} = \cfrac{1}{i + \cfrac{2(1 - i)}{2}}$$

$$= \frac{1}{i + 1 - i} = 1.$$

Problems

1. Express in the form $a + ib$

(a) $\left(\dfrac{1 + i\sqrt{3}}{2}\right)^2$;

(b) $(1 - i)^4$;

(c) $(\cos\varphi + i\sin\varphi)^2$;

(d) $\dfrac{1}{\cos\varphi - i\sin\varphi}$;

(e) $\dfrac{1}{1 + (7 + 8i)} + \dfrac{1}{1 + \dfrac{1}{7 + 8i}}$.

2. What are the values of z for which the complex number

$$\frac{1}{1 + z} + \frac{1}{1 + 1/z}$$

is undefined?

3. Let $s = 1 + z + z^2 + \cdots + z^n$, where z is a complex number, $z \neq 1$. Find a closed formula for s by forming the expression $s - zs$.

2.2   Geometrical Interpretation

The discussion of §2.1 will have satisfied the student that complex numbers can be manipulated exactly like real numbers. At the moment this result may seem a mere formal curiosity. However, the significance of complex numbers in analysis is based on the fact that the algebraic operations discussed in §2.1 admit simple and beautiful geometric interpretations.

A complex number $z = a + ib$ can be represented geometrically in either of two ways:

(*i*) We can associate with z the point with coordinates (a, b) in the (x, y)-plane. If the points of an (x, y)-plane are thought of as complex numbers, that plane is called *complex plane*. Its x-axis is the locus of complex numbers that are real; it is therefore called the *real axis*. Its y-axis carries the points with real part zero; it is called the *imaginary axis* of the complex plane.

(*ii*) We can associate with a complex number z the two-dimensional vector with components a and b. We think of this vector as a *free* vector, i.e., a vector that may be shifted around arbitrarily as long as its direction remains unchanged.

Both interpretations can be combined into one by attaching the *vector z* to the origin of the plane. It then becomes the radius vector pointing from the origin to the *point z*.

EXAMPLE

6. To the four complex numbers $z = 1$, $z = i$, $z = -1$, $z = -i$ there correspond points (Fig. 2.2c) or radius vectors (Fig. 2.2d).

If complex numbers are interpreted as vectors, addition of two complex numbers amounts to ordinary addition of the corresponding vectors. For indeed, the sum of two vectors is obtained by forming the sum of corresponding components, just as in the case of complex numbers (see Fig. 2.2e).

The difference $z_1 - z_2$ of two complex numbers can be interpreted as the difference of the corresponding vectors. As is well known, the difference of two vectors z_1 and z_2 can be defined by adding to z_1 the vector $-z_2$, i.e., the vector which has the same length as z_2, but direction opposite to z_2. If we attach both vectors z_1 and z_2 to the same point, then the difference $z_1 - z_2$ corresponds to the vector pointing from the head of z_2 to the head of z_1 (see Fig. 2.2f).

Multiplication of a complex number by a real number can be similarly interpreted. Let c be a real number, and let $z = a + ib$. Writing $c = c + i0$, we obtain from the multiplication rule

$$cz = (c + i0)(a + ib) = ca + icb.$$

The vector cz thus has the same direction as z, if $c > 0$, and it has the opposite direction, if $c < 0$. If $c = 0$, cz is the zero vector. The length of cz is always $|c|$ times the length of z.

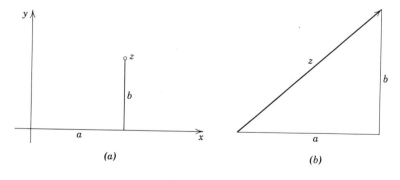

(a)   *(b)*

Figure 2.2

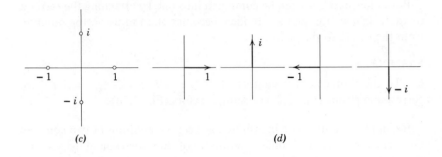

(c)   (d)

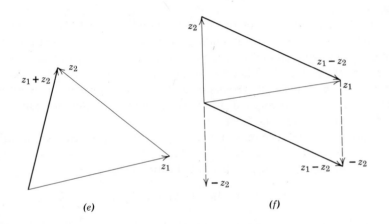

(e)   (f)

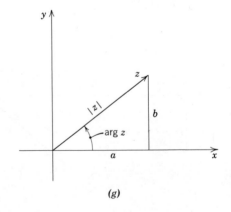

(g)

Figure 2.2

The question of how to interpret the product of two *arbitrary* complex numbers now arises. In order to arrive at a suitable interpretation, we require the notions of the *absolute value* and of the *argument* of a complex number.

The *absolute value* or *modulus* of a complex number $z = a + ib$ is denoted by $|z|$ and is defined as the length of the vector representing z, or equivalently, as the distance from the origin of the point in the complex plane representing z. It follows from Pythagoras' theorem that

$$(2\text{-}8) \qquad\qquad |z| = \sqrt{a^2 + b^2}.$$

The absolute value of a complex number z is zero if and only if $z = 0$. Otherwise the absolute value is positive.

The argument of a complex number is defined only when $z \neq 0$. It is denoted by arg z and denotes any angle subtended by the vector z and the positive real axis. The angle is counted positively if a counterclockwise rotation of the positive real axis would be required to make its direction coincide with the direction of the vector z (see Fig. 2.2g).

The argument of a complex number is not uniquely determined. If φ is a value of arg z, then every angle $\varphi + 2\pi k$ (k an arbitrary integer) is also a possible value of arg z. Thus, the argument of a complex number is determined only† up to multiples of 2π.

To compute an argument φ of $z = a + ib \neq 0$ we proceed as follows. Letting $|z| = r$, we have from figure 2.2g

$$(2\text{-}9) \qquad\qquad a = r \cos \varphi, \qquad b = r \sin \varphi.$$

We can use the first of these equations to determine the absolute value of φ by means of the relations

$$\cos \varphi = \frac{a}{r}, \qquad |\varphi| \leq \pi.$$

If $b \neq 0$, the sign of φ is the same as the sign of b, as follows from the relations

$$\text{sign } \varphi = \text{sign (sin } \varphi) = \text{sign } \frac{b}{r} = \text{sign } b,$$

since $r > 0$. If $b = 0$, then $a = r$ or $a = -r$. In the former case, $\varphi = 0$. In the latter case, $\varphi = \pm \pi$. Since π and $-\pi$ differ by 2π, both values are equally admissible values of arg z.

---

† The more advanced development of the theory of complex numbers shows that it would be unwise to restrict the argument artificially to an interval of length 2π by stipulating a condition such as $-\pi < \varphi \leq \pi$ or $0 \leq \varphi < 2\pi$.

EXAMPLES

7. Let $z = 4 + i3$. From (2-8)

$$|z| = \sqrt{4^2 + 3^2} = \sqrt{25} = 5.$$

To compute $\varphi = \arg z$, we first use the relation

$$\cos \varphi = \tfrac{4}{5} \text{ to find } |\varphi| = 36° \, 52'.$$

Since Im $z = 3 > 0$, it follows that $\varphi = 36° \, 52'$.
8. For $z = -4 - i3$ we likewise find $|z| = 5$, but in view of $\cos \varphi = -\tfrac{4}{5}$ we now have $|\varphi| = 143° \, 8'$ and, since Im $z = -3 < 0$, $\varphi = -143° \, 8'$.
9. The absolute value of a complex number with imaginary part zero coincides with the absolute value (as defined for real numbers) of its real part. Its argument is 0 (or $2\pi k$), if the real part is positive, and π (or $\pi + 2\pi k$), if the real part is negative.

Relation (2-9) shows that every complex number $z \neq 0$ can be represented in a unique manner as the product of a real, positive number and of a complex number with absolute value 1. This representation is given by

(2-10) $z = r(\cos \varphi + i \sin \varphi)$,

where $r = |z|$, and where φ denotes any value of arg z. The representation (2-10) is called the *polar representation* or *polar form* of the complex number z.

The polar representation enables us to give a geometric interpretation of the product and the quotient of two complex numbers. Let

$$z_1 = r_1(\cos \varphi_1 + i \sin \varphi_1)$$
$$z_2 = r_2(\cos \varphi_2 + i \sin \varphi_2)$$

be any two nonzero complex numbers. Multiplying by the algebraic rules of §2.1 we find

$$z_1 z_2 = r_1 r_2[\cos \varphi_1 \cos \varphi_2 - \sin \varphi_1 \sin \varphi_2 + i(\cos \varphi_1 \sin \varphi_2 + \sin \varphi_1 \cos \varphi_2)]$$

or, by virtue of the addition theorems of the trigonometric functions,

$$z_1 z_2 = r_1 r_2[\cos (\varphi_1 + \varphi_2) + i \sin (\varphi_1 + \varphi_2)].$$

This, however, is the polar representation of the complex number with absolute value $r_1 r_2$ and argument $\varphi_1 + \varphi_2$. We thus have proved the relations

(2-11) $|z_1 z_2| = |z_1| |z_2|$, $\arg (z_1 z_2) = \arg z_1 + \arg z_2$.

The second of these is to be interpreted in the following sense: The sum of any two admissible values of $\arg z_1$ and $\arg z_2$ yields an admissible value of $\arg (z_1 z_2)$.

EXAMPLES

10. Let z_2 be a positive real number.   We then have $|z_2| = z_2$, $\varphi_2 = 0$. Multiplication of z_1 by z_2 amounts to a dilatation or contraction of the vector z_1, according to whether $z_2 > 1$ or $0 < z_2 < 1$.

11. If z_2 is a complex number of absolute value 1, multiplication by z_2 amounts to a counterclockwise rotation by $\arg z_2$.   Special case: Multiplication by i amounts to a rotation by $\pi/2$.

It has already been noted that for real numbers complex multiplication reduces to ordinary real multiplication.   Inasmuch the argument of a negative real number may be taken as $\pm \pi$, the rules (2-11) can be thought of as a generalization of the rule "minus $\times$ minus $=$ plus" of high school algebra.

We now compute the quotient of two complex numbers given in polar form.   Using (2-7) we find

$$\frac{z_2}{z_1} = \frac{r_1 r_2 (\cos \varphi_1 \cos \varphi_2 + \sin \varphi_1 \sin \varphi_2)}{r_1^2}$$

$$+ i \frac{r_1 r_2 (\cos \varphi_1 \sin \varphi_2 - \cos \varphi_2 \sin \varphi_1)}{r_1^2}$$

$$= \frac{r_2}{r_1} [\cos (\varphi_2 - \varphi_1) + i \sin (\varphi_2 - \varphi_1)].$$

The expression on the right is the polar form of the complex number with absolute value r_2/r_1 and argument $\varphi_2 - \varphi_1$.   We thus have the formulae

(2-12) $\left| \dfrac{z_2}{z_1} \right| = \dfrac{|z_2|}{|z_1|}$, $\arg \left(\dfrac{z_2}{z_1} \right) = \arg z_2 - \arg z_1.$

An important special case arises when $z_2 = 1$.   Since $|1| = 1$, $\arg 1 = 0$ we then have

(2-13) $\left| \dfrac{1}{z} \right| = \dfrac{1}{|z|}$, $\arg \left(\dfrac{1}{z} \right) = -\arg z.$

The construction of the reciprocal $1/z$ for a given z thus involves two interchangeable steps: (i) Multiply z by $1/|z|^2$; (ii) Reverse the sign of the argument of the number thus obtained.   The second of these steps— reversal of the sign of the argument—geometrically amounts to a reflection of the number on the real axis.   As this operation of reflection occurs frequently in other contexts, too, a special notation has been devised for it.

If $z = a + ib$ is any complex number, reflection on the real axis yields the number $a - ib$. This number is called the complex conjugate of z and is usually denoted by $\bar{z}$. Evidently,

(2-14) $|\bar{z}| = |z|, \quad \arg \bar{z} = -\arg z.$

The real and imaginary parts of a complex number can be expressed in terms of a complex number and its conjugate. By adding and subtracting the relations $z = a + ib$, $\bar{z} = a - ib$ we obtain

$$z + \bar{z} = 2a, \quad z - \bar{z} = 2ib,$$

hence

(2-15) $a = \operatorname{Re} z = \dfrac{z + \bar{z}}{2}, \quad b = \operatorname{Im} z = \dfrac{z - \bar{z}}{2i}.$

From

$$z\bar{z} = (a + ib)(a - ib) = a^2 + b^2 = |z|^2$$

there follows the relation

(2-16) $|z| = \sqrt{z\bar{z}}.$

The following rules of computation are easily verified:

$$\overline{z_1 + z_2} = \bar{z}_1 + \bar{z}_2$$

(2-17)

$$\overline{z_1 z_2} = \bar{z}_1 \bar{z}_2.$$

These rules say, in effect, that in forming the complex conjugate of a sum or a product it is immaterial whether we take the complex conjugate of the result of the algebraic operation, or whether we perform the algebraic operation on the complex conjugate quantities. In the case of the reciprocal we have

$$\frac{1}{\bar{z}} = \overline{\left(\frac{1}{z}\right)}.$$

We shall use the rules just established to give a purely analytical proof of the so-called *triangle inequalities*. These inequalities say that in an arbitrary triangle the length of one edge is at most equal to the sum, and at least equal to the difference, of the lengths of the two other sides. If we represent the sides of the triangle by the vectors z_1, z_2, and $z_1 + z_2$ (see Fig. 2.2e), the triangle inequalities are given by

(2-18) $\big||z_1| - |z_2|\big| \leq |z_1 + z_2| \leq |z_1| + |z_2|.$

To prove this analytically, we note that by (2-16) and (2-17)

$$(2\text{-}19) \qquad |z_1 + z_2|^2 = (z_1 + z_2)\overline{(z_1 + z_2)}$$
$$= (z_1 + z_2)(\bar{z}_1 + \bar{z}_2)$$
$$= z_1\bar{z}_1 + z_1\bar{z}_2 + z_2\bar{z}_1 + z_2\bar{z}_2$$
$$= |z_1|^2 + z_1\bar{z}_2 + z_2\bar{z}_1 + |z_2|^2.$$

In view of $\bar{z}_1 z_2 = \bar{z}_1\bar{\bar{z}}_2 = \overline{z_1\bar{z}_2}$ we have by (2-15)

$$z_1\bar{z}_2 + \bar{z}_1 z_2 = z_1\bar{z}_2 + \overline{z_1\bar{z}_2} = 2 \operatorname{Re} z_1\bar{z}_2.$$

Now for an arbitrary complex number z,

$$|\operatorname{Re} z| \leq |z|.$$

Setting $z = z_1\bar{z}_2$, we thus have by (2-11) and (2-14)

$$|\operatorname{Re} z_1\bar{z}_2| \leq |z_1\bar{z}_2| = |z_1| |\bar{z}_2| = |z_1| |z_2|.$$

Thus there follows from (2-19)

$$|z_1 + z_2|^2 \leq |z_1|^2 + |z_2|^2 + 2|z_1| |z_2|$$
$$= (|z_1| + |z_2|)^2$$

and hence by taking square roots,

$$|z_1 + z_2| \leq |z_1| + |z_2|.$$

On the other hand, since $\operatorname{Re} z \geq -|z|$, we also have

$$|z_1 + z_2|^2 \geq |z_1|^2 + |z_2|^2 - 2|z_1| |z_2|$$
$$= (|z_1| - |z_2|)^2$$

and hence, by taking square roots,

$$|z_1 + z_2| \geq ||z_1| - |z_2||.$$

This completes the proof of (2-18).

Problems

4. Compute the absolute value and all arguments of the complex numbers

$$\text{(a)} \ -i2, \qquad \text{(b)} \ -6 + i8, \qquad \text{(c)} \ \frac{1+i}{2}.$$

5. Let $\omega = -\frac{1}{2} + i(\sqrt{3}/2)$. Show algebraically that $\omega^3 = 1$. What do you conclude about $\arg \omega$? What about $|\omega|$? Verify your conclusions by calculating $|\omega|$ and $\arg \omega$.

6. Show that the arguments φ of the complex number $z = a + ib$ $(a \neq 0)$ are given by

$$\varphi = \text{arc tan } \frac{b}{a} + 2\pi k, \quad \text{if } a > 0,$$

$$\varphi = \text{arc tan } \frac{b}{a} + (2k + 1)\pi, \quad \text{if } a < 0,$$

where k denotes an arbitrary integer, and where arctan x denotes the principal value (lying between $-\pi/2$ and $\pi/2$) of the arcus tangens function. What is arg z if $a = 0$?

7. Using induction, prove that for arbitrary complex numbers $z_1, z_2, \ldots, z_n$

$$|z_1 + z_2 + \cdots + z_n| \leqq |z_1| + |z_2| + \cdots + |z_n|.$$

8. Determine the equation of the locus of those points $z = x + iy$ of the complex plane with the following property: The ratio of the distances of z from the points $+1$ and -1 has the constant value k. (Hint: The distances can be expressed in the form $|z - 1|$ and $|z + 1|$.)

9. What is i^n for arbitrary integral n?

2.3   Powers and Roots

Applying the multiplication formula (2-11) to a product of n equal factors $z = r(\cos \varphi + i \sin \varphi)$, we obtain

(2-20) $[r(\cos \varphi + i \sin \varphi)]^n = r^n(\cos n\varphi + \sin n\varphi).$

Setting $r = 1$, there results Moivre's formula

(2-21) $(\cos \varphi + i \sin \varphi)^n = \cos n\varphi + \sin n\varphi.$

We expand the expression on the left by the binomial theorem and observe that $i^2 = -1, i^3 = -i, i^4 = 1, \ldots$ Equating real and imaginary parts on both sides of (2-21), we obtain

$$\cos^n \varphi - \binom{n}{2} \cos^{n-2} \varphi \sin^2 \varphi + \binom{n}{4} \cos^{n-4} \varphi \sin^4 \varphi - \cdots = \cos n\varphi,$$

$$\binom{n}{1} \cos^{n-1} \varphi \sin \varphi - \binom{n}{3} \cos^{n-3} \varphi \sin^3 \varphi + \cdots = \sin n\varphi.$$

Here the symbol $\binom{n}{k}$ denotes the *binomial coefficient* defined by

$$\binom{n}{k} = \frac{n!}{k!(n-k)!}, \quad k = 0, 1, \ldots, n.$$

EXAMPLE

12. For $n = 3$ we obtain

$$\cos 3\varphi = \cos^3 \varphi - \binom{3}{2} \cos \varphi \sin^2 \varphi = \cos^3 \varphi - 3 \cos \varphi \sin^2 \varphi,$$

$$\sin 3\varphi = \binom{3}{1} \cos^2 \varphi \sin \varphi - \binom{3}{3} \sin^3 \varphi = 3 \cos^2 \varphi \sin \varphi - \sin^3 \varphi.$$

We are now ready to tackle the problem of computing the nth roots of a complex number. In the real domain we call nth root of a real number a any number x satisfying the equation

$$x^n = a.$$

Analogously we shall call nth root of the complex number w any number $z = x + iy$ satisfying the equation

(2-22) $z^n = w.$

If $w = 0$ this clearly has the only solution $z = 0$. In order to determine all possible solutions z for $w \neq 0$, we write

$$w = \rho(\cos \alpha + i \sin \alpha)$$

and seek to determine the polar form $z = r(\cos \varphi + i \sin \varphi)$. By (2-20),

$$z^n = r^n(\cos n\varphi + i \sin n\varphi),$$

and condition (2-22) assumes the form

(2-23) $r^n(\cos n\varphi + i \sin n\varphi) = \rho(\cos \alpha + i \sin \alpha).$

This condition evidently will be satisfied if

$$r^n = \rho, \qquad \text{i.e., } r = \sqrt[n]{\rho}$$

and

$$n\varphi = \alpha, \qquad \text{i.e., } \varphi = \frac{\alpha}{n}.$$

Thus the number

$$z = \sqrt[n]{\rho} \left(\cos \frac{\alpha}{n} + i \sin \frac{\alpha}{n} \right)$$

certainly is an nth root of the complex number $\rho(\cos \alpha + i \sin \alpha)$. Is it the only one? The absolute value of a nonzero complex number being positive, $\sqrt[n]{\rho}$ evidently is the only possible absolute value of z. However, other values of the argument are possible. We recall that the argument of a complex number is determined only up to multiples of 2π. Thus condition (2-23) is also satisfied if

$$n\varphi = \alpha + k2\pi,$$

where k is an arbitrary integer. We thus obtain an infinity of possible values of φ:

$$\varphi = \frac{\alpha}{n} + \frac{k2\pi}{n}, \qquad k = 0, \pm 1, \pm 2, \ldots .$$

Not all these values yield different solutions of (2-23), however. The same solution z is defined by two values of φ that differ merely by an integral multiple of 2π. This will occur as soon as the corresponding values of k differ by an integral multiple of n. Different solutions of (2-23) thus are obtained only by selecting the values $k = 0, 1, \ldots, n - 1$. We summarize this result as follows:

> **Theorem 2.3** The equation $z^n = w$, where $w \neq 0$ and n is a positive integer, has exactly n solutions. They are given by
>
> $$z = r(\cos \varphi + i \sin \varphi),$$
>
> where
>
> $$r = \sqrt[n]{|w|} \quad \text{and} \quad \varphi = \frac{\arg w}{n} + k\frac{2\pi}{n}, \qquad k = 0, 1, 2, \ldots, n - 1.$$

Geometrically, these solutions all lie on a circle of radius r. The argument of one solution is $1/n$ times the argument (*any* argument) of w; the remaining solutions divide the circumference of the circle in n equal parts.

EXAMPLES

13. To compute all solutions of $z^4 = i$. Since $|i| = 1$, all solutions have absolute value one. In view of $\arg i = \pi/2$, one solution has the argument $\pi/8$; the remaining solutions divide the unit circle into four equal parts (see Fig. 2.3).

14. In the real domain the number 1 has either one or two nth roots, according to whether n is odd or even. In the complex domain the number $1 = 1 + i0$ must have n nth roots. Since $\sqrt[n]{1} = 1$, all roots have absolute value 1. Since $\arg 1 = 0$, the arguments of these n nth roots of unity are given by $2\pi k/n$, $k = 0, 1, \ldots, n - 1$. Special example: For $n = 3$ we obtain the following three third roots of unity:

$$\cos \frac{2\pi 0}{3} + i \sin \frac{2\pi 0}{3} = 1$$

$$\cos \frac{2\pi}{3} + i \sin \frac{2\pi}{3} = -\frac{1}{2} + i\frac{\sqrt{3}}{2}$$

$$\cos \frac{4\pi}{3} + i \sin \frac{4\pi}{3} = -\frac{1}{2} - i\frac{\sqrt{3}}{2}.$$

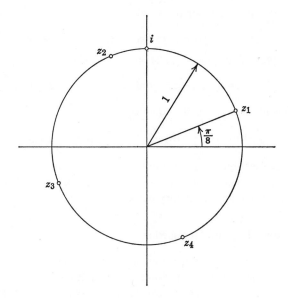

Figure 2.3

For n even one root of unity (namely the one obtained for $k = n/2$) has the argument

$$\frac{2\pi}{n} \frac{n}{2} = \pi,$$

and thus is real and negative, i.e., we obtain two real nth roots of unity. For n odd the equation

$$\frac{2\pi}{n} k = \pi$$

is not satisfied for any integer k, thus there is no negative root of unity. The example shows how by introducing complex numbers cumbersome distinctions of special cases can be avoided or dealt with from a unified point of view.

Problems

10. Determine all solutions of the equation

$$1 + z + z^2 + \cdots + z^n = 0.$$

(Use problem 3.)

11. Find closed expressions for the sums

$$A_n = 1 + r \cos \varphi + r^2 \cos 2\varphi + \cdots + r^n \cos n\varphi,$$

$$B_n = r \sin \varphi + r^2 \sin 2\varphi + \cdots + r^n \sin n\varphi.$$

(Hint: Let $z = r(\cos \varphi + i \sin \varphi)$ and consider $A_n + iB_n$.)

12. Determine all solutions of the equations

    (a) $z^4 = -4$,   (b) $z^6 = i$,   (c) $z^3 = 3\sqrt{3}$.

13. Express the function $\cos 4\varphi$ as a polynomial in $\cos \varphi$.

14. Express the two values of $\sqrt{x + iy}$ in the form $a + ib$. (Hint: Square and compare real and imaginary parts.)

2.4   The Complex Exponential Function

Beginning with this section we shall study certain important *functions* of a complex variable. In the real domain, a function is defined if to every real number of a certain set of real numbers (called the *domain* of the function) there is made to correspond (for instance, by an algebraic formula) a number of another set (called the *range* of the function). A complex function is defined in the same manner, with the difference that now both domain and range of the function are, in general, sets of complex numbers.

We begin by defining the exponential function for complex values of the variable. In the real domain the exponential function can be defined by the exponential series,

$$e^x = 1 + \frac{x}{1!} + \frac{x^2}{2!} + \frac{x^3}{3!} + \cdots.$$

Replacing x by iy, where y is real, we obtain formally

$$e^{iy} = 1 + \frac{iy}{1!} + \frac{(iy)^2}{2!} + \frac{(iy)^3}{3!} + \cdots.$$

Using the fact that $i^2 = -1$, $i^3 = -i$, $i^4 = 1, \ldots$, this may be written

$$e^{iy} = 1 + \frac{iy}{1!} - \frac{y^2}{2!} - \frac{iy^3}{3!} + \cdots,$$

or, collecting real and imaginary terms,

$$e^{iy} = 1 - \frac{y^2}{2!} + \frac{y^4}{4!} - \cdots + i\left(\frac{y}{1!} - \frac{y^3}{3!} + \frac{y^5}{3!} - \cdots\right).$$

The two series on the right are recognized as the Maclaurin expansions of $\cos y$ and $\sin y$. Thus we obtain

(2-24) $$e^{iy} = \cos y + i \sin y.$$

The above computations are purely formal in the sense that they lack analytic justification. (It has not been defined, for instance, what is meant by the sum of an infinite series with complex terms, nor do we know whether it is permissible to rearrange the terms in such a series.) However, nothing keeps us from adopting equation (2-24) as the *definition* of the exponential function e^z for purely imaginary values of z. It will soon be obvious that this definition is reasonable for a number of reasons.

We notice, first of all, that the polar form of a complex number $z \neq 0$, where $r = |z|$ and $\varphi = \arg z$, can be written thus:

$$z = re^{i\varphi}.$$

We also note the following properties of the function e^{iy}, which are analogous to properties of the real exponential function:

(a) $$e^{i0} = \cos 0 + i \sin 0 = 1 + i0 = 1.$$

For the one number which is real as well as purely imaginary, the new definition (2-24) is compatible with the real definition.

(b) The real exponential function satisfies the so-called addition theorem

$$e^{x_1}e^{x_2} = e^{x_1 + x_2}$$

for arbitrary real numbers x_1 and x_2. In analogy, the following identity holds for arbitrary real y_1 and y_2:

(2-25) $$e^{iy_1}e^{iy_2} = e^{i(y_1 + y_2)}.$$

The proof follows from the multiplication rule (2-11) in view of the fact that e^{iy} is a complex number with absolute value 1 and argument y.

The following two properties of the complex exponential function have no direct analog in the realm of real numbers.

(c) We have $e^{2\pi i} = \cos 2\pi + i \sin 2\pi$ and hence

$$e^{2\pi i} = 1.$$

(This formula connects several important numbers of analysis.)

(d) It follows from (2-25) and (c) that

$$e^{i(y + 2\pi)} = e^{iy}e^{2\pi i} = e^{iy}$$

for arbitrary y. This relation expresses the fact that the exponential function is *periodic* with period $2\pi i$.

Above, we have defined e^{iy} in terms of $\cos y$ and $\sin y$. We now shall show how to express the trigonometric functions in terms of the exponential function. Replacing in (2-24) y by $-y$, we get

(2-26) $e^{-iy} = \cos(-y) + i\sin(-y) = \cos y - i\sin y.$

Adding and subtracting (2-24) and (2-26) and solving for $\cos y$ and $\sin y$, we obtain Euler's formulas

(2-27) $\cos y = \dfrac{e^{iy} + e^{-iy}}{2}, \qquad \sin y = \dfrac{e^{iy} - e^{-iy}}{2i}.$

EXAMPLE

15. Moivre's formulas enable us to express $\cos n\varphi$ and $\sin n\varphi$ in terms of powers of $\cos\varphi$ and $\sin\varphi$. Here we are concerned with the converse problem: To express a power of $\cos\varphi$ or $\sin\varphi$ as a linear combination of the functions $\cos\varphi$, $\cos 2\varphi$, ..., $\sin\varphi$, $\sin 2\varphi$, ... We explain the method by considering the function $\cos^4\varphi$. By Euler's formula,

$$(\cos\varphi)^4 = \left(\frac{e^{i\varphi} + e^{-i\varphi}}{2}\right)^4$$

or, using the binomial theorem, since $(e^{i\varphi})^k = e^{ik\varphi}$,

$$(\cos\varphi)^4 = \tfrac{1}{16}(e^{4i\varphi} + 4e^{2i\varphi} + 6 + 4e^{-2i\varphi} + e^{-4i\varphi}).$$

Collecting the terms with the same absolute value of the exponents and applying Euler's formula again, we find

$$(\cos\varphi)^4 = \tfrac{1}{16}(e^{4i\varphi} + e^{-4i\varphi}) + \tfrac{1}{4}(e^{2i\varphi} + e^{-2i\varphi}) + \tfrac{3}{8}$$
$$= \tfrac{1}{8}\cos 4\varphi + \tfrac{1}{2}\cos 2\varphi + \tfrac{3}{8}.$$

There remains the problem of defining e^z for an arbitrary complex number $z = x + iy$. We adopt the following definition:

(2-28) $e^{x+iy} = e^x e^{iy} = e^x(\cos y + i\sin y).$

With this definition, the addition theorem

(2-29) $e^{z_1} e^{z_2} = e^{z_1 + z_2}$

holds for arbitrary complex numbers $z_1 = x_1 + iy_1$ and $z_2 = x_2 + iy_2$. Proof:

$$\begin{aligned}
e^{z_1} e^{z_2} &= e^{x_1} e^{x_2} e^{iy_1} e^{iy_2} \quad &\text{(by (2-28))}\\
&= e^{x_1 + x_2} e^{i(y_1 + y_2)} \quad &\text{(by (2-25))}\\
&= e^{x_1 + x_2 + i(y_1 + y_2)} \quad &\text{(by (2-28))}\\
&= e^{z_1 + z_2}.
\end{aligned}$$

As a consequence of the addition theorem we find that for any positive integer n

$$(e^z)^n = e^{nz};$$

furthermore, setting $z_1 = z$ and $z_2 = -z$ and observing property (a) we obtain

$$e^z e^{-z} = e^{z-z} = e^0 = 1.$$

Consequently the relation

$$e^{-z} = \frac{1}{e^z},$$

familiar for real values of z, holds true also for arbitrary complex values. It shows, among other things, that the complex exponential function never assumes the value zero.

Problems

15. Express $\sin^4 \varphi \cos^3 \varphi$ in terms of the functions $\cos \varphi$, $\cos 2\varphi$, ...

16. Prove that

$$\int_0^{2\pi} (\cos \varphi)^{2n} \, d\varphi = 2\pi \frac{\binom{2n}{n}}{2^{2n}}.$$

17. The logarithm $y = \log x$ of a real, positive number x is defined as the unique solution of the equation $e^y = x$. How would you define the logarithm of a complex number w? Does your definition lead to a unique value of $\log w$? What are the possible value(s) of $\log(-1)$?

18. If t is regarded as the time, the point $z = z(t) = Re^{it}$ travels on a circle of radius R. What is the locus of the points

(a) $z = ae^{it} + be^{-it}$ $(a, b$ real$)$;

(b) $z = \cos 2t + i \cos t$ $(0 \leq t \leq \pi)$?

19. Prove that $e^{\bar{z}} = \overline{e^z}$ for all complex z.

2.5 Polynomials

Let $a_0, a_1, a_2, \ldots, a_n$ be $n + 1$ arbitrary complex numbers, $a_0 \neq 0$. The complex-valued function p defined by

$$(2\text{-}30) \qquad p(z) = a_0 z^n + a_1 z^{n-1} + a_2 z^{n-2} + \cdots + a_n$$

is called a *polynomial* of degree n. The constants $a_0, \ldots, a_n$ are called the *coefficients* of the polynomial. The number a_0 is called the *leading* coefficient. A polynomial is called *real* if all its coefficients are real.

Any real or complex number z for which

$$p(z) = 0$$

is called a *zero* of the polynomial p. We know that a real polynomial does not necessarily have real zeros; it suffices to recall the case of polynomials of degree 2. In the complex case, however, the situation is different.

Theorem 2.5a   Every polynomial of degree $n \geq 1$ has at least one zero.

This is the so-called fundamental theorem of algebra, which was first stated and proved (in five different ways) by C. F. Gauss (1777–1855). The proofs of Gauss were either algebraically involved, or they used advanced methods. Today the modern tools of analysis make it possible to give a relatively simple proof that is accessible to any student of advanced calculus (see e.g., Landau [1951], p. 233). However, a presentation of this proof would lead us too far astray from the main themes of this book and it is therefore omitted.

Taking the fundamental theorem for granted, we now discuss several of its applications.

Theorem 2.5b   Let p be a polynomial of degree $n \geq 1$, and let $p(z_1) = 0$. Then there exists a polynomial p_1 of degree $n - 1$ such that

$$p(z) = (z - z_1)p_1(z)$$

identically in z.

Proof.   Let the polynomial p be given by (2-30). In view of $p(z_1) = 0$ we have

$$p(z) = p(z) - p(z_1)$$
$$= a_0(z^n - z_1^n) + a_1(z^{n-1} - z_1^{n-1}) + \cdots + a_{n-1}(z - z_1).$$

We now make use of the identities

$$z^k - z_1^k = (z - z_1)(z^{k-1} + z^{k-2}z_1 + z^{k-3}z_1^2 + \cdots + z_1^{k-1})$$

$(k = 1, 2, \ldots, n)$. Factoring out $z - z_1$, we obtain

$$p(z) = (z - z_1)[a_0(z^{n-1} + z^{n-2}z_1 + \cdots + zz_1^{n-2} + z_1^{n-1})$$
$$+ a_1(z^{n-2} + z^{n-3}z_1 + \cdots + z_1^{n-2})$$
$$+ \cdots$$
$$+ a_{n-2}(z + z_1) + a_{n-1}].$$

Calling the expression in brackets $p_1(z)$, we find upon rearranging that

$$p_1(z) = a_0 z^{n-1} + (a_0 z_1 + a_1) z^{n-2} + \cdots$$
$$+ (a_0 z_1^{n-1} + a_1 z_1^{n-2} + \cdots + a_{n-1});$$

p_1 thus is again a polynomial; since $a_0 \neq 0$, the degree of p_1 is $n - 1$. This completes the proof of theorem 2.5b.

Let now p be a polynomial of degree $n > 1$. By the fundamental theorem, it has at least one zero, z_1 say. By theorem 2.5b, p can thus be represented in the form $(z - z_1)p_1(z)$, where p_1 is of degree $n - 1 \geq 1$. Again by the fundamental theorem, p_1 has at least one zero, say z_2, and thus is representable in the form $(z - z_2)p_2(z)$, where p_2 is a polynomial of degree $n - 2$. The process evidently can be continued until we arrive at a polynomial p_n of degree zero. Since the leading coefficient of every p_k is a_0, it follows that $p_n(z) = a_0 \neq 0$. We thus find the following representation for the polynomial p:

$$(2\text{-}31) \qquad p(z) = (z - z_1)(z - z_2)\ldots(z - z_n)a_0.$$

We thus have represented $p(z)$ as a product of polynomials of degree 1, or *linear* polynomials, of the form $z - z_k$. The numbers z_k $(k = 1, 2, \ldots, n)$ evidently all are zeros of p. Any number different from all the z_k's cannot be a zero, since a product of nonzero complex numbers never vanishes. It is not asserted that the numbers z_k are all different. However, we have proved the following: *A polynomial of degree $n \geq 1$ has, at most, n distinct zeros.*

An important consequence of this statement is as follows: *If two polynomials p and q, both known to have degrees $\leq n$, assume the same values at $n + 1$ points, then they are identical.* For the proof, consider the polynomial $p - q$. Its degree is at most n, and yet its value is zero at the $n + 1$ points where p and q agree. This can only be if $p - q$ vanishes identically.

Above, we have in the relation (2-31) represented a polynomial of degree n as a product of linear factors. It is conceivable that this representation depends on the order in which the zeros $z_1, z_2, \ldots, z_n$ have been split off. However, we now shall show that *for a given polynomial p the representation (2-31) is unique* (apart, obviously, from the order in which the factors appear).

Our assertion is trivial if p has n distinct zeros, for in this case each zero must appear in the representation (2-31), and since each must occur, each can occur only once. It is quite possible, however, that a polynomial of degree $n > 1$ has less than n distinct zeros, and that as a consequence the numbers $z_1, z_2, \ldots, z_n$ are not all different from each other.

EXAMPLE

16. The polynomial $p(z) = z^4 - 4z^3 + 6z^2 - 4z + 1$ has the representation

$$p(z) = (z - 1)^4 = (z - 1)(z - 1)(z - 1)(z - 1).$$

There is only one zero, $z = 1$; in the representation (2-31) we have $z_1 = z_2 = z_3 = z_4 = 1$.

In order to prove the uniqueness of the representation (2-31) even in the case of repeated zeros, assume p has another representation of the same form, say

(2-32) $$p(z) = (z - z_1')(z - z_2')\ldots(z - z_n')b_0.$$

From equations (2-31) and (2-32) we obtain the identity

$$(z - z_1)(z - z_2)\ldots(z - z_n)a_0 = (z - z_1')(z - z_2')\ldots(z - z_n')b_0.$$

The expression on the left is zero for $z = z_1$. Hence the expression on the right, too, must vanish when $z = z_1$. This is possible only if one of the numbers z_k', say z_1', equals z_1. Cancelling the factor $z - z_1$, we obtain

$$(z - z_2)\ldots(z - z_n)a_0 = (z - z_2')\ldots(z - z_n')b_0.$$

Because we have divided by $z - z_1$, this identity is at the moment proved for $z \neq z_1$ only; however, since both sides are polynomials (of degree $n - 1$), and since the set of all points $\neq z_1$ comprises more than $n - 1$ points, the identity must in fact hold for all z. Now we can continue as above: One of the numbers $z_2', \ldots, z_n'$, say z_2', must be equal to z_2 (even though z_2 could be the same as z_1). Splitting off the factor $z - z_2'$ and continuing in the same vein, we find that for a suitable numbering of the z_k'

$$z_k = z_k', \qquad k = 1, 2, \ldots, n.$$

We thus have proved:

Theorem 2.5c   A polynomial $p(z) = a_0 z^n + a_1 z^{n-1} + \cdots + a_n$ of degree n can be represented in a unique manner (up to the order of the factors) in the form

$$p(z) = (z - z_1)(z - z_2)\ldots(z - z_n)a_0.$$

Each zero of p occurs at least once among the numbers $z_1, z_2, \ldots, z_n$.

If a zero z of p occurs precisely k times among the numbers $z_1, z_2, \ldots, z_n$, the zero is said to have *multiplicity* k. A zero of multiplicity one is also called a *simple* zero. The zeros alone do not completely determine a polynomial, not even up to a constant factor, but the zeros together with their multiplicities do.

EXAMPLE

17. The two third degree polynomials

$$p(z) = (z - 1)(z + 1)^2 = z^3 + z^2 - z - 1$$

$$q(z) = (z - 1)^2(z + 1) = z^3 - z^2 - z + 1$$

both have the zeros $z = 1$ and $z = -1$, and no other zeros. In the case of p, $z = 1$ has multiplicity 1 and $z = -1$ has multiplicity 2. In the case of q the multiplicities are reversed.

Multiplying out the factors in (2-31) and comparing the coefficients with those of (2-30) we obtain relations between the numbers z_k and the coefficients of the polynomial. These relations are known as Vieta's formulas. The simplest of these formulas are those obtained by comparing the coefficients of z^{n-1} and by comparing the constant terms:

(2-33)
$$z_1 z_2 \ldots z_n = (-1)^n \frac{a_n}{a_0}$$

(2-34)
$$z_1 + z_2 + \cdots + z_n = -\frac{a_1}{a_0}.$$

The fundamental theorem of algebra guarantees the *existence* of at least one zero of any polynomial of positive degree, and thus indirectly of the representation (2-31) in terms of linear factors. A completely different problem is the actual *computation* of these zeros. As pointed out already in §1.3, explicit formulas (involving root operations) can be given only if the degree does not exceed four, and are frequently impractical already when the degree is three. For polynomials of degree > 4 the zeros can be found, in general, only by algorithmic techniques. However, for some special polynomials of arbitrary degree the zeros can be found explicitly.

EXAMPLES

18. Let $p(z) = z^n - 1$. According to §2.3 the zeros of this polynomial are the nth roots of unity. Since $a_0 = 1$, the representation in terms of linear factors is thus given by

$$z^n - 1 = (z - 1)(z - e^{i\varphi})(z - e^{2i\varphi}) \ldots (z - e^{(n-1)i\varphi}),$$

where $\varphi = 2\pi/n$.

19.   To determine the representation of the polynomial

$$p(z) = z^n + z^{n-1} + z^{n-2} + \cdots + 1$$

in terms of linear factors.   We find

$$zp(z) = z^{n+1} + z^n + z^{n-1} + \cdots + z,$$

and upon subtracting $p(z)$ and dividing by $z - 1$,

$$p(z) = \frac{z^{n+1} - 1}{z - 1} \qquad (z \neq 1).$$

The zeros of p thus are the $(n + 1)$st roots of unity, with the exception of $z = 1$.   For instance for $n = 3$,

$$1 + z + z^2 + z^3 = (z - i)(z + 1)(z + i).$$

   The above theory refers to arbitrary polynomials with complex coefficients.   Real polynomials are contained in this class as a special case. We now shall state two theorems that are true only for real polynomials.

   Theorem 2.5d   If z is a zero of the real polynomial p, then $\bar{z}$ is a zero also.

Proof.   Let

$$p(z) = a_0 z^n + a_1 z^{n-1} + \cdots + a_n$$

be the given real polynomial.   If

$$0 = a_0 z^n + a_1 z^{n-1} + \cdots + a_n,$$

then by repeated applications of the relations (2-17)

$$\begin{aligned}
0 = \bar{0} &= \overline{a_0 z^n + a_1 z^{n-1} + \cdots + a_n} \\
&= \overline{a_0 z^n} + \overline{a_1 z^{n-1}} + \cdots + \overline{a_n} \\
&= \overline{a_0}\,\overline{z^n} + \overline{a_1}\,\overline{z^{n-1}} + \cdots + \overline{a_n}.
\end{aligned}$$

However, since the a_k are real, $\overline{a_k} = a_k$ $(k = 0, 1, \ldots, n)$; furthermore, $\overline{z^k} = (\bar{z})^k$.   We thus have

$$0 = a_0 \bar{z}^n + a_1 \bar{z}^{n-1} + \cdots + a_n,$$

i.e., the number $\bar{z}$ is a zero of p.

   In addition to the statement of theorem 2.5d, it is also true that the multiplicities of z and $\bar{z}$ are the same.   For a proof see problem 26 in §2.6.   As a consequence, we find that in the factored representation of a real polynomial each factor $z - z_k$ involving a nonreal z_k can be grouped with a factor $z - z_{k+1}$ where $z_{k+1} = \overline{z_k}$. Multiplying out such a pair of complex conjugate factors we obtain

$$(z - z_k)(z - \overline{z_k}) = z^2 - (z_k + \overline{z_k})z + z_k\overline{z_k},$$

a quadratic polynomial with the real coefficients $-(z_k + \overline{z_k}) = -2\,\mathrm{Re}\,z_k$ and $z_k\overline{z_k} = |z_k|^2$. We thus have obtained:

Theorem 2.5e Any real polynomial of degree $n \geq 1$ can be represented in a unique manner as a product of real linear factors and of real quadratic factors with complex conjugate zeros.

EXAMPLES

20.
$$z^3 + z^2 + z + 1 = (z - i)(z + 1)(z + i)$$
$$= (z^2 + 1)(z + 1).$$

21. Let

$$p(z) = z^6 - 1 = (z - 1)\left(z - \frac{1 + i\sqrt{3}}{2}\right)\left(z - \frac{-1 + i\sqrt{3}}{2}\right)$$
$$\times (z + 1)\left(z - \frac{-1 - i\sqrt{3}}{2}\right)\left(z - \frac{1 - i\sqrt{3}}{2}\right).$$

The representation by real factors is

$$p(z) = (z - 1)(z^2 - z + 1)(z^2 + z + 1)(z + 1).$$

Problems

20. Prove: A real polynomial of odd degree has at least one real zero.

21. Applying Vieta's formulas to $z^n - 1$, prove that

$$1 + \cos\frac{2\pi}{n} + \cos\left(2\,\frac{2\pi}{n}\right) + \cdots + \cos\left((n-1)\frac{2\pi}{n}\right) = 0,$$

$$\sin\frac{2\pi}{n} + \sin\left(2\,\frac{2\pi}{n}\right) + \cdots + \sin\left((n-1)\frac{2\pi}{n}\right) = 0.$$

22. Find all zeros of the following polynomials: (a) $z^4 + 6z^2 + 25$, (b) $z^4 - 6z^2 + 25$, (c) $z^8 + 14z^4 + 625$. (Use problem 14.)

23. The real polynomial $p(z) = z^4 + a_1z^3 + a_2z^2 + a_3z + a_4$ is known to have the zeros $1 + i$ and $-1 - i$. Determine the coefficients $a_1, \ldots, a_4$.

2.6 Multiplicity and Derivative

In this section we wish to point out the following connection between the multiplicity of a zero of a polynomial p and the values of the derivatives of p at the zero: If the multiplicity of a zero z of a polynomial p is k, then

(2-35)
$$\begin{cases} p(z) = p'(z) = \cdots = p^{(k-1)}(z) = 0, \\ p^{(k)}(z) \neq 0. \end{cases}$$

The multiplicity of a zero of a polynomial thus can be determined without completely factoring the polynomial simply by evaluating the derivatives of the polynomial at the zero.

For real zeros of real polynomials the relations (2-35) are a straightforward consequence of elementary differentiation rules of calculus. If x_1 is a real zero of multiplicity k, we need only to observe that by theorem 2.5c we can write

(2-36)
$$p(x) = (x - x_1)^k q(x),$$

where q is a polynomial such that $q(x_1) \neq 0$. Differentiating (2-36) by means of the Leibnitz formula for differentiation of a product (see Kaplan [1953], p. 19), we get precisely the relations (2-35).

For complex zeros and complex polynomials, however, there arises first of all the question of what is meant by the derivatives $p', p'', \ldots$. (After all, the variables considered in the definition of the derivative in calculus are always real.)   It is possible to extend the limit definition of the derivative to the complex domain in such a manner that the calculus proof of (2-35) retains its validity.   However, some of the issues involved in the theory of complex differentiation are fairly sophisticated, and their discussion is best left to a course in complex analysis.   Fortunately, as far as polynomials are concerned, it is possible to discuss derivatives in a purely algebraic manner, without considering limits.

If p is a polynomial of degree n with real or complex coefficients,

(2-37)
$$p(z) = a_0 z^n + a_1 z^{n-1} + \cdots + a_n,$$

we *define* the derivative of first order of p to be the polynomial of degree $n - 1$,

(2-38) $p'(z) = na_0 z^{n-1} + (n - 1)a_1 z^{n-2} + \cdots + a_{n-1}.$

Derivatives of higher orders are defined recursively, for instance

$p''(z) = (p')'(z) = n(n - 1)a_0 z^{n-2} + (n - 1)(n - 2)a_1 z^{n-3} + \cdots + 2a_{n-2}.$

EXAMPLES†

22.   Using binomial coefficients, the kth derivative of the polynomial (2-37) can be expressed as follows:

$$p^{(k)}(z) = k! \left[\binom{n}{k} a_0 z^{n-k} + \binom{n-1}{k} a_1 z^{n-k-1} + \cdots + \binom{k}{k} a_{n-k} \right].$$

---

† Readers unfamiliar with binomial coefficients should at this point consult §3.5 for the definition and basic properties of these coefficients.

23. If p denotes the polynomial (2-37), then clearly

$$p^{(n)}(z) = n!a_0.$$

24. We wish to calculate the kth derivative of the special polynomial

$$p(z) = (z - a)^n$$

$$= z^n - \binom{n}{1}az^{n-1} + \binom{n}{2}a^2z^{n-2} - \cdots + (-1)^n\binom{n}{n}a^n.$$

By example 22, we have

$$p^{(k)}(z) = k!\left[\binom{n}{k}z^{n-k} - \binom{n-1}{k}\binom{n}{1}az^{n-k-1}\right.$$

$$\left. + \binom{n-2}{k}\binom{n}{2}a^2z^{n-k-2} - \cdots + (-1)^{n-k}\binom{k}{k}\binom{n}{n-k}a^{n-k}\right].$$

The products of binomial coefficients occurring here can be recombined as follows:

$$\binom{n-m}{k}\binom{n}{m} = \frac{(n-m)!}{k!(n-m-k)!}\frac{n!}{m!(n-m)!}$$

$$= \frac{n!}{k!(n-k)!}\frac{(n-k)!}{m!(n-k-m)!}$$

$$= \binom{n}{k}\binom{n-k}{m}.$$

Thus we have

$$p^{(k)}(z) = k!\binom{n}{k}\left[\binom{n-k}{0}z^{n-k} - \binom{n-k}{1}az^{n-k-1}\right.$$

$$\left. + \binom{n-k}{2}a^2z^{n-k-2} - \cdots + (-1)^{n-k}\binom{n-k}{n-k}a^{n-k}\right].$$

By the binomial theorem, the expression in brackets reduces to $(z - a)^{n-k}$. We thus find

$$p^{(k)}(z) = k!\binom{n}{k}(z - a)^{n-k}.$$

This result is familiar, of course, when z and a are real.

It is clear that the derivative defined by equation (2-38) satisfies some of the laws that we expect from real calculus. Thus, if c is an arbitrary constant, we have

$$(cp)' = cp';$$

furthermore, if p and q are any two polynomials, then

$$(p + q)' = p' + q'.$$

It is somewhat less obvious that the familiar rule for the differentiation of a product also holds in the complex case:

(2-39) $$(pq)' = p'q + pq'.$$

In order to prove (2-39), let p be given by (2-37) and q by

$$q(z) = b_0 z^m + b_1 z^{m-1} + \cdots + b_m.$$

We agree to set $a_k = 0$ for $k > n$ and $b_k = 0$ for $k > m$. The product of the two polynomials then can be written

$$p(z)q(z) = c_0 z^{n+m} + c_1 z^{n+m-1} + \cdots + c_{m+n},$$

where

$$c_k = a_0 b_k + a_1 b_{k-1} + \cdots + a_{k-1} b_1 + a_k b_0,$$

$k = 0, 1, 2, \ldots, m + n$. The coefficient of $z^{m+n-1-k}$ in the derivative of pq then is $(m + n - k)c_k$. If, on the other hand, we form the expression $p'q + pq'$ directly, we find

$$p'(z)q(z) + p(z)q'(z)$$

$$= (na_0 z^{n-1} + (n - 1)a_1 z^{n-2} + \cdots + a_{n-1})(b_0 z^m + b_1 z^{m-1} + \cdots + b_m)$$

$$+ (a_0 z^n + a_1 z^{n-1} + \cdots + a_n)(mb_0 z^{m-1} + (m-1)b_1 z^{m-2} + \cdots + b_{m-1}).$$

Collecting the coefficients of all terms combining to $z^{m+n-1-k}$, we find

$$na_0 b_k + (n - 1)a_1 b_{k-1} + \cdots + (n - k)a_k b_0$$
$$+ a_0(m - k)b_k + a_1(m - k + 1)b_{k-1} + \cdots + a_k mb_0$$
$$= (n + m - k)(a_0 b_k + a_1 b_{k-1} + \cdots + a_k b_0).$$

Since this agrees with $(m + n - k)c_k$, our assertion is proved.

As in real calculus, we now obtain by induction from (2-39) the Leibnitz formula for the kth derivative of a product of two polynomials:

(2-40) $$(pq)^{(k)} = \binom{k}{0} p^{(k)} q + \binom{k}{1} p^{(k-1)} q' + \cdots + \binom{k}{k} pq^{(k)}.$$

We now are ready to prove:

Theorem 2.6 Let p be a polynomial of positive degree, and let $z = z_1$ be a zero of multiplicity m of p. Then

$$p(z_1) = p'(z_1) = \cdots = p^{(m-1)}(z_1) = 0, \qquad p^{(m)}(z_1) \neq 0.$$

Proof. By theorem 2.5c, the polynomial p can be written in the form

(2-41) $$p(z) = (z - z_1)^m q(z),$$

where q is a polynomial such that $q(z_1) \neq 0$. We now form the kth derivative of (2-41) by the Leibnitz formula (2-40). By example 24 we find

$$p^{(k)}(z) = \binom{k}{0} k! \binom{m}{k} (z - z_1)^{m-k} q(z)$$

$$+ \binom{k}{1} (k-1)! \binom{m}{k-1} (z - z_1)^{m-k+1} q'(z)$$

$$+ \cdots$$

$$+ \binom{k}{k} (z - z_1)^m q^{(k)}(z).$$

Here we set $z = z_1$ and distinguish two cases. If $k < m$, each term contains a positive power of $z - z_1$ and thus vanishes for $z = z_1$. If $k = m$, all terms except the first contain positive powers of $z - z_1$. The first term reduces to a nonzero constant times $q(z)$, which is $\neq 0$ for $z = z_1$. Thus theorem 2.6 is proved.

Problems

24 The polynomial

$$p(z) = z^5 - z^4 - 8z^3 + 20z^2 - 17z + 5$$

has the zero $z = 1$. Determine its multiplicity.

25. What are the multiplicities of the zeros at $z = \pm i$ of the polynomial

$$p(z) = z^5 + z^4 + 2z^3 + 2z^2 + z + 1 ?$$

26. Prove: Complex conjugate zeros of real polynomials have equal multiplicities. (Hint: The derivatives of a real polynomial are real polynomials. Now use the theorems 2.5d and 2.6.)

27. Using the definition (2-38) of the derivative, prove Taylor's theorem for arbitrary polynomials. That is, show that if p is a polynomial of degree n, then for arbitrary z and h

$$p(z + h) = p(z) + \frac{h}{1!} p'(z) + \frac{h^2}{2!} p''(z) + \cdots + \frac{h^n}{n!} p^{(n)}(z).$$

(Use the representation of example 22 for the kth derivative of p.)

Recommended Reading

A more thorough treatment of complex numbers and polynomials will be found in Birkhoff and MacLane [1953].

chapter 3 difference equations

Many algorithms of numerical analysis consist in determining solutions of *difference equations*. In addition, difference equations play an important part in other branches of pure and applied mathematics such as combinatorial analysis, the theory of probability, and mathematical economics.

Many aspects of the theory of difference equations are similar to certain aspects of the theory of differential equations. We thus begin this chapter with a brief review of the idea of a *differential equation*.

3.1 Differential Equations

Suppose $f = f(x, y, z)$ is a real valued function defined for all x in an interval $I = [a, b]$, and for all y and z lying in certain sets of real numbers S_0 and S_1. The sets S_0 and S_1 may depend on x. The following problem is called a *differential equation* of the first order: To find a function $y = y(x)$, differentiable for $x \in I$, such that, for all x in I,

(*i*) $$y(x) \in S_0, \qquad y'(x) \in S_1;$$

(*ii*) $$f(x, y(x), y'(x)) = 0.$$

The essential condition here is (*ii*); condition (*i*) is merely imposed in order that condition (*ii*) makes sense. The problem thus defined is symbolically denoted by

(3-1) $$f(x, y, y') = 0;$$

any function $y = y(x)$ satisfying the conditions (*i*) and (*ii*) is called a *solution* of the differential equation (3-1).

EXAMPLES

1. Let I, S_0, S_1 all be the sets of all real numbers, and let

$$f(x, y, z) = z - ky$$

where k is a constant. Every function

$$y(x) = Ce^{kx},$$

where C is an arbitrary constant, is a solution of the resulting differential equation

$$y' = ky.$$

2. Let I be the set of all reals, and let $S_0 = S_1 = [-1, 1]$. If

$$f(x, y, z) = z^2 - (1 - y^2)$$

there results the differential equation

$$y'^2 = 1 - y^2,$$

solutions of which are given by the functions

$$y(x) = \sin(x - a),$$

where a is again arbitrary.

More generally, we consider differential equations of order N, where N is a positive integer. Let $f = f(x, y_0, y_1, \ldots, y_N)$ be a real valued function defined for $x \in I$ and for all $y_0, y_1, \ldots, y_N$ in certain sets of real numbers $S_0, S_1, \ldots, S_N$. The problem of finding a function $y = y(x)$ defined for $x \in I$, having N derivatives on I, and satisfying the conditions

(i) $$y^{(k)}(x) \in S_k, \quad k = 0, 1, \ldots, N;$$

(ii) $$f(x, y(x), y'(x), \ldots, y^{(N)}(x)) = 0$$

for all x in I, is called a *differential equation of order* N, and is symbolically denoted by

$$f(x, y, y', \ldots, y^{(N)}) = 0.$$

EXAMPLE

3. Let $N = 2$, and let I, S_0, and S_2 be the set of all real numbers. If

$$f(x, y_0, y_1, y_2) = y_0 + y_2,$$

every function of the form

$$y(x) = A \cos x + B \sin x$$

where A and B are constants, is a solution of the resulting differential equation $y'' + y = 0$.

The problems studied in the theory of differential equations not only concern the analytic representation of some or all of their solutions, but also the general behavior of these solutions, especially when x tends to infinity.

3.2 Difference Equations

The fundamental problem in the theory of difference equations is in many ways similar to that of the theory of differential equations, with the exception that the mathematical object sought here is a *sequence* rather than a *function*. From the abstract point of view, a sequence is merely a function defined on a set of integers. More concretely, a sequence s is defined by associating with every member of a set I of integers a certain (real) number. Traditionally, the number associated with the integer n is denoted by a symbol such as s_n and not by the usual functional notation $s(n)$.

In almost all applications, the domain of definition of a sequence is a set of *consecutive* integers, i.e., the set of all integers contained between two fixed limits, one or both of which may be infinite. Frequently I is the set of all nonnegative integers.

Sequences can be denoted by a single letter such as s, much as functions are denoted by single letters such as f. More commonly, however, the sequence with the general element s_n is written either as $\{s_0, s_1, s_2, \ldots\}$ or simply as $\{s_n\}$.

A sequence can be defined by giving an explicit formula for s_n, such as

$$s_n = \frac{1}{n^2 + 1}, \qquad n \geqq 27.$$

More often than not, however, the elements of a sequence are defined only implicitly.   Difference equations are among the most important tools for defining sequences.

We first explain what is meant by a *difference equation of order* 1.   Let $f = \{f_n(y, z)\}$ be a sequence of functions defined on a set I of consecutive integers and for all y and z belonging to some set of real numbers S.   The problem is to find a sequence $\{x_n\}$ of real numbers defined on a set containing I so that the following conditions are satisfied for all $n \in I$:

(i) $\qquad\qquad\qquad\qquad x_n \in S, \qquad x_{n-1} \in S;$

(ii) $\qquad\qquad\qquad\qquad f_n(x_n, x_{n-1}) = 0.$

A sequence $\{x_n\}$ satisfying these conditions is called a *solution* of the difference equation symbolized by equation (ii).

EXAMPLES

4. Let I be the set of all integers.   The choice $f_n(y, z) = y - z - 1$ yields the difference equation

$$x_n - x_{n-1} = 1.$$

Obvious solutions are given by $x_n = n + c$ for every constant c.

5. Let I be the set of nonnegative integers, $f_n(y, z) = y - z - n$. There results the difference equation

$$x_n - x_{n-1} = n,$$

having (among others) the solution

$$x_n = \frac{n(n + 1)}{2}.$$

6. Let I be the set of all integers, and let $f_n(y, z) = y - qz$ where q is a nonzero constant. We obtain the difference equation

$$x_n = qx_{n-1}.$$

A solution satisfying $x_0 = 1$ is given by $x_n = q^n$.

The difference equation of order N, where N is a positive integer, is defined in a similar manner. Let $f = \{f_n(y_0, y_1, \ldots, y_N)\}$ be a sequence, defined on a set I of consecutive integers, whose elements are functions defined for y_k ($k = 0, 1, \ldots, N$) in a set of real numbers S. The problem is to find a sequence $\{x_n\}$, defined on a set containing I and satisfying the following conditions for all $n \in I$:

(*i*) $\qquad\qquad x_n \in S, \; x_{n-1} \in S, \ldots, x_{n-N} \in S;$

(*ii*) $\qquad\qquad f_n(x_n, x_{n-1}, \ldots, x_{n-N}) = 0.$

A sequence $\{x_n\}$ with these properties is again called a *solution* of the difference equation symbolized by (*ii*).

EXAMPLES

7. Let I be the set of all integers, $f_n(y_0, y_1, y_2) = y_0 - 2 \cos \varphi \, y_1 + y_2$, where φ is real. A solution of the resulting difference equation

$$x_n - 2 \cos \varphi \, x_{n-1} + x_{n-2} = 0$$

is given by $x_n = \cos (n\varphi)$.

8. Let I be the set of nonnegative integers, $f_n(y_0, y_1, y_2) = y_0 - y_1 - y_2$. Can you find a solution of the difference equation

$$x_n - x_{n-1} - x_{n-2} = 0?$$

As in the case of differential equations, the problems studied in the theory of difference equations not only concern the analytic representation of some or all solutions, but also the general behavior of these solutions, especially when n tends to infinity.

A differential equation usually has many solutions. In order to pin down a solution, we have to specify some additional property of it, such as its value, and perhaps the values of some of its derivatives, at a given point. The problem of finding a solution of the differential equation under these side conditions is called an *initial value problem*. Similarly, in order to pin down the solution of a difference equation, we have to prescribe the values of some elements of the solution $\{x_n\}$. For instance, in example 8 above, we may require that $x_0 = x_1 = 1$.

The solution of an initial value problem for a difference equation is, in a way, a trivial matter. We assume that the equation

$$f_n(y_0, y_1, \ldots, y_N) = 0$$

can be solved for y_0,

$$y_0 = g_n(y_1, y_2, \ldots, y_N).$$

The difference equation can then be written as a recurrence relation for the elements of the sequence $\{x_n\}$,

(3-2) $$x_n = g_n(x_{n-1}, x_{n-2}, \ldots, x_{n-N}).$$

Once N consecutive elements of the solution are known, further elements can be obtained by successively evaluating the functions g_n. In this manner we find for the difference equation of example 8 the solution

$$\{1, 1, 2, 3, 5, 8, 13, \ldots\}.$$

There remains the problem, however, of finding an explicit formula for x_n, and of determining *all* solutions of a given difference equation.

3.3   Linear Difference Equations of Order One

A difference equation is called *linear*, if, for each $n \in I$, the function f_n is a linear function of $y_0, y_1, \ldots, y_N$. The coefficients in that linear function may, however, depend on n. That is, for certain sequences $\{a_{0,n}\}$, $\{a_{1,n}\}, \ldots, \{a_{N,n}\}$, and $\{b_n\}$ we have

(3-3) $$f_n(y_0, y_1, \ldots, y_N) = a_{0,n}y_0 + a_{1,n}y_1 + \cdots + a_{N,n}y_N + b_n.$$

EXAMPLES

9. The difference equations considered in the examples 4 through 8 are linear.

10. The difference equation

$$x_n + 5nx_{n-1} + n^2x_{n-2} = 2$$

is linear.

11. The difference equation

$$x_n - 2x_{n-1}^2 = 0$$

is not linear.

A linear difference equation is called *homogeneous*, if $f_n(0, 0, \ldots, 0) = 0$ for all $n \in I$, i.e., if the sequence $\{b_n\}$ has zero elements only. The linear difference equations considered in the examples 6, 7, and 8 are homogeneous. The equation of example 10 is not homogeneous. Likewise, the equation of example 11 is not homogeneous, because it is not linear.

Postponing the discussion of linear difference equations of order $N > 1$ until chapter 6, we shall in the present section consider linear difference equations of order 1 only. Assuming that $a_{0,n} \neq 0$, $n \in I$, we may divide through by $a_{0,n}$ and write any such difference equation in the form (3-2), viz.,

$$(3\text{-}4) \qquad x_n = a_n x_{n-1} + b_n.$$

We shall assume that I is the set of positive integers and shall try to find a closed expression for the solution of (3-4) satisfying an arbitrary initial condition $x_0 = c$.

We first consider the homogeneous equation

$$(3\text{-}5) \qquad x_n = a_n x_{n-1}.$$

It is seen immediately that the solution satisfying $x_0 = c$ is given by

$$(3\text{-}6) \qquad x_n = c\pi_n,$$

where

$$(3\text{-}7) \qquad \pi_0 = 1, \; \pi_n = a_1 a_2 \ldots a_n, \qquad n = 1, 2, \ldots.$$

To find the solution of the nonhomogeneous equation (3-4) satisfying $x_0 = c$, we use the method of variation of constants familiar from the theory of ordinary differential equations. Thus we set

$$(3\text{-}8) \qquad x_n = c_n \pi_n,$$

where the sequence $\{c_n\}$ is to be determined. In view of $x_0 = c_0 \pi_0 = c_0$ the initial condition yields $c_0 = c$. Substituting (3-8) into (3-4) we find

$$c_n \pi_n = a_n c_{n-1} \pi_{n-1} + b_n = c_{n-1} \pi_n + b_n.$$

We assume that $a_n \neq 0$, $n = 1, 2, \ldots$, which implies that $\pi_n \neq 0$. The last relation then yields

$$c_n = c_{n-1} + \frac{b_n}{\pi_n},$$

and it follows that

$$c_n = c_0 + (c_1 - c_0) + (c_2 - c_1) + \cdots + (c_n - c_{n-1})$$

$$= c + \frac{b_1}{\pi_1} + \frac{b_2}{\pi_2} + \cdots + \frac{b_n}{\pi_n}.$$

For later reference we summarize the above as follows:

Theorem 3.3 Let $a_n \neq 0$, $n = 1, 2, \ldots$. The solution of the difference equation

$$x_n = a_n x_{n-1} + b_n$$

satisfying $x_0 = c$ is given by

(3-9) $$x_n = \pi_n \left(c + \frac{b_1}{\pi_1} + \frac{b_2}{\pi_2} + \cdots + \frac{b_n}{\pi_n} \right), \qquad n = 0, 1, \ldots,$$

where $\pi_0 = 1$, $\pi_n = a_1 a_2 \ldots a_n$, $n = 1, 2, \ldots$.

We observe that this solution can be considered as being composed of the sequence $\{c\pi_n\}$, which is the solution of the *homogeneous* equation having initial value c, and of the sequence

$$x_n = \pi_n \left(\frac{b_1}{\pi_1} + \frac{b_2}{\pi_2} + \cdots + \frac{b_n}{\pi_n} \right), \qquad n = 1, 2, \ldots,$$

which is the solution of the nonhomogeneous equation having initial value zero. The principle of superposition familiar from the theory of differential equations thus prevails also for difference equations.

Whether or not the solution (3-9) can be expressed in simple form depends on the possibility of expressing the products π_n and the sums c_n in simple form, much as in the case of differential equations the simplicity of the solution depends on the possibility to evaluate certain integrals.

Problems

1. Let $-1 < a < 1$. Solve the difference equation

$$x_0 = 1, \qquad x_n = ax_{n-1} + 1$$

and determine the limit of the sequence $\{x_n\}$ as $n \rightarrow \infty$.

2. Let z be an arbitrary real number, $z \neq 0$. Solve the difference equation

$$x_0 = 1, \qquad x_n = \frac{n}{z} x_{n-1} + 1.$$

Show that

$$\frac{z^n}{n!} x_n \rightarrow e^z, \qquad n \rightarrow \infty.$$

3. Let the infinite series

$$\sum_{n=0}^{\infty} b_n$$

be convergent, and let its sum be s. Show that

$$s = \lim_{n \rightarrow \infty} x_n,$$

where $\{x_n\}$ denotes the solution of the difference equation

$$x_0 = b_0, \qquad x_n = x_{n-1} + b_n.$$

4. It is shown in calculus that

$$\prod_{n=1}^{\infty} \left(1 - \frac{z^2}{n^2}\right) = \frac{\sin \pi z}{\pi z}.$$

Show that the infinite product can be obtained as $\lim_{n \to \infty} x_n$, where $\{x_n\}$ satisfies the linear and homogeneous difference equation

$$x_0 = 1, \qquad x_n = \left(1 - \frac{z^2}{n^2}\right) x_{n-1}, \qquad n = 1, 2, \ldots .$$

3.4   Horner's Scheme

Several applications of theorem 3.3 will be made in later chapters.   At this point we wish to show how difference equations can be used to calculate the values of a polynomial.

Let $b_0, b_1, \ldots, b_N$ be $N + 1$ given constants, and let z be a given real number.   We define

Algorithm 3.4   Calculate the numbers $x_0, x_1, \ldots, x_N$ recursively by the relations

(3-10) $x_0 = b_0, \qquad x_n = zx_{n-1} + b_n, \qquad n = 1, 2, \ldots, N.$

Evidently, the finite sequence $\{x_n\}$ defined in algorithm 3.4 is the solution satisfying $x_0 = b_0$ of a difference equation (3-4), where $a_n = z$, $n = 1, 2, \ldots, N$.   In this special case, $\pi_n = z^n$ and hence, by virtue of theorem 3.3,

$$x_n = b_0 z^n + b_1 z^{n-1} + \cdots + b_n$$

and in particular for $n = N$,

$$x_N = b_0 z^N + b_1 z^{N-1} + \cdots + b_N.$$

We thus have obtained:

Theorem 3.4   The numbers x_n produced by algorithm 3.4 equal the values of the polynomials p_n defined by

$$p_n(x) = b_0 x^n + b_1 x^{n-1} + \cdots + b_n \qquad (n = 0, 1, \ldots, N)$$

at $x = z$.

Algorithm 3.4 may thus be regarded as a method for evaluating a polynomial.   This method is generally known as *Horner's rule*.   It requires n multiplications and n additions for the evaluation of a polynomial of degree n, whereas the "straightforward" method—building up the powers of x recursively by $x^n = x \cdot x^{n-1}$ and subsequently multiplying by the b_n—normally requires $2n - 1$ multiplications and n additions. Schematically, Horner's method may be indicated thus:

$$
\begin{array}{cccc}
b_0 & b_1 & b_2 \cdots & b_N \\
\downarrow & \downarrow & \downarrow & \downarrow \\
x_0 \Rightarrow & x_1 \Rightarrow & x_2 \cdots \Rightarrow & x_N
\end{array}
$$

(The symbol $\Rightarrow$ indicates multiplication by z.)

Scheme 3.4

EXAMPLE

12. To evaluate the polynomial

$$p(x) = 7x^4 + 5x^3 - 2x^2 + 8$$

at $x = 0.5$. Scheme 3.4 yields (note that the coefficient $b_3 = 0$ may not be suppressed!)

$$
\begin{array}{ccccc}
7 & 5 & -2 & 0 & 8 \\
7 & 8.5 & 2.25 & 1.125 & 8.5625.
\end{array}
$$

It follows that $p(0.5) = 8.5625$.

3.5 Binomial Coefficients

We wish to obtain a generalization of algorithm 3.4 and recall to this end some facts about binomial coefficients.

The binomial coefficients

$$\binom{n}{m} \qquad \text{(to be read: ``n over m'')}$$

are for integral n and m, $n \geq 0$, $0 \leq m \leq n$, defined by the expansion

$$(1 + x)^n = \binom{n}{0} + \binom{n}{1}x + \binom{n}{2}x^2 + \cdots + \binom{n}{n}x^n.$$

It is well known that

$$(3\text{-}11) \qquad \binom{n}{m} = \frac{n(n-1)(n-2)\ldots(n-m+1)}{1 \cdot 2 \cdot 3 \ldots m} = \frac{n!}{m!(n-m)!}$$

From (3-11) it follows immediately that

$$(3\text{-}12) \qquad \binom{n}{n-m} = \binom{n}{m}.$$

The identity

$$(3\text{-}13) \qquad \binom{n}{m} + \binom{n}{m+1} = \binom{n+1}{m+1}$$

likewise is easily proved.   It states that if the binomial coefficients are arranged in a triangular array known as Pascal's triangle,

$$
\begin{array}{ccccccccc}
 & & & & 1 & & & & \\
 & & & 1 & & 1 & & & \\
 & & 1 & & 2 & & 1 & & \\
 & 1 & & 3 & & 3 & & 1 & \\
1 & & 4 & & 6 & & 4 & & 1 \\
\end{array}
$$

$$\cdots$$

then each entry in this array is equal to the sum of the two entries immediately above it.   We also shall require the following slightly less obvious property of the binomial coefficients.

Theorem 3.5   For any two nonnegative integers k and $n \geq k$ the following identity holds:

$$\binom{k}{k} + \binom{k+1}{k} + \binom{k+2}{k} + \cdots + \binom{n}{k} = \binom{n+1}{k+1}.$$

Proof.   Keeping k fixed, we use induction with respect to n.   Set

$$x_n = \binom{k}{k} + \binom{k+1}{k} + \cdots + \binom{n}{k}.$$

Assuming that for some integer $n - 1 \geq k$

$$(3\text{-}14) \qquad\qquad x_{n-1} = \binom{n}{k+1},$$

which is certainly true for $n - 1 = k$, we have, using (3-13),

$$x_n = x_{n-1} + \binom{n}{k} = \binom{n}{k+1} + \binom{n}{k} = \binom{n+1}{k+1},$$

proving (3-14) with n increased by one, which suffices to establish the formula for all positive integers n.

EXAMPLE

13.   For $k = 2$, $n = 5$ we get

$$\binom{2}{2} + \binom{3}{2} + \binom{4}{2} + \binom{5}{2} = 1 + 3 + 6 + 10 = 20 = \binom{6}{3}.$$

Problems

5. Use the definition of the binomial coefficients to prove that

$$\binom{n}{0} + \binom{n}{1} + \binom{n}{2} + \cdots + \binom{n}{n} = 2^n.$$

6. Obtain an alternate proof of theorem 3.5 by differentiating the identity

$$(x - 1)(1 + x + x^2 + \cdots + x^n) = x^{n+1} - 1$$

$k + 1$ times and setting $x = 1$.

7. Show by means of theorem 3.3 that the difference initial value problem

$$y_0 = 1, \qquad y_n = ny_{n-1} + n! \binom{k + n}{n}$$

has the solution

$$y_n = n! \left[\binom{k}{0} + \binom{k + 1}{1} + \binom{k + 2}{2} + \cdots + \binom{k + n}{n} \right].$$

Show that the same initial value problem is also solved by

$$y_n = n! \binom{k + n + 1}{n}$$

and hence obtain yet another proof of theorem 3.5.   (Use (3-12).)

3.6   Evaluation of the Derivatives of a Polynomial

One is frequently required not only to find the value of a polynomial, but also the values of one or several derivatives at a given point. For instance, if the polynomial

$$p(x) = b_0 x^N + b_1 x^{N-1} + \cdots + b_N$$

is to be expanded in powers of a new variable $h = x - z$, then we have by Taylor's theorem (see Taylor [1959], p. 471)

$$p(x) = p(z + h) = c_0 h^N + c_1 h^{N-1} + \cdots + c_N,$$

where

(3-15) $$c_{N-k} = \frac{1}{k!} p^{(k)}(z), \qquad k = 0, 1, \ldots, N.$$

These coefficients could be evaluated, of course, by differentiating p analytically and evaluating the resulting polynomials by Horner's algorithm 3.4. It turns out, however, that there is a far shorter way. To this end we consider the following extension of algorithm 3.4.

Algorithm 3.6   Let the sequence generated in algorithm 3.4 now be denoted by $\{x_n^{(0)}\}$. For $k = 1, 2, \ldots, N$, generate the sequences $\{x_n^{(k)}\}$ recursively by the difference equation

(3-16) $$x_0^{(k)} = x_0^{(k-1)}, \qquad x_n^{(k)} = zx_{n-1}^{(k)} + x_n^{(k-1)},$$
$$n = 0, 1, \ldots, N - k.$$

Each of the sequences $\{x_n^{(k)}\}$ is generated from the preceding sequence

$\{x_n^{(k-1)}\}$ just as the sequence $\{x_n^{(0)}\}$ was generated from the sequence $\{b_n\}$. However, the new sequence always terminates one step before the old sequence. Scheme 3.6 indicates the resulting two-dimensional array for $N = 4$.

$$x_0^{(0)} \quad x_1^{(0)} \quad x_2^{(0)} \quad x_3^{(0)} \quad \underline{x_4^{(0)}}$$
$$\downarrow$$
$$x_0^{(1)} \quad x_1^{(1)} \Rightarrow x_2^{(1)} \quad \underline{x_3^{(1)}}$$
$$x_0^{(2)} \quad x_1^{(2)} \quad \underline{x_2^{(2)}}$$
$$x_0^{(3)} \quad \underline{x_1^{(3)}}$$
$$\underline{x_0^{(4)}}$$

(The symbol $\Rightarrow$ indicates multiplication by z.)

Scheme 3.6

The relevance of this scheme to the problem stated above is based on the following fact:

Theorem 3.6   If p_n denotes the polynomial

$$p_n(x) = b_0 x^n + b_1 x^{n-1} + \cdots + b_n \quad (n = 0, 1, \ldots, N)$$

then the numbers $x_n^{(k)}$ determined in algorithm 3.6 satisfy

(3-17)
$$x_n^{(k)} = \frac{1}{k!} p_{n+k}^{(k)}(z),$$

$$n = 1, 2, \ldots, N - k; \quad k = 0, 1, \ldots, N.$$

Thus, the coefficients c_{N-k} required in (3-15) are just the entries in the lower diagonal (underlined in scheme 3.6) in the array of numbers generated by algorithm 3.6.

Proof.   Theorem 3.4 asserts the truth of formula (3-17) for $k = 0$. We use induction with respect to k to prove it for arbitrary k. Assuming that (3-17) is true for a certain $k \geqq 0$, we can write the difference equation (3-16) defining the sequence $\{x_n^{(k+1)}\}$ as follows:

$$x_0^{(k+1)} = \frac{1}{k!} p_k^{(k)}(z), \, x_n^{(k+1)} = zx_{n-1}^{(k+1)} + \frac{1}{k!} p_{n+k}^{(k)}(z).$$

Expressing the solution of this equation by means of formula (3-9), where

$$a_n = z, \qquad \pi_n = z^n, \qquad b_n = \frac{1}{k!} p_{n+k}^{(k)}(z),$$

we obtain

$$x_n^{(k+1)} = \frac{z^n}{k!} \{p_k^{(k)}(z) + z^{-1} p_{k+1}^{(k)}(z) + z^{-2} p_{k+2}^{(k)}(z) + \cdots + z^{-n} p_{n+k}^{(k)}(z)\}.$$

Here we express the polynomial derivatives by means of the formula of example 22, chapter 2:

$$\frac{1}{k!}p_m^{(k)}(z) = \binom{m}{k}b_0 z^{m-k} + \binom{m-1}{k}b_1 z^{m-k-1} + \cdots + \binom{k}{k}b_{m-k}.$$

This yields the expression

$$x_n^{(k+1)} = z^n\left\{\binom{k}{k}b_0 + z^{-1}\left[\binom{k+1}{k}b_0 z + \binom{k}{k}b_1\right] + \cdots \right.$$

$$\left. + z^{-n}\left[\binom{n+k}{k}b_0 z^n + \binom{n+k-1}{k}b_1 z^{n-1} + \cdots + \binom{k}{k}b_n\right]\right\}.$$

By rearranging the sum by collecting terms involving like powers of z this can be transformed into

$$x_n^{(k+1)} = \left[\binom{k}{k} + \binom{k+1}{k} + \cdots + \binom{n}{k}\right]b_0 z^n$$

$$+ \left[\binom{k}{k} + \binom{k+1}{k} + \cdots + \binom{n-1}{k}\right]b_1 z^{n-1} + \cdots + \binom{k}{k}b_n.$$

By using the identity of theorem 3.5, this simplifies into

$$x_n^{(k+1)} = \binom{n+k+1}{k+1}b_0 z^n + \binom{n+k}{k+1}b_1 z^{n-1} + \cdots + \binom{k+1}{k+1}b_n$$

$$= \frac{1}{(k+1)!}p_{n+k+1}^{(k+1)}(z).$$

EXAMPLE

14. To compute the Taylor expansion of the polynomial

$$p(x) = 7x^4 + 5x^3 - 2x^2 + 8$$

at $x = 0.5$. Continuing the scheme started in example 12, we obtain

7	5	−2	0	8
7	8.5	2.25	1.125	8.5625
7	12.0	8.25	5.25	
7	15.5	16.0		
7	19.0			
7				

We thus have

$$7x^4 + 5x^3 - 2x^2 + 8 = 7(x - 0.5)^4 + 19(x - 0.5)^3$$
$$+ 16(x - 0.5)^2$$
$$+ 5.25(x - 0.5) + 8.5625.$$

Problems

8. Find the Taylor expansion of the polynomial

$$p(x) = 4x^3 - 5x + 1$$

   at $x = -0.3$.

9. Calculate $p(x)$, $p'(x)$, $p''(x)$ for $x = 1.5$, where

$$p(x) = 2x^5 - 7x^4 + 3x^3 - 6x^2 + 4x - 5.$$

PART ONE

SOLUTION OF EQUATIONS

chapter 4 iteration

In this chapter we shall begin the study of the problem of solving (non-linear) equations in one and several variables. The method of iteration is the prototype of all numerical methods for attacking this problem. It is based on the solution of a certain nonlinear difference equation of the first order. The principle of iteration is of great importance also in certain branches of theoretical mathematics such as functional analysis and the theory of differential equations.

4.1 Definition and Hypotheses

We are here concerned with the problem of solving equations of the form

$$(4\text{-}1) \qquad x = f(x),$$

where f is a given function. The method of iteration consists in executing the following algorithm:

> **Algorithm 4.1** Choose x_0 arbitrarily, and generate the sequence $\{x_n\}$ recursively from the relation
>
> $$(4\text{-}2) \qquad x_n = f(x_{n-1}), \qquad n = 1, 2, \dots.$$

At the outset, we cannot even be sure that this algorithm is well defined. (It could be that f is undefined at some point $f(x_n)$.) However, let us assume that f is defined on some closed finite interval $I = [a, b]$, and *that the values of f lie in the same interval.* Geometrically this means that the graph of the function $y = f(x)$ is contained in the square $a \leq x \leq b$, $a \leq y \leq b$ (see Fig. 4.1a).

Under this assumption, if $x_0 \in I$, we can say that all elements of the sequence $\{x_n\}$ are in I. For if some $x_n \in I$ with $n \geq 0$, then also $x_{n+1} = f(x_n) \in I$, since f has its values in I.

A glance at figure 4.1a shows that the above hypotheses are not sufficient

61

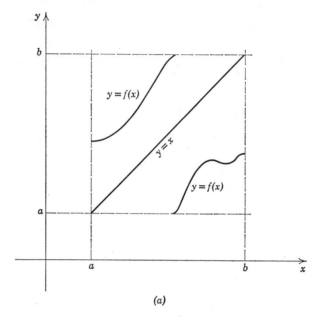

(a)

Figure 4.1

to guarantee that equation (4-1) has a solution. The graph of the function $y = f(x)$ need not intersect the graph of the function $y = x$. However, if we assume the function f to be *continuous*, then the graph shows that the equation has at least one solution. For the graph of $y = f(x)$ originates somewhere on the vertical straight line segment joining the points (a, a) and (a, b), and it ends somewhere on the straight line segment joining (b, a) and (b, b). Since the graph is now continuous, it must intersect the straight line $y = x$ ($a \leq x \leq b$), perhaps at an endpoint. If s is the abscissa of the point of intersection, then

$$y = s \quad \text{and} \quad y = f(s)$$

at that point, hence $s = f(s)$, i.e., the number s is a solution of (4-1) (see Fig. 4.1b).

The above intuitive consideration can be couched in purely analytical terms, as follows. Consider the function g defined by $g(x) = x - f(x)$, $a \leq x \leq b$. This function is continuous on the interval $[a, b]$; moreover, since $f(a) \geq a$, $f(b) \leq b$, it satisfies $g(a) \leq 0$, $g(b) \geq 0$. By the intermediate value theorem of calculus (see Taylor [1959], p. 240) it assumes all values between $g(a)$ and $g(b)$ somewhere in the interval $[a, b]$. Therefore it must assume the value zero, say at $x = s$. This implies $0 = s - f(s)$, or $s = f(s)$. Thus the number s is the desired solution.

If an element of the sequence $\{x_n\}$ defined by algorithm 4.1 is equal to s, then all later elements will also be equal to s.   For this reason, any solution s of $x = f(x)$ is frequently called a *fixed point* of the iteration defined by the function f.

The assumptions made so far do not preclude the possibility of the existence of several, or even infinitely many, fixed points in the interval $[a, b]$ (see Fig. 4.1c).

If we wish to be sure that there is not more than one solution, we must make some assumption guaranteeing that the function f does not vary too rapidly.   We could assume, for instance, that f is differentiable, and that its derivative f' satisfies

$$(4\text{-}3) \qquad |f'(x)| \leq L, \qquad a \leq x \leq b,$$

where L is some constant so that $L < 1$.   It turns out, however, that the following weaker assumption is sufficient to guarantee uniqueness: *There exists a constant $L < 1$ so that for any two points x_1 and x_2 in I the following inequality holds*:

$$(4\text{-}4) \qquad |f(x_1) - f(x_2)| \leq L|x_1 - x_2|.$$

Any condition of the form (4-4) (whether $L < 1$ or not) is called a *Lipschitz condition*, and the constant L is called the *Lipschitz constant*.   Let us first show that condition (4-3) implies condition (4-4) with the same

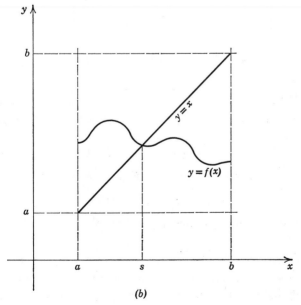

(b)

Figure 4.1

value of L. By the mean value theorem of differential calculus (see Taylor [1959], p. 240),

$$f(x_1) - f(x_2) = f'(x^*)(x_1 - x_2),$$

where x^* is a suitable point between x_1 and x_2. Relation (4-4) now follows readily by taking absolute values and using (4-3).

We now show that the Lipschitz condition (4-4) with $L < 1$ implies that equation (4-1) has at most one solution. Assume that there are two

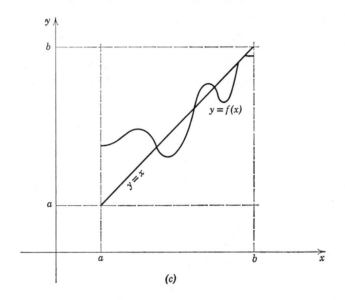

(c)

Figure 4.1

solutions s_1 and s_2, for instance. This means that the following relations both are true:

$$s_1 = f(s_1), \qquad s_2 = f(s_2).$$

Subtracting the second relation from the first, we obtain

$$s_1 - s_2 = f(s_1) - f(s_2).$$

Taking absolute values and applying (4-4) to the difference on the right, we get

$$|s_1 - s_2| = |f(s_1) - f(s_2)| \leq L|s_1 - s_2|.$$

If $s_1 \neq s_2$, we can divide by $|s_1 - s_2|$ and obtain $1 \leq L$, contradicting the assumption that $L < 1$. Thus $s_1 = s_2$, i.e., any two solutions of (4-1) are identical.

Problems

1. Show that the following functions satisfy Lipschitz conditions with $L < 1$:
   (a) $f(x) = 5 - \frac{1}{4} \cos 3x,$ $0 \leq x \leq 2\pi/3$;
   (b) $f(x) = 2 + \frac{1}{2}|x|,$ $-1 \leq x \leq 1$;
   (c) $f(x) = x^{-1},$ $2 \leq x \leq 3.$
2. Let m be any real number, and let $|\varepsilon| < 1$.   Show that the equation

$$x = m - \varepsilon \sin x$$

   has a unique solution in the interval $[m - \pi, m + \pi]$.
3. Prove that any function satisfying a Lipschitz condition on an interval I is continuous at every point of I.
4. Construct an example showing that not every continuous function satisfies a Lipschitz condition.

4.2   Convergence of the Iteration Method

Having disposed of the theoretical preliminaries required to establish existence and uniqueness of the solution of (4-1), we now turn to the practical question of determining this solution by means of algorithm 4.1. Rather surprisingly it turns out that the assumptions which were made in §4.1 to guarantee existence and uniqueness suffice also to establish the fact that the sequence $\{x_n\}$ generated by algorithm 4.1 converges to the solution s.

Theorem 4.2   Let $I = [a, b]$ be a closed finite interval, and let the function f satisfy the following conditions:

(*i*) f is continuous on I;
(*ii*) $f(x) \in I$ for all $x \in I$;
(*iii*) f satisfies the Lipschitz condition (4-4) with a Lipschitz constant $L < 1$.

Then for any choice of $x_0 \in I$ the sequence defined by algorithm 4.1 converges to the unique solution s of the equation $x = f(x)$.

Proof.   That the hypotheses of theorem 4.2 guarantee the existence of a unique solution s of $x = f(x)$ has already been shown in §4.1.   To prove convergence of the sequence $\{x_n\}$, we shall estimate the difference $x_n - s$. By definition,

$$x_n - s = f(x_{n-1}) - s = f(x_{n-1}) - f(s)$$

and hence, by the Lipschitz condition,

$$|x_n - s| \leq L|x_{n-1} - s|.$$

Applying the same inequality repeatedly, we find

(4-5)
$$|x_n - s| = L^n |x_0 - s|.$$

Since $0 \leqq L < 1$,

$$\lim_{n \to \infty} L^n = 0,$$

and it follows that

$$\lim_{n \to \infty} |x_n - s| = 0,$$

which means the same as

$$\lim_{n \to \infty} x_n = s,$$

completing the proof.

The convergence of the sequence $\{x_n\}$ is illustrated very suggestively by plotting the points (x_0, x_0), (x_0, x_1), (x_1, x_1), (x_1, x_2), (x_2, x_2), (x_2, x_3), ... in a graph of the function $y = f(x)$ (see Fig. 4.2). It is evident from the figure that convergence cannot take place in general if condition (4-4) does not hold with some $L < 1$.

EXAMPLE

1. It is desired to find a solution of the equation

$$x = e^{-x}.$$

For $0 \leqq x \leqq 1$, the values of the function $f(x) = e^{-x}$ lie in the interval

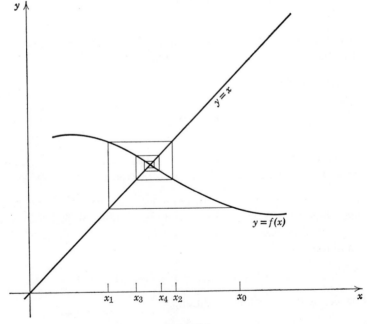

Figure 4.2

$[e^{-1}, 1]$, and thus a fortiori in $[0, 1]$.   For x_1, x_2 in this interval, we have by the mean value theorem

$$|f(x_1) - f(x_2)| = |f'(x^*)| |x_1 - x_2|,$$

where $f'(x^*) = -e^{-x^*}$.   Since the maximum of $|f'(x)|$ for $x \in [0, 1]$ is 1, (4-4) holds only with $L = 1$, violating the condition that $L < 1$.   However, let us consider the smaller interval $I = [\frac{1}{2}, \log 2]$.   Since

$$\tfrac{1}{2} = e^{-\log 2} \leqq e^{-x} \leqq e^{-1/2} < \log 2$$

for $\frac{1}{2} \leqq x \leqq \log 2$, I is again mapped into I by the function f, and (4-4) now holds with $L = |f'(\frac{1}{2})| = e^{-1/2} = 0.606531$.   Beginning with $x_0 = 0.5$, the first few values of the sequence $\{x_n\}$ are as follows:

Table 4.2

n	x_n	$f(x_n)$	x_n'
0	0.500000	0.606531	0.567624
1	0.606531	0.545239	0.567299
2	0.545239	0.579703	0.567193
3	0.579703	0.560065	0.567159
4	0.560065	0.571172	0.567148
5	0.571172	0.564863	0.567145
6	0.564863	0.568438	0.567144
7	0.568438	0.566410	0.567144
8	0.566410	0.567560	
9	0.567560	0.566907	
10	0.566907	0.567278	
11	0.567278	0.567067	
12	0.567067	0.567187	
13	0.567187	0.567119	
14	0.567119		

Problems

5. Kepler's equation

$$m = x - E \sin x,$$

where m and E are given and x is sought, plays a considerable role in dynamical astronomy.   Solve the equation iteratively for $m = 0.8$, $E = 0.2$, by writing it in the form

$$x = m + E \sin x$$

and starting with $x_0 = m$.

6. The solution of Kepler's equation can also be represented analytically in the form

$$x = m + \sum_{n=1}^{\infty} \frac{2}{n} J_n(nE) \sin nm,$$

where J_n denotes the Bessel function of order n. Using tables of Bessel functions, check the result obtained in problem 5 and compare the amount of labor required.

7. Find the only positive solution of the equation

$$x^3 - x^2 - x - 1 = 0$$

by iteration, writing it in the form

$$x = 1 + \frac{1}{x} + \frac{1}{x^2}.$$

(Begin with $x_0 = 1$.)

8. Assume that the function f, in addition to the hypotheses of theorem 4.2, is differentiable and satisfies $f'(x) < 0$, $x \in I$. If $x_0 < s$, prove both analytically and by considering the graph of f that

$$x_0 < x_2 < x_4 < \cdots < s < \cdots < x_5 < x_3 < x_1.$$

9. Suppose the function g is defined and differentiable on the interval $[0, 1]$, and suppose that $g(0) < 0 < g(1)$, $0 < a \leq g'(x) \leq b$, where a and b are constants. Show that there exists a constant M such that the solution of the equation $g(x) = 0$ can be found by applying iteration to the function

$$f(x) = x + Mg(x).$$

10. What is the value of

$$s = \sqrt{2 + \sqrt{2 + \sqrt{2 + \cdots}}}\,?$$

(Hint: The number s may be considered as the limit of the sequence $\{x_n\}$ generated by algorithm 4.1, where $f(x) = \sqrt{2 + x}$, $x_0 = 0$. Show that f satisfies, in a suitable interval, the hypotheses of theorem 4.2.)

4.3 The Error after a Finite Number of Steps

No computing process can be carried on indefinitely, and in any practical application algorithm 4.1 must be artificially terminated after having computed, say, the element x_n. We are interested in finding a bound for the quantity $|x_n - s|$, that is, for the error of x_n considered as an approximation to the solution s. This bound should depend only on quantities that are known *a priori* and should not depend on a knowledge of the solution itself.   (This is why a result such as (4-5) does not satisfy our purpose.)

To establish such a bound, we require the following auxiliary result: For $n = 0, 1, 2, \ldots,$

$$(4\text{-}6) \qquad |x_{n+1} - x_n| \leq L^n |x_1 - x_0|.$$

Evidently this is true for $n = 0$.   Assuming the truth of (4-6) for $n = k - 1$, where k is an integer > 0, we have

$$\begin{aligned}
|x_{k+1} - x_k| &= |f(x_k) - f(x_{k-1})| \\
&\leq L|x_k - x_{k-1}| \\
&\leq L \cdot L^{k-1}|x_1 - x_0| \\
&= L^k|x_1 - x_0|,
\end{aligned}$$

establishing (4-6) for $n = k$.   The truth of (4-6) for all positive integers n now follows by the principle of mathematical induction.

Let n now be a fixed positive integer, and let $m > n$.   We shall find a bound for $x_m - x_n$.   Writing

$$x_m - x_n = (x_m - x_{m-1}) + (x_{m-1} - x_{m-2}) + \cdots + (x_{n+1} - x_n)$$

and applying the triangle inequality (see Taylor [1959], p. 443) we get

$$|x_m - x_n| \leq |x_m - x_{m-1}| + |x_{m-1} - x_{m-2}| + \cdots + |x_{n+1} - x_n|$$

and using (4-6) to estimate each term on the right,

$$|x_m - x_n| \leq (L^{m-1} + \cdots + L^n)|x_1 - x_0|.$$

Now, since $|L| < 1$,

$$\begin{aligned}
L^{m-1} + \cdots + L^n &= L^n(1 + L + \cdots + L^{m-n-1}) \\
&\leq L^n(1 + L + L^2 + \cdots) \\
&= L^n \frac{1}{1 - L}
\end{aligned}$$

by virtue of the familiar formula for the sum of the geometric series.   We thus have

$$|x_m - x_n| \leq \frac{L^n}{1 - L}|x_1 - x_0|.$$

In this relation let $m \to \infty$ while keeping n fixed.   By virtue of $x_m \to s$ we obtain

Corollary 4.3   Under the conditions of theorem 4.2 the error of the nth approximation x_n defined by algorithm 4.1 is bounded as follows:

$$(4\text{-}7) \qquad \cdot|x_n - s| \leq \frac{L^n}{1 - L}|x_1 - x_0|.$$

EXAMPLE

2. The error of the final element x_{14} produced in example 1 is less than

$$\frac{L^{14}}{1-L}|x_1 - x_0| = \frac{e^{-7}}{0.393469} \times 0.106531 = 0.000247.$$

Actually, since $f' < 0$, it follows from problem 8 that the solution satisfies

$$0.567119 < s < 0.567187.$$

Problem

11. For Kepler's equation considered in problem 5, estimate the error of x_{10}.

4.4 Accelerating Convergence

Let us now assume that the function f, in addition to satisfying the hypotheses of theorem 4.2, is continuously differentiable throughout the interval I, and that the derivative f' is never zero.   This means that f is either monotonically strictly increasing or monotonically strictly decreasing throughout the whole interval I.   If $x_0 \neq s$, it is evident from a graph that under this condition no x_n can be equal to the exact solution s, i.e., the iteration process cannot terminate in a finite number of steps.   An analytical proof of this fact is as follows: Assume the contrary, namely, that $f(x_n) = x_n$ for some n.   If n is the first index for which this happens, then

$$x_n = f(x_{n-1}) = f(x_n), \qquad x_{n-1} \neq x_n,$$

and hence, by the mean value theorem (see Taylor [1959], p. 240),

$$0 = f(x_{n-1}) - f(x_n) = f'(x^*)(x_{n-1} - x_n),$$

where x^* is between x_{n-1} and x_n.   Since $x_{n-1} - x_n \neq 0$, it follows that $f'(x^*) = 0$, contradicting the hypotheses that f' never vanishes.

The above implies that the error

$$d_n = x_n - s$$

is never zero.   We now ask: Does the limit

$$\lim_{n \to \infty} \frac{d_{n+1}}{d_n}$$

exist, and if it does, what is its value?

Using the mean value theorem once more, we have

$$\begin{aligned} d_{n+1} &= x_{n+1} - s \\ &= f(x_n) - s \\ &= f(s + d_n) - f(s) \\ &= f'(s + \theta_n d_n) d_n \end{aligned}$$

where $0 < \theta_n < 1$. Let us define ε_n by

$$f'(s + \theta_n d_n) = f'(s) + \varepsilon_n.$$

We then have

(4-8) $$d_{n+1} = (f'(s) + \varepsilon_n) d_n$$

and, since $\varepsilon_n \to 0$ for $n \to \infty$ by virtue of the continuity of f',

(4-9) $$\lim_{n \to \infty} \frac{d_{n+1}}{d_n} = f'(s).$$

This equation shows that the error at the $(n + 1)st$ step is approximately equal to $f'(s)$ times the error at the nth step. As s is unknown, the limiting ratio $f'(s)$ of two consecutive errors is, of course, unknown, and all we really know is that the ratio of two consecutive errors approaches some unknown limit. We now shall show, however, how to obtain a significant improvement of the convergence of the iteration method by a judicious use of this incomplete information.

Let us heuristically proceed under the assumption that (4-8) is exact with $\varepsilon_n = 0$ for finite values of n. We then have, writing $f'(s) = A$ for brevity,

$$x_{n+1} - s = A(x_n - s)$$
$$x_{n+2} - s = A(x_{n+1} - s).$$

It is an easy matter now to eliminate the unknown quantity A and solve for s. Subtracting the first equation from the second, we obtain

$$x_{n+2} - x_{n+1} = A(x_{n+1} - x_n),$$

hence

$$A = \frac{x_{n+2} - x_{n+1}}{x_{n+1} - x_n}.$$

Solving the first equation for s and substituting for A, we get

$$s = \frac{1}{1 - A} (x_{n+1} - A x_n)$$

$$= x_n + \frac{1}{1 - A} (x_{n+1} - x_n)$$

$$= x_n - \frac{(x_{n+1} - x_n)^2}{x_{n+2} - 2x_{n+1} + x_n}.$$

If our assumption that $\varepsilon_n = 0$ were correct, we thus could obtain the exact solution from any three consecutive iterates x_n, x_{n+1}, x_{n+2}. In reality, of course, the ε_n are not exactly zero: They are, however, ultimately small compared to $f'(s)$. We thus may hope that for n large the quantities

$$(4\text{-}10) \qquad x_n' = x_n - \frac{(x_{n+1} - x_n)^2}{x_{n+2} - 2x_{n+1} + x_n}$$

yield a better approximation to s than the quantities x_n. We are thus led to investigate the properties of the following algorithm:

Algorithm 4.4 (Aitken's Δ^2-method) Given a sequence of numbers $\{x_n\}$, generate from it a new sequence $\{x_n'\}$ by means of formula (4-10).

Formula (4-10) can be simplified somewhat by means of the difference operator Δ. If $\{x_n\}$ is any sequence, we write

$$\Delta x_n = x_{n+1} - x_n, \qquad n = 0, 1, 2, \ldots.$$

Higher powers of the operator Δ are defined recursively. For instance,

$$\Delta^2 x_n = \Delta(\Delta x_n) = \Delta x_{n+1} - \Delta x_n$$
$$= x_{n+2} - 2x_{n+1} + x_n.$$

It is now evident that formula (4-10) can be written thus:

$$(4\text{-}11) \qquad x_n' = x_n - \frac{(\Delta x_n)^2}{\Delta^2 x_n}.$$

This notation, together with the fact that algorithm 4.4 was discussed by Aitken [1926], explains its traditional name.

EXAMPLE

3. The last column of table 4.2 (see example 1) contains the accelerated Aitken values of the sequence $\{x_n\}$. It is seen that after six steps the sequence $\{x_n'\}$ has converged to the number of digits given, whereas the original sequence $\{x_n\}$ has not even nearly converged after fourteen steps.

Problem

12. The convergence behavior typified by (4-8) is sometimes characterized by the statement that "the number of correct decimal digits in x_n grows at a fixed rate." Give a quantitative interpretation of this statement, allowing you to answer the following question: As $n \to \infty$, how many steps are necessary (on the average) to reduce the error by a factor $\frac{1}{10}$?

4.5   Aitken's Δ^2-Method

After having discovered algorithm 4.4 in a heuristic manner, we now shall present a rigorous analysis of it. This analysis applies to arbitrary sequences having certain convergence properties and is not confined to sequences arising through iteration. The basic result is as follows.

Theorem 4.5 Let $\{x_n\}$ be any sequence converging to the limit s such that the quantities $d_n = x_n - s$ satisfy $d_n \neq 0$,

(4-12) $$d_{n+1} = (A + \varepsilon_n)\, d_n,$$

where A is a constant, $|A| < 1$, and $\varepsilon_n \to 0$ for $n \to \infty$. Then the sequence $\{x_n'\}$ derived from $\{x_n\}$ by means of algorithm 4.4 is defined for n sufficiently large and converges to s faster than the sequence $\{x_n\}$ in the sense that

(4-13) $$\frac{x_n' - s}{x_n - s} \to 0, \qquad n \to \infty.$$

Proof. Applying (4-12) twice, we have

$$d_{n+2} = (A + \varepsilon_{n+1})(A + \varepsilon_n)\, d_n.$$

Hence

$$\begin{aligned} \varDelta^2 x_n &= x_{n+2} - 2x_{n+1} + x_n \\ &= d_{n+2} - 2d_{n+1} + d_n \\ &= [(A - 1)^2 + \varepsilon_n']\, d_n, \end{aligned}$$

where

$$\varepsilon_n' = A(\varepsilon_n + \varepsilon_{n+1}) - 2\varepsilon_n + \varepsilon_n \varepsilon_{n+1}.$$

By virtue of $\varepsilon_n \to 0$ it follows that also

(4-14) $$\varepsilon_n' \to 0, \qquad n \to \infty.$$

We conclude that $(A - 1)^2 + \varepsilon_n' \neq 0$ for all sufficiently large n, $n > n_0$, say. It follows that $\varDelta^2 x_n \neq 0$ for $n > n_0$; hence the sequence $\{x_n'\}$ is defined for $n > n_0$. We have

$$\varDelta x_n = \varDelta d_n = (A + \varepsilon_n - 1)\, d_n$$

and hence, subtracting s from (4-11),

$$\begin{aligned} x_n' - s &= d_n - \frac{(\varDelta x_n)^2}{\varDelta^2 x_n} \\[2mm] &= d_n - \frac{(A - 1 + \varepsilon_n)^2\, d_n}{(A - 1)^2 + \varepsilon_n'} \\[2mm] &= \frac{\varepsilon_n' - 2\varepsilon_n(A - 1) - \varepsilon_n^2}{(A - 1)^2 + \varepsilon_n'}\, d_n. \end{aligned}$$

The hypothesis that $\varepsilon_n \to 0$ and (4-14) now insure that

$$\frac{x_n' - s}{d_n} = \frac{\varepsilon_n' - 2\varepsilon_n(A - 1) - \varepsilon_n^2}{(A - 1)^2 + \varepsilon_n'} \to 0,$$

as desired.

The result of theorem 4.5 is immediately applicable to sequences generated by iteration, as shown by corollary 4.5.

Corollary 4.5   Let the function f, in addition to the hypotheses of theorem 4.2, have a continuous derivative on I, and assume that $f' \neq 0$. Provided that $x_0 \neq s$, the sequence $\{x_n\}$ generated by algorithm 4.1 satisfies the hypotheses of theorem 4.5, and algorithm 4.4 thus results in a speedup of convergence.

Proof.   As was shown at the beginning of §4.4, the hypotheses of the corollary suffice not only to show that $d_n \neq 0$, $n = 0, 1, 2, \ldots$, but also that (4-8) holds, which is precisely what is needed in theorem 4.5.

Although originally motivated by the iterative procedure, the effectiveness of the Δ^2-acceleration is in no way confined to sequences generated by iteration. Some other instances where it may be applied are given in problems 15, 16, and in chapter 7.

Problems

13. Apply algorithm 4.4 to the sequences obtained in problems 5 and 7.

14. Let x_n be the nth partial sum of the infinite series $\sum_{k=0}^{\infty} a_k$,

$$x_n = \sum_{k=0}^{n} a_k.$$

Show that algorithm 4.4 in this case yields the formula

$$x_n' = x_n + \frac{a_{n+1}^2}{a_{n+1} - a_{n+2}}.$$

15. Show that the hypotheses of theorem 4.5 are satisfied if the terms a_n of the series considered in problem 14 satisfy

$$a_n = (a + \varepsilon_n)z^n,$$

where a is a constant, $|z| < 1$, and where the sequence $\{\varepsilon_n\}$ tends to zero. (Hint: Introduce the quantities

$$\delta_n = \sup_{k \geq n} |\varepsilon_k|.)$$

16. Evaluate the sum

$$\sum_{n=0}^{\infty} \frac{1}{\cosh(n \log 2)}$$

to 5 decimal places.   (Apply problem 14.)

17*. Assume the quantities ε_n in theorem 4.5 are such that the limit

(4-15)
$$\lim_{n \to \infty} \frac{\varepsilon_{n+1}}{\varepsilon_n} = B$$

exists.   Show that the convergence of the sequence $\{x_n'\}$ can be sped up even further by applying algorithm 4.4 once more.

18*. Show that condition (4-15) is satisfied if the sequence $\{x_n\}$ is generated by algorithm 4.1, provided that the second derivative f'' of f exists, is continuous, and satisfies $f''(s) \neq 0$.

4.6 Quadratic Convergence

In §4.4 we had assumed that $f'(x) \neq 0$ on the interval I, and thus in particular that $f'(s) \neq 0$. We then obtained a convergence behavior characterized by relation (4-8). This is known as *linear* convergence. (The number of correct decimal places is approximately a linear function of the number of iterations performed.) Let us now investigate the asymptotic behavior of the error if $f'(s) = 0$. We first note that if this is known to be the case, then it is not necessary to verify all the hypotheses of theorem 4.2.

> **Theorem 4.6** Let I be an interval (finite or infinite), and let the function f be defined on I and satisfy the following conditions:
>
> (*i*) f and f' are continuous on I;
> (*ii*) the equation $x = f(x)$ has a solution s, located in the interior of I, such that $f'(s) = 0$.
>
> Then there exists a number $d > 0$ such that algorithm 4.1 converges to s for any choice of x_0 satisfying $|x_0 - s| \leq d$.

The conclusion of the theorem can be expressed by saying that the algorithm *always* converges when the starting point is "sufficiently close" to the solution.

Proof. Let I_d denote the interval $[s - d, s + d]$. Since s is in the interior of I, I_d is contained in I if d is sufficiently small, $d \leq d_0$ say. Let L be given, $0 < L < 1$. By the continuity of f', there exists d satisfying $0 < d \leq d_0$ such that $|f'(x) - f'(s)| = |f'(x)| \leq L$ for $x \in I_d$. An application of the mean value theorem now shows that for $x \in I_d$,

$$|f(x) - s| = |f(x) - f(s)| \leq L|x - s| \leq Ld < d;$$

thus, the values taken by f in I_d lie in I_d. The hypotheses of theorem 4.2 are thus satisfied for the interval I_d, and as a consequence algorithm 4.1 converges.

Let us now assume, in addition to the hypotheses of theorem 4.6, that f'' exists, is continuous, and does not vanish on I_d. As at the beginning of §4.4, we then can show that if $x_0 \neq s$, no x_n will be accidentally equal to s, and that the iteration algorithm cannot yield the exact solution in a finite number of steps.

Using Taylor's theorem with remainder (see Taylor [1959], p. 476) we find the following expression for the error $d_{n+1} = x_{n+1} - s$:

$$
\begin{aligned}
d_{n+1} &= x_{n+1} - s \\
&= f(x_n) - f(s) \\
&= f'(s) d_n + \tfrac{1}{2} f''(s + \theta_n d_n) d_n^2.
\end{aligned}
$$

Here θ_n denotes, as usual, an unspecified number between zero and one. By virtue of our assumption that $f'(s) = 0$ the above expression simplifies, and we get

(4-16) $$d_{n+1} = \tfrac{1}{2} f''(s + \theta_n d_n) d_n^2.$$

Since $d_n \neq 0$ and $d_n \to 0$ for $n \to \infty$ it follows that

(4-17) $$\lim_{n \to \infty} \frac{d_{n+1}}{d_n^2} = \tfrac{1}{2} f''(s).$$

Relation (4-16) states the remarkable fact that if $f'(s) = 0$, the error at the $(n + 1)st$ step is proportional to the *square* of the error at the nth step. This type of convergence behavior is known as *quadratic convergence*. It is frequently, if somewhat vaguely, described by stating that the number of correct decimal places is *doubled* at each step.

EXAMPLE

4. Let $a > 0$, and let $f(x) = \tfrac{1}{2}(x + a/x)$ for $x > 0$. The equation $x = f(x)$ has the solution $x = \sqrt{a}$. It is easily seen that $f'(\sqrt{a}) = 0$, and that $f''(x) > 0$, $x > 0$. It thus follows that the sequence defined by

$$x_{n+1} = \frac{1}{2}\left(x_n + \frac{a}{x_n}\right)$$

converges to $\sqrt{a}$ quadratically, provided that x_0 is sufficiently close to $\sqrt{a}$. (Actually, it can be shown that the sequence converges for every choice of $x_0 > 0$, see §4.9.) This algorithm for calculating $\sqrt{a}$ is a special case of Newton's method, which is discussed in the next section.

Problems

19. For what values of the constant M in problem 9 does the sequence defined by $x_n = f(x_{n-1})$ converge quadratically to the solution s?
20. Let the function f have a nonvanishing third derivative, and assume that $s = f(s)$, $f'(s) = f''(s) = 0$. Show that in this case the limit of d_{n+1}/d_n^3 exists for x_0 sufficiently close to s. (The convergence is called cubic in this case.)
21*. Let

$$f(x) = \frac{x^3 + bx}{cx^2 + d}.$$

Determine the constants b, c, d in such a manner that the sequence defined by algorithm 4.1 converges cubically to $\sqrt{z}$. Use the algorithm thus obtained to calculate $\sqrt{10}$ to ten decimal places, starting with $x_0 = 3$.

4.7 Newton's Method

The reader might well be under the impression that the discussion of quadratic convergence in the last section is without much practical value, since for an equation of the form $x = f(x)$ the condition $f'(s) = 0$ will be satisfied only by accident. However, it will now be shown that, at least for differentiable functions f, the basic iteration procedure of algorithm 4.1 can always be reformulated in such a way that it becomes quadratically convergent.

Let the function F be defined and twice continuously differentiable on the interval $I = [a, b]$, let $F'(x) \neq 0$ for $x \in I$, and let the equation

$$(4\text{-}18) \qquad F(x) = 0$$

have the solution (necessarily the *only* solution) $x = s$, where s lies in the interior of I. We have already observed that this solution can be found by applying iteration to the function

$$f(x) = x + MF(x),$$

where M is a constant that must satisfy certain inequalities (see problem 9). Unless we are lucky, the convergence of the iteration sequence thus generated is linear. We now ask: Can we determine a function h (depending on f, but easily calculable) so that the iteration sequence generated by the function

$$f(x) = x + h(x)F(x)$$

converges to s quadratically?

In view of theorem 4.6 the sole condition to be satisfied by f (in addition to the obviously satisfied condition $f(s) = s$) is $f'(s) = 0$. In view of

$$f'(x) = 1 + h'(x)F(x) + h(x)F'(x)$$

this yields

$$h'(s)F(s) + h(s)F'(s) = -1$$

or, since $F(s) = 0$,

$$h(s) = -\frac{1}{F'(s)}.$$

A simple way to satisfy this condition (we do not claim it is the only way) is to choose

$$h(x) = -\frac{1}{F'(x)}.$$

We are thus led to the following algorithm:

Algorithm 4.7   Choose x_0, and determine the sequence $\{x_n\}$ from the recurrence relation

$$(4\text{-}19) \qquad x_{n+1} = x_n - \frac{F(x_n)}{F'(x_n)}, \qquad n = 0, 1, 2, \ldots.$$

Algorithm 4.7 is known as *Newton's method* or as the *Newton-Raphson method*. The convergence of Newton's method for starting values x_0 sufficiently close to s follows immediately from theorem 4.6, since the iteration function $f(x) = x - F(x)/F'(x)$ was specifically constructed so as to satisfy $f'(s) = 0$.   If we assume, in addition to the hypotheses made above, that F''' exists and is continuous, then it easily follows that f'' is continuous, and hence that the convergence of the sequence defined by (4-19), if it takes place at all, is quadratic.

Formula (4-19) has a very simple graphical interpretation.   We approximate the graph of the function F by its tangent at the point x_n, that is, $F(x)$ is replaced by

$$F(x_n) + (x - x_n)F'(x_n).$$

Setting this expression equal to zero and solving for x, we find equation (4-19) (see Fig. 4.7).   Intuitively appealing as they may be, considerations such as these tell us nothing about the nature of convergence, nor are they easily extended to the case of systems of equations.

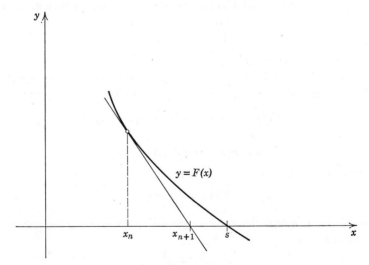

Figure 4.7

4.8   A Non-local Convergence Theorem for Newton's Method

The results proved in §4.7 are still unsatisfactory inasmuch convergence is proved only for starting values "sufficiently close" to the solution s. Since the exact solution is unknown, the question how to find the first approximation x_0 remains unanswered. We now shall prove a result which explicitly specifies an interval in which the first approximation may be taken. The one new hypothesis which must be made is that F'' does not change sign in the interval under consideration.

Theorem 4.8   Let the function F be defined and twice continuously differentiable on the closed finite interval $[a, b]$, and let the following conditions be satisfied:

(i) $F(a)F(b) < 0$;
(ii) $F'(x) \neq 0$, $x \in [a, b]$;
(iii) $F''(x)$ is either ≥ 0 or ≤ 0 for all $x \in [a, b]$;
(iv) If c denotes that endpoint of $[a, b]$ at which $|F'(x)|$ is smaller, then

$$\left| \frac{F(c)}{F'(c)} \right| \leq b - a.$$

Then Newton's method converges to the (only) solution s of $F(x) = 0$ for any choice of x_0 in $[a, b]$.

Some explanation of the hypotheses is in order. Condition (i) merely states that $F(a)$ and $F(b)$ have different signs, and hence that the equation $F(x) = 0$ has at least one solution in (a, b). By virtue of condition (ii) there is only one solution. Condition (iii) states that the graph of F is either concave from above or concave from below. Condition (iv) states that the tangent to the curve $y = F(x)$ at that endpoint where $|F'(x)|$ is smallest intersects the x-axis within the interval $[a, b]$.

Proof.   Theorem 4.8 covers the following four different situations:

(a) $F(a) < 0$, $F(b) > 0$, $F''(x) \leq 0$ $(c = b)$;
(b) $F(a) > 0$, $F(b) < 0$, $F''(x) \geq 0$ $(c = b)$;
(c) $F(a) < 0$, $F(b) > 0$, $F''(x) \geq 0$ $(c = a)$;
(d) $F(a) > 0$, $F(b) < 0$, $F''(x) \leq 0$ $(c = a)$.

The cases (b) and (d) are readily reduced to the cases (a) and (c), respectively, by considering the function $-F$ in place of F. (This change does not change the sequence $\{x_n\}$.) Case (c) is reduced to case (a) by replacing x by $-x$. (This changes the sequence $\{x_n\}$ into $\{-x_n\}$, and the solution s to $-s$.) It thus suffices to prove the theorem in case (a). Here the graph of F looks as given in figure 4.8.

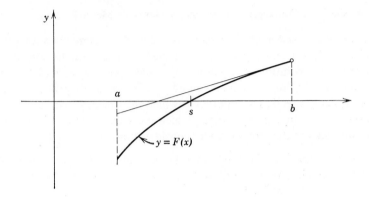

Figure 4.8

Let s be the unique solution of $F(x) = 0$. We first assume that $a \leqq x_0 \leqq s$. By virtue of $F(x_0) \leqq 0$, it is clear that

$$x_1 = x_0 - \frac{F(x_0)}{F'(x_0)} \geqq x_0.$$

We assert that $x_n \leqq s$, $x_{n+1} \geqq x_n$ for all values of n. This being true for $n = 0$, it suffices to perform the induction step from n to $n + 1$. If $x_n \leqq s$, then by the mean value theorem

$$-F(x_n) = F(s) - F(x_n) = (s - x_n)F'(x_n^*)$$

where $x_n \leqq x_n^* \leqq s$. By virtue of $F''(x) \leqq 0$, F' is decreasing, hence $F'(x_n^*) \leqq F'(x_n)$

$$-F(x_n) \leqq (s - x_n)F'(x_n)$$

and

$$x_{n+1} = x_n - \frac{F(x_n)}{F'(x_n)} \leqq x_n + (s - x_n) = s.$$

Consequently, $F(x_{n+1}) \leqq 0$ and $x_{n+2} = x_{n+1} - F(x_{n+1})/F'(x_{n+1}) \geqq x_{n+1}$, completing the induction step.

Since every bounded monotonic sequence has a limit (see Taylor [1959], p. 453) it follows that $\lim_{n \to \infty} x_n$ exists and is $\leqq s$. Denoting this limit by q and letting $n \to \infty$ in the relation

$$x_{n+1} = x_n - \frac{F(x_n)}{F'(x_n)}$$

it follows by the continuity of F and F' that

$$q = q - \frac{F(q)}{F'(q)}$$

and hence that $F(q) = 0$, implying that $q = s$.

Consider now the case where $s < x_0 \leqq b$. Using once more the mean value theorem, we have $F(x_0) = F'(x_0^*)(x_0 - s)$, where $s < x_0^* < x_0$, and hence, since F' is decreasing, $F(x_0) \geqq (x_0 - s)F'(x_0)$. It follows that

$$x_1 = x_0 - \frac{F(x_0)}{F'(x_0)} \leqq x_0 - (x_0 - s) = s.$$

On the other hand $F(x_0) = F(b) - (b - x_0)F'(x_0')$ where $x_0 \leqq x_0' \leqq b$, hence

$$F(x_0) \leqq F(b) - (b - x_0)F'(b).$$

By virtue of condition (iv) of the theorem we thus have

$$x_1 = x_0 - \frac{F(x_0)}{F'(x_0)} \geqq x_0 - \frac{F(x_0)}{F'(b)} \geqq x_0 - \frac{F(b)}{F'(b)} + (b - x_0)$$

$$= x_0 - (b - a) + (b - x_0) = a.$$

Hence $a \leqq x_1 \leqq s$, and it follows by what has been proved above for the case $a \leqq x_0 \leqq s$ that the sequence $\{x_n\}$ converges to s. Thus the proof of theorem 4.8 is complete.

The literature on Newton's method is extensive. Convergence can be proved under various sets of conditions other than those of theorem 4.8, and it is also possible to give bounds for the error after a finite number of steps (see Ostrowski [1960]). The theorem given above, however, covers some of the most important special cases.

4.9   Some Special Cases of Newton's Method

We now shall give some examples for the application of theorem 4.8.

(i) *Determination of square roots.* Let c be a given number, $c > 0$, and let

$$F(x) = x^2 - c \qquad (x > 0).$$

We wish to solve the equation $F(x) = 0$, i.e., to compute $x = \sqrt{c}$. Newton's method takes the form

$$x_{n+1} = x_n - \frac{F(x_n)}{F'(x_n)}$$

$$= x_n - \frac{x_n^2 - c}{2x_n},$$

or

(4-20)
$$x_{n+1} = \frac{1}{2}\left(x_n + \frac{c}{x_n}\right).$$

We are thus led to the algorithm considered in example 4. Since $F'(x) > 0$, $F''(x) > 0$ for $x > 0$, we are in case (d) of theorem 4.8. In any interval $[a, b]$ with $0 < a < \sqrt{c} < b$ the smallest value of the slope occurs at $x = a$, and it is easily seen that condition (*iv*) is satisfied for every $b \geq \frac{1}{2}(a + c/a)$. Thus it follows that the sequence defined by (4-20) converges to $\sqrt{c}$ for every choice of $x_0 > 0$.

Formula (4-20) states that the new approximation is always the arithmetic mean of the old approximation and of the result of dividing c by the old approximation. For work on a desk computer it is not necessary to work with full accuracy from the beginning since every new value can be regarded as new starting value of the iteration.

EXAMPLE

5. $c = 10$, $x_0 = 3$ yields

$$
\begin{array}{ll}
x_0 = 3 & c/x_0 = 3.3 \\
x_1 = 3.15 & c/x_1 = 3.1746 \\
x_2 = 3.1622 & c/x_2 = 3.16225532 \\
x_3 = 3.16227766 & c/x_3 = 3.16227766
\end{array}
$$

Since $x_4 = x_3$ to the number of digits given, the last value is accepted as final.

(*ii*) *Finding roots of arbitrary order.* If $F(x) = x^k - c$, where $c > 0$ and k is any positive integer, Newton's method yields the formula

$$
x_{n+1} = x_n - \frac{x_n^k - c}{k x_n^{k-1}}
$$

or

(4-21)
$$
x_{n+1} = \left(1 - \frac{1}{k}\right)x_n + \frac{1}{k}cx_n^{1-k}
$$

Again the conditions of theorem 4.8 are satisfied for every interval $[a, b]$ if $0 < a < \sqrt[k]{c}$ and b is sufficiently large, and the sequence defined by (4-21) converges to $\sqrt[k]{c}$ for arbitrary $x_0 > 0$.

(*iii*) *Finding reciprocals without division.* For a given $c > 0$ we wish to determine the number

$$
s = \frac{1}{c}.
$$

This may be regarded as the solution of the equation

$$
F(x) = \frac{1}{x} - c = 0.
$$

Newton's method yields

$$x_{n+1} = x_n - \frac{\dfrac{1}{x_n} - c}{-\dfrac{1}{x_n^2}}$$

$$= x_n + (1 - cx_n)x_n$$

or

(4-22) $$x_{n+1} = x_n(2 - cx_n).$$

No divisions are necessary to compute the sequence $\{x_n\}$. Since

$$F'(x) = -\frac{1}{x^2} < 0, \qquad F''(x) = \frac{2}{x^3} > 0$$

for $x > 0$, we are in case (b) of theorem 4.8, and convergence of the algorithm is assured if we can find an interval $[a, b]$ so that $a < c^{-1} < b$ and

$$\frac{F(b)}{F'(b)} = b(bc - 1) \leqq b - a.$$

The last inequality is satisfied if

$$b = \frac{1 + \sqrt{1 - ac}}{c}.$$

Since $a > 0$ may be made arbitrarily small, this means that the sequence defined by (4-22) converges to c^{-1} for any choice of x_0 such that

$$0 < x_0 < 2c^{-1}.$$

EXAMPLE

6. To calculate e^{-1}, where $e = 2.7182183$. Starting with $x_0 = 0.3$, we find

$x_0 = 0.3$	$2 - ex_0 = 1.1846$
$x_1 = 0.355$	$2 - ex_1 = 1.0350106$
$x_2 = 0.367429$	$2 - ex_2 = 1.0012244$
$x_3 = 0.36787889$	$2 - ex_3 = 1.00000150$
$x_4 = 0.36787994$	$2 - ex_4 = 1.00000000.$

The quadratic nature of the convergence is quite evident in this example (doubling of the number of zeros in the second column at each step).

Problems

22. Show that the function $F(x) = x - e^{-x}$ satisfies the conditions of theorem 4.8 in the interval $[0, 1]$. Hence solve the equation by Newton's method, starting with $x_0 = 0.5$.

23. Calculate $\sqrt{\pi}$ to nine decimal places.

24. Determine a numerical value of $\frac{1}{3}$ without division, beginning a Newton iteration with $x_0 = 0.3$.

25. Show that the numbers defined by (4-20) satisfy

$$\frac{x_n - \sqrt{c}}{x_n + \sqrt{c}} = \left(\frac{x_0 - \sqrt{c}}{x_0 + \sqrt{c}}\right)^{2^n}, \qquad n = 0, 1, 2, \ldots.$$

Hence verify that the sequence $\{x_n\}$ converges to $\sqrt{c}$ quadratically for arbitrary choices of $x_0 > 0$.

26. If (4-22) is started with $x_0 = 1$, show that

$$x_n = \frac{1 - (1 - c)^{2^n}}{c}.$$

Deduce that the sequence thus generated converges for $0 < c < 2$, and that the convergence is again quadratic.

27*. How do we have to choose the function h in the formula

$$f(x) = x - \frac{F(x)}{F'(x)} + h(x)\left(\frac{F(x)}{F'(x)}\right)^2$$

such that the iteration defined by f converges cubically to a solution of $F(x) = 0$?

4.10   Newton's Method Applied to Polynomials

Newton's method is especially well suited to the problem of determining the zeros of a polynomial in view of the simple algorithm that is available for calculating both the value of a polynomial and that of its first derivative.   Let

$$p(x) = a_0 x^N + a_1 x^{N-1} + \cdots + a_N$$

be a given polynomial of degree N.   If z is any (real or complex) number, and if the constants $b_0, b_1, \ldots, b_N$ and $c_0, c_1, \ldots, c_{N-1}$ are determined from the recurrence relations† of algorithm 3.6,

(4-23) $b_0 = a_0, \qquad b_n = zb_{n-1} + a_n \qquad (n = 1, \ldots, N)$

(4-24) $c_0 = b_0, \qquad c_n = zc_{n-1} + b_n \qquad (n = 1, \ldots, N-1)$

then we have by a special case of theorem 3.6

$$b_N = p(z), \qquad c_{N-1} = p'(z).$$

The quantities $p(z)$ and $p'(z)$ required in each step of Newton's method can

---

† We now write c_n in place of x_n to avoid confusion with the variable x.

thus be calculated very easily.   The computation proceeds as indicated in scheme 4.10.

$$a_0 \quad a_1 \quad a_2 \cdots \quad a_{N-1} \quad a_N$$

$$\downarrow \quad \downarrow \quad \downarrow \qquad \downarrow \qquad \downarrow$$

$$b_0 \Rightarrow b_1 \Rightarrow b_2 \cdots \Rightarrow b_{N-1} \Rightarrow \underline{b_N}$$

$$\downarrow \quad \downarrow \quad \downarrow \qquad \downarrow$$

$$c_0 \Rightarrow c_1 \Rightarrow c_2 \cdots \Rightarrow \underline{c_{N-1}}$$

Scheme 4.10

$\rightarrow$ indicates addition, $\Rightarrow$ multiplication by z and addition.

EXAMPLE

7.   To determine a zero of the polynomial

$$p(x) = x^3 - x^2 + 2x + 5$$

near $x = -1$.   (In table 4.10, the coefficients a_n, b_n, c_n are arranged in columns rather than in rows.)

Table 4.10

	a_n	b_n	c_n	p/p'
	1	1	1	
$x_0 = -1$	-1	-2	-3	
	2	4	7	0.142857
	5	1		
	1	1	1	
$x_1 = -1.142857$	-1	-2.142857	-3.285714	-0.010305
	2	4.448979	8.204076	
	5	-0.084546		
	1	1	1	
$x_2 = -1.129807$	-1	-2.129807	-3.259614	
	2	4.406241	8.089006	0.002691
	5	0.021764		
	1	1	previous	
$x_3 = -1.132498$	-1	-2.132498	value of p'	
	2	4.415050	retained!	-0.000004
	5	-0.000035		
	1	1		
$x_4 = -1.132494$	-1	-2.132494		
	2	4.415037		
	5	-0.000003		

If z is a zero of the polynomial p, then, according to theorem 2.5b, p can be represented in the form

(4-25) $$p(x) = (x - z)q(x),$$

where

$$q(x) = b_0 x^{N-1} + b_1 x^{N-2} + \cdots + b_{N-1}$$

is a certain polynomial of degree $N - 1$. Multiplying the two polynomials on the right of equation (4-25) and comparing coefficients of like powers of x, we find

$$a_0 = b_0,$$

$$a_{n+1} = -zb_n + b_{n-1}, \qquad n = 1, 2, \ldots, N.$$

Solving for b_{n+1} and replacing $n + 1$ by n, we find that the second relation is identical with (4-23). Thus we have:

Theorem 4.10 If z is a zero of the polynomial p, the coefficients $b_0, b_1, \ldots, b_{N-1}$ defined by equation (4-23) are identical with the coefficients of the polynomial $(x - z)^{-1}p(x)$.

EXAMPLE

8. Accepting $x_4 = -1.132494$ as a zero of the polynomial p considered in example 7, we have

$$q(x) = \frac{p(x)}{x + 1.132494} = x^2 - 2.132494x + 4.415037.$$

Thus, having found a zero of a polynomial p of degree N, the polynomial q whose zeros are identical with those of p save the one which has already been determined is easily constructed. The remaining zeros of p can now be found more easily as the zeros of q. The process of passing from p to q is known as *deflating p*. By successive deflations, the zeros of a given polynomial thus can be found by working with polynomials of successively lower degrees.

Unfortunately, the above process is subject to accumulation of round-off errors in view of the fact that the data of the problem (in this case, the coefficients of the given polynomial) enters the computation only at the first step. It is thus advisable to recheck each zero found by using it once more as a starting value for a Newton's process applied to the full (undeflated) polynomial p.

We mention without proof that local convergence of Newton's method can also be established for complex zeros and for polynomials with complex coefficients. However, for the determination of pairs of conjugate complex zeros of polynomials with real coefficients a more efficient method is available (see §5.6).

Problems

28. Determine the unique positive zero of the polynomial

$$p(x) = x^3 - x^2 - x - 1$$

by Newton's method.

29. Find the zero near 0.9 of the polynomial

$$p(x) = 4x^6 - 5x^5 + 4x^4 - 3x^3 + 7x^2 - 7x + 1.$$

30. Prove the relation $c_{N-1} = p'(z)$ by differentiating the recurrence relation (4-24) and observing that the b_n are functions of z.

4.11   Some Modifications of Newton's Method

Newton's method requires the evaluation of the derivative F' of the given function F.   While in most textbook problems this requirement is trivial, this may not be the case in more involved situations, for instance if the function F itself is the result of a complicated computation.   It therefore seems worthwhile to discuss some methods for solving $F(x) = 0$ that do not require the evaluation of F' and nevertheless retain some of the favorable convergence properties of Newton's method.

(*i*) *Whittaker's method* (Whittaker and Robinson [1928]).   The simplest way to avoid the computation of F' is to replace $F'(x_n)$ in (4-19) by a constant value, say m.   The resulting formula

(4-26) $$x_{n+1} = x_n - \frac{F(x_n)}{m}$$

then defines, for a certain range of values of m, a linearly converging procedure, unless we happen to pick $m = F'(s)$.   If the estimate of m is good, convergence may nevertheless be quite rapid.   Especially in the final stages of Newton's process it is usually not necessary to recompute F' at each step.

(*ii*) *Regula falsi.*   Here the value of the derivative $F'(x_n)$ is approximated by the difference quotient

$$\frac{F(x_n) - F(x_{n-1})}{x_n - x_{n-1}}$$

formed with the two preceding approximations.   There results the formula

(4-27) $$x_{n+1} = x_n - \frac{(x_n - x_{n-1})F(x_n)}{F(x_n) - F(x_{n-1})}.$$

This is identical with

$$x_{n+1} = \frac{x_{n-1}F(x_n) - x_n F(x_{n-1})}{F(x_n) - F(x_{n-1})}.$$

but for numerical purposes the "incremental" form (4-27) is preferable.

The algorithm suggested by (4-27) is known as the *regula falsi*. It is defined by a difference equation of order 2 and thus is not covered by the general theory given at the beginning of this chapter. A more detailed investigation (see Ostrowski [1960], p. 17) shows that the degree of convergence of the regula falsi lies somewhere between that of Newton's method and of ordinary iteration.

(*iii*) *Muller's method* (Muller [1956]). The regula falsi can be obtained by approximating the graph of the function F by the straight line passing through the points $(x_{n-1}, F(x_{n-1}))$ and $(x_n, F(x_n))$. The point of intersection of this line with the x-axis defines the new approximation x_{n+1} (see Fig. 4.11).

Instead of approximating F by a linear function, it seems natural to try to obtain more rapid convergence by approximating F by a polynomial p of degree $k > 1$ coinciding with F at the points $x_n, x_{n-1}, \ldots, x_{n-k}$, and to determine x_{n+1} as one of the zeros of p. Muller has made a detailed study of the case $k = 2$ and found that this choice of k yields very satisfactory results. Since the construction of p depends on the theory of the interpolating polynomial, we postpone the derivation of Muller's algorithm to chapter 10.

(*iv*) *Newton's method in the case $F'(s) = 0$.* Newton's algorithm was derived in §4.7 under the assumption that $F'(x) \neq 0$, implying in particular that $F'(s) \neq 0$. Let us now consider the general situation where

$$F(s) = F'(s) = \cdots = F^{(m-1)}(s) = 0, \; F^{(m)}(s) \neq 0,$$

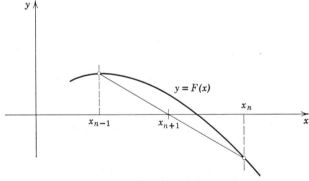

Figure 4.11

where $m \geq 1$.   If we set $x = s + h$, the iteration function for Newton's process,

$$f(x) = x - \frac{F(x)}{F'(x)}$$

can be expanded in powers of h to give

$$f(s + h) = s + h - \frac{(m!)^{-1}F^{(m)}(s)h^m + 0(h^{m+1})}{[(m-1)!]^{-1}F^{(m)}(s)h^{m-1} + 0(h^m)}$$

$$= s + h - \frac{1}{m}h + 0(h^2).$$

From this we find

$$f'(s) = \lim_{h \to 0} \frac{f(s + h) - f(s)}{h} = 1 - \frac{1}{m}.$$

Thus, if $m \neq 1$, $f'(s) \neq 0$, and the convergence fails to be quadratic. However, the above analysis shows how to modify the iteration function in order to achieve quadratic convergence.   If we set

$$f(x) = \begin{cases} x - \dfrac{mF(x)}{F'(x)}, & x \neq s \\ s, & x = s \end{cases}$$

then a computation similar to the one performed above shows that $f'(s) = 0$.   By theorem 4.6, the sequence defined by

(4-28) $$x_{n+1} = x_n - \frac{mF(x_n)}{F'(x_n)}, \qquad n = 0, 1, \ldots$$

converges to s quadratically, provided that x_0 is sufficiently close to s. For $m = 1$ this algorithm reduces to the ordinary Newton process.

Admittedly the above discussion is somewhat academic, because only rarely in practice we have an a priori knowledge of the fact that $F'(x) = 0$ at a solution of $F(x) = 0$.   However, formula (4-28) has also been used in the early stages of Newton's process if two solutions of $F(x) = 0$ are very close together.   In this case m was chosen in a heuristic fashion somewhere between 1 and 2 (see Forsythe [1958], p. 234).

Problems

31. Show that if the regula falsi is applied to the equation $x^2 = 1$, then, assuming that the errors $d_n = x_n - 1$ tend to zero, the stronger relation

$$\lim_{n \to \infty} \frac{d_{n+1}}{d_n d_{n-1}} = \frac{1}{2}$$

holds.

32. Give a closed formula for x_n if the equation $x^2 = 0$ is solved (a) by the ordinary Newton's method, (b) by the modification of Newton's method given by (4-28), and discuss the result. (Assume $x_0 = 1$ in both cases.)

33. Show that if the equation $[F(x)]^m = 0$ is solved by (4-28), there results the ordinary Newton's method for solving $F(x) = 0$.

4.12   The Diagonal Aitken Procedure

We continue to discuss the problem of achieving quadratic (i.e., Newton-like) convergence by methods not requiring the evaluation of any derivatives. We recall that none of the substitutes for Newton's method offered in the preceding section quite achieved quadratic convergence. Returning to equations written in the form

$$(4-29) \qquad\qquad x = f(x)$$

we now shall show that true quadratic convergence can be achieved without derivative evaluation by a modification of the Aitken acceleration procedure discussed in §4.4.

This modification proceeds as follows. We start out, as we do in ordinary iteration, by choosing an initial value x_0 and calculating $x_1 = f(x_0)$, $x_2 = f(x_1)$. Aitken's formula (4-11) is now applied to x_0, x_1, x_2, yielding

$$(4-30) \qquad\qquad x_0' = x_0 - \frac{(x_1 - x_0)^2}{x_2 - 2x_1 + x_0}.$$

The number $x_0' = x_0^{(1)}$ is used as a new starting value for two more iterations. Having calculated $x_1^{(1)} = f(x_0^{(1)})$, $x_2^{(1)} = f(x_1^{(1)})$, we apply Aitken's formula again, obtaining an accelerated value $x_0^{(2)}$, which in turn is used to start a new iteration, etc. If a denominator in Aitken's formula happens to be zero, we set $x_0^{(k+1)} = x_0^{(k)}$, thus in effect terminating the iteration. Schematically, the algorithm is described by the following table:

$$\begin{matrix} x_0^{(0)} \\ x_1^{(0)} \\ x_2^{(0)} \end{matrix} \Big\} \nearrow \begin{matrix} x_0^{(1)} \\ x_1^{(1)} \\ x_2^{(1)} \end{matrix} \Big\} \nearrow \begin{matrix} x_0^{(2)} \\ x_1^{(2)} \\ x_2^{(2)} \end{matrix} \Big\} \nearrow \cdots$$

Scheme 4.12

Since $x_1 = f(x_0)$, $x_2 = f(x_1) = f(f(x_0))$, the values $x^{(k+1)}$ can be thought of as being generated from $x^{(k)}$ in a single iteration step by means of a function F defined in terms of f as follows. We set

$$N(x) = f(f(x)) - 2f(x) + x$$

and put

$$(4\text{-}31) \qquad F(x) = \begin{cases} x - \dfrac{[f(x) - x]^2}{N(x)}, & N(x) \neq 0, \\ x, & N(x) = 0. \end{cases}$$

Thus we are led to the following formal statement of the new procedure:

Algorithm 4.12 Choose $x^{(0)}$, and determine the sequence $\{x^{(k)}\}$ recursively by

$$(4\text{-}32) \qquad x^{(k+1)} = F(x^{(k)}), \qquad k = 0, 1, 2, \ldots,$$

where F is defined by equation (4-31).

This algorithm is sometimes known as Steffensen's iteration, as it was first proposed by Steffensen [1933].

The questions arise whether Steffensen's iteration is well defined, whether the sequence $\{x^{(k)}\}$ converges to a solution of $x = f(x)$, and whether the degree of convergence is higher than linear. We shall give affirmative answers under the following hypotheses: The equation $x = f(x)$ has a solution $x = s$, and the function f is three times continuously differentiable in a neighborhood of $x = s$ and satisfies $f'(s) \neq 1$. It will be shown that these hypotheses imply that F satisfies $F(s) = s$, is twice continuously differentiable in a neighborhood of $x = s$, and satisfies $F'(s) = 0$. Quadratic convergence of the sequence $\{x^{(k)}\}$ for all $x^{(0)}$ sufficiently close to s then follows as a consequence of theorem 4.6.

As direct differentiation of F turns out to be cumbersome, we introduce an auxiliary function g by setting

$$g(h) = \begin{cases} \dfrac{f(s + h) - s}{h}, & h \neq 0, \\ f'(s), & h = 0. \end{cases}$$

The function g is still at least twice continuously differentiable near $h = 0$. The definition of g implies

$$f(s + h) = s + hg(h),$$
$$f(f(s + h)) = f(s + hg(h))$$
$$= s + hg(h)g(hg(h)).$$

If $x = s + h$, it follows that

$$N(x) = f(f(x)) - 2f(x) + x = hG(h),$$

where

$$G(h) = 1 - 2g(h) + g(h)g(hg(h)).$$

Like g, the function G is twice continuously differentiable near $h = 0$; furthermore, since $g(0) = f'(s) \neq 1$,

(4-33) $$G(0) = [g(0) - 1]^2 \neq 0,$$

showing that $G(h) \neq 0$ for $|h|$ sufficiently small. Hence $N(x) \neq 0$ for $x \neq s$, $|x - s|$ sufficiently small, and by (4-31),

$$F(x) = s + h - h\frac{[g(h) - 1]^2}{G(h)}.$$

This representation of F also holds for $h = 0$; it shows that F, too, is twice continuously differentiable, and that $F(s) = s$. Furthermore,

$$F'(s) = \lim_{h \to 0} \frac{F(s + h) - s}{h}$$

$$= \lim_{h \to 0} \left\{ 1 - \frac{[g(h) - 1]^2}{G(h)} \right\}$$

$$= 1 - \frac{[g(0) - 1]^2}{G(0)} = 0,$$

by virtue of (4-33).

EXAMPLE

9. We apply Steffensen iteration to the equation $x = e^{-x}$ considered in example 1. Starting with $x^{(0)} = 0.5$, we obtain the following values (arranged in the manner of scheme 4.12)

0.500000	0.567624	0.567144	0.567143
0.606531	0.566871	0.567143	
0.545239	0.567298	0.567143.	

The values in the top row are seen to converge very rapidly.

Problems

34. Apply Steffensen iteration to the solution of Kepler's equation given in problem 5, where $m = 1$, $E = 0.8$.

35. Apply algorithm 4.12 to the equation

$$x = \frac{10}{x},$$

starting with $x = 3$. Compare the sequence $\{x^{(k)}\}$ with the sequence obtained by Newton's method for computing $\sqrt{10}$.

36. Give a somewhat simpler proof of the quadratic convergence of Steffensen's iteration by assuming that f can be expanded in powers of $h = x - s$.

37. Show that if, in addition to the hypotheses stated above, $f'(s) = 0$, then also $F''(s) = 0$. Thus in this case Steffensen's iteration converges at least cubically (see Householder [1953], p. 128).

4.13   A Non-local Convergence Theorem for Steffensen's Method†

As was the case with the corresponding result concerning Newton's method, the convergence statement concerning Steffensen's iteration proved in the last section is unsatisfactory because it guarantees convergence only for choices of $x^{(0)}$ "sufficiently close" to the solution s. We now shall prove a result which guarantees convergence no matter where the iteration is started. The extra hypothesis which will be added concerns the signs of f' and of f''.

Theorem 4.13   Let I denote the semi-infinite interval (a, ∞), and let the function f satisfy the following conditions:

(*i*) f is defined and twice continuously differentiable on I;
(*ii*) $f(x) > a,\ x \in I$;
(*iii*) $f'(x) < 0,\ x \in I$;
(*iv*) $f''(x) > 0,\ x \in I$.

Then algorithm 4.12 defines a sequence which converges to the (unique) solution s of $x = f(x)$ for any choice of the starting value $x^{(0)}$ in I.

Proof.   By virtue of the hypotheses, the graph of the function f looks as indicated in figure 4.13 (where $a = 0$); it shows that the equation $x = f(x)$ has a unique solution s.

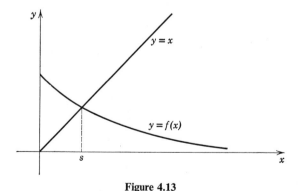

Figure 4.13

---

† This section may be omitted at first reading.

Let $x > s$.   Then, by virtue of (iii), $f(x) < s$, and $f(f(x)) > s$.   Hence, by application of the mean value theorem,

$$f(x) - s = f(x) - f(s) = f'(t_1)(x - s),$$

where $s < t_1 < x$.   Furthermore,

$$f(f(x)) - s = f(f(x)) - f(s)$$
$$= f'(t_2)(f(x) - s)$$

where $f(x) < t_2 < s$.   We thus find

$$f(x) - x = f(x) - s - (x - s)$$
$$= [f'(t_1) - 1](x - s)$$

and

$$f(f(x)) - f(x) = f(f(x)) - s - [f(x) - s]$$
$$= [f'(t_2) - 1][f(x) - s]$$
$$= [f'(t_2) - 1]f'(t_1)(x - s).$$

Thus the expression N in equation (4-31) is given by

$$f(f(x)) - 2f(x) + x = f(f(x)) - f(x) - [f(x) - x]$$
$$= \{[f'(t_2) - 1]f'(t_1) - [f'(t_1) - 1]\}(x - s).$$

The expression inside the braces may be written

$$[f'(t_1) - 1]^2 + [f'(t_2) - f'(t_1)]f'(t_1).$$

This is positive, since by (iv), $f'(t_2) < f'(t_1)$ and hence by (iii),

$$[f'(t_2) - f'(t_1)]f'(t_1) > 0.$$

Hence $N(x) \neq 0$, and the first definition of F applies for all $x > s$.   Since N is a continuous function of x, it follows that F is continuous at each point $x > s$.   (Continuity at $x = s$ has already been established in the preceding section under wider assumptions.)

Using the above work, we find for $x > s$

$$F(x) - s = (1 - Q)(x - s),$$

where

(4-34) $$Q = \frac{[f'(t_1) - 1]^2}{[f'(t_1) - 1]^2 + [f'(t_2) - f'(t_1)]f'(t_1)}$$

By the above, $0 < Q < 1$, and it follows that

$$0 < F(x) - s < x - s$$

or

$$s \leq F(x) < x.$$

Thus for $x^{(0)} \geq s$, the sequence $\{x^{(k)}\}$ is decreasing and bounded below by *s*. It therefore must have a limit $q \geq s$. By the continuity of F, $F(q) = q$. This is possible only for $q = s$. Thus the relation

$$\lim_{k \to \infty} x^{(k)} = s$$

has been proved for the case $x^{(0)} > s$.

If $x^{(0)} = s$, then all $x^{(k)} = s$, and the result is trivial. Thus let now $x < s$. In that situation, $f(x) > s, f(f(x)) < s$. The above computations remain valid, with the difference that now

$$x < t_1 < s, \qquad s < t_2 < f(x).$$

We find again $N(x) \neq 0$; however, since the order of t_1 and t_2 is now reversed, the numerator in (4-34) now exceeds the denominator, and we have $Q > 1$. It follows that $F(x) > s$. Thus if $x^{(0)} < s$, then $x^{(1)} = F(x^{(0)}) > s$, and the sequence $\{x^{(k)}\}$ decreases from $k = 1$ onward. Convergence to s now follows as in the case $x^{(0)} > s$.

EXAMPLES

10. The hypotheses of theorem 4.13 are satisfied for $f(x) = e^{-x}$, $x > 0$. Steffensen iteration thus will produce the solution of $x = e^{-x}$ for any choice of the starting value $x^{(0)} > 0$.

11. Let $c > 0$, and put $f(x) = c/x$ for $x > 0$. Again this function satisfies the hypotheses of theorem 4.13, and the Steffensen iteration converges to the solution $s = \sqrt{c}$ for any choice of $x^{(0)} > 0$. Since $f(f(x)) = x$, the function F is given by

$$F(x) = x - \frac{\left(\frac{c}{x} - x\right)^2}{x - 2\frac{c}{x} + x} = \frac{1}{2}\left(x + \frac{c}{x}\right),$$

and the iteration function is identical with that given by Newton's method.

Problems

38. Prove results analogous to theorem 4.13 for the cases

(a) $\qquad 0 \leq f'(x) < 1, \qquad f''(x) > 0$;
(b) $\qquad f'(x) > 1, \qquad f''(x) > 0.$

39. Show that the function $f(x) = A + B/x^n$ satisfies the hypotheses of theorem 4.13 for $x \geq A > 0$, $B > 0$. Hence find the solution $s > 1$ of the equation $x^3 = x^2 + 1$.

Recommended Reading

The theory of iteration of functions of one variable is treated very thoroughly in the books by Householder [1953] and Ostrowski [1960]. Householder also gives numerous references to earlier work.

Research Problem

Find bounds (exhibiting the quadratic nature of the convergence) for the error of the approximations generated by Steffensen's iteration. (Similar bounds are known for Newton's method, see for example, Ostrowski [1960].)

chapter 5 iteration for systems of equations

In this chapter we will show how to apply the basic iteration algorithm discussed in chapter 4 and its modifications to obtain solutions of systems of equations. The equations envisaged here are nonlinear. Inasmuch as linear equations are a special case of nonlinear ones, the algorithms discussed here are naturally applicable to linear systems. However, for the reasons mentioned in the preface, the many important algorithms that are especially designed for the solution of linear systems of equations are not treated in this book.

5.1 Notation

The algorithms discussed in this chapter are, in principle, applicable to problems involving any number of equations and unknowns. However, for greater concreteness, and also in order to avoid cumbersome notation, we shall consider explicitly only the case of two equations with two unknowns. These equations will usually be written in the form

$$(5\text{-}1) \qquad \begin{cases} x = f(x, y) \\ y = g(x, y), \end{cases}$$

where f and g are certain functions of the point (x, y) that are defined in suitable regions of the plane. Each of the two equations $x - f(x, y) = 0$ and $y - g(x, y) = 0$ defines, in general, a curve in the (x, y) plane. The problem of solving the system of equations (5-1) is equivalent to the problem of finding the point or points of intersection of these curves. It will usually be assumed, and in some cases also proved, that such a point of intersection exists. Its coordinates will be denoted by s and t. The quantities s and t then satisfy the relations

$$s = f(s, t)$$
$$t = g(s, t).$$

EXAMPLE

1. Let $f(x, y) = x^2 + y^2$, $g(x, y) = x^2 - y^2$. The system (4-1) reads in this case

$$x = x^2 + y^2$$
$$y = x^2 - y^2.$$

The equation $x^2 + y^2 - x = 0$ defines a circle centered at $(\frac{1}{2}, 0)$, the equation $x^2 - y^2 - y = 0$ a hyperbola centered at $(0, -\frac{1}{2})$. Both the circle and the hyperbola pass through the origin, thus our system has the obvious solution $s = t = 0$. But an inspection of the graphs shows that there must be another solution near $x = 0.8$, $y = 0.4$.

It will be convenient to employ *vector notation*. Thus we not only shall be able to simplify the writing of our equations, but also to aid the understanding of the theoretical analysis and even the programming of our algorithms. We represent the coordinates of the point (x, y) by the column vector

$$\mathbf{x} = \begin{pmatrix} x \\ y \end{pmatrix}.$$

The functions f and g then become functions of the vector $\mathbf{x}$, whose value for a particular $\mathbf{x}$ we denote by $f(\mathbf{x})$ and $g(\mathbf{x})$. If we denote by $\mathbf{f}$ the column vector with components f and g, the system (5-1) can be written more simply as follows:

(5-2) $\mathbf{x} = \mathbf{f}(\mathbf{x}).$

The fact that the vector

$$\mathbf{s} = \begin{pmatrix} s \\ t \end{pmatrix}$$

is a solution of equation (5-1) is expressed in the form

$$\mathbf{s} = \mathbf{f}(\mathbf{s}).$$

A vector analog of the absolute value of a number (or *scalar*, as numbers are called in this context) will be required. Clearly, the *length* of a vector is such an analog. If $\mathbf{x} = (x, y)$, we write

(5-3) $\|\mathbf{x}\| = \sqrt{x^2 + y^2}$

The quantity $\|\mathbf{x}\|$ is called the Euclidean norm of the vector $\mathbf{x}$. It is nonnegative, and zero only if $\mathbf{x}$ is the zero vector $\mathbf{0} = (0, 0)$. Furthermore, if the sum of two vectors and the product of a vector and a scalar are defined in the natural way, we have

(5-4) $\|c\mathbf{x}\| = |c| \, \|\mathbf{x}\|$

(5-5) $\|\mathbf{x}_1 + \mathbf{x}_2\| \leq \|\mathbf{x}_1\| + \|\mathbf{x}_2\|.$

Relation (5-5) is called the *triangle inequality*. If x_1 and x_2 are two vectors, then $\|x_1 - x_2\|$ is the distance of the points whose coordinates are the components of x_1 and x_2.

We shall have occasion to consider sequences of vectors $\{x_n\}$. Such a sequence is said to converge to a vector v, if

$$\|x_n - v\| \to 0 \qquad \text{for } n \to \infty.$$

The following criterion due to Cauchy is necessary and sufficient for the convergence of the sequence $\{x_n\}$ to some vector v (see Buck [1956], p. 13). Given any number $\varepsilon > 0$, there exists an integer N so that for all $n > N$ and all $m > N$

$$\|x_m - x_n\| < \varepsilon.$$

5.2 A Theorem on Contracting Maps

Iteration in several variables is defined as in the case of one variable.

Algorithm 5.2 Choose a vector x_0, and calculate the sequence of vectors $\{x_n\}$ recursively by

(5-6) $$x_n = f(x_{n-1}), \qquad n = 1, 2, 3, \ldots.$$

As in the scalar situation, there arise the questions whether the sequence $\{x_n\}$ is well defined, whether it converges, and whether its limit necessarily is a solution of the equation

(5-7) $$x = f(x).$$

All these questions are answered by the following result:

Theorem 5.2 Let R denote the rectangular region $a \le x \le b$, $c \le y \le d$, and let the functions f and g satisfy the following conditions:

(*i*) f and g are defined and continuous on R;

(*ii*) For each $x \in R$, the point $(f(x), g(x))$ also lies in R;

(*iii*) There exists a constant $L < 1$ such that for any two points x_1 and x_2 in R the following inequality holds:

(5-8) $$\|f(x_1) - f(x_2)\| \le L\|x_1 - x_2\|.$$

Then the following statements are true:

(a) Equation (5-7) has precisely one solution s in R;

(b) for any choice of x_0 in R, the sequence $\{x_n\}$ given by algorithm 5.2 is defined and converges to s;

(c) for any $n = 1, 2, \ldots$, the following inequality holds:

(5-9) $$\|x_n - s\| \le \frac{L^n}{1 - L}\|x_1 - x_0\|.$$

Condition (5-8) is again called a *Lipschitz condition.*   It here expresses the fact that the mapping $\mathbf{x} \rightarrow \mathbf{f(x)}$ diminishes the distance between any two points in R at least by the factor L.   For this reason a mapping with the property (5-8) is called a *contraction mapping.*

It should be noted that already statement (a) is not as trivial now as in the case of one variable, where the existence of a solution could be inferred from an inspection of the graph of the function f.   Statement (b) guarantees convergence of the algorithm, while (c) gives an upper bound for the error after n steps.

Proof of theorem 5.2.   The proof will be accomplished in several stages. From condition (*ii*) it is evident that the sequence $\{\mathbf{x}_n\}$ is defined, and that its elements lie in R.   Proceeding exactly as in the proof of (4-6) (§4.3) we now can show that (*iii*) implies

(5-10) $$\|\mathbf{x}_{n+1} - \mathbf{x}_n\| \leqq L^n \|\mathbf{x}_1 - \mathbf{x}_0\|, \qquad n = 0, 1, 2, \ldots.$$

Now let n be a fixed positive integer, and let $m > n$.   With a view towards applying the Cauchy criterion, we shall find a bound for $\|\mathbf{x}_m - \mathbf{x}_n\|$. Writing

$$\mathbf{x}_m - \mathbf{x}_n = (\mathbf{x}_{n+1} - \mathbf{x}_n) + (\mathbf{x}_{n+2} - \mathbf{x}_{n+1}) + \cdots + (\mathbf{x}_m - \mathbf{x}_{m-1})$$

and applying the triangle inequality (5-5), we obtain

$$\|\mathbf{x}_m - \mathbf{x}_n\| \leqq \|\mathbf{x}_{n+1} - \mathbf{x}_n\| + \|\mathbf{x}_{n+2} - \mathbf{x}_{n+1}\| + \cdots + \|\mathbf{x}_m - \mathbf{x}_{m-1}\|$$

and, using (5-10) to estimate each term on the right,

(5-11) $$\|\mathbf{x}_m - \mathbf{x}_n\| \leqq (L^n + L^{n+1} + \cdots + L^{m-1})\|\mathbf{x}_1 - \mathbf{x}_0\|$$
$$\leqq \frac{L^n}{1 - L} \|\mathbf{x}_1 - \mathbf{x}_0\|,$$

by virtue of $0 \leqq L < 1$.   Since the expression on the right does not depend on m and tends to zero as $n \rightarrow \infty$, we have established that the sequence $\{\mathbf{x}_n\}$ satisfies the Cauchy criterion.   It thus has a limit $\mathbf{s}$.   Since R is compact, $\mathbf{s} \in R$.

We next show that $\mathbf{s}$ is a solution of (5-7).   By virtue of the continuity of f and g,

(5-12) $$\lim_{n \rightarrow \infty} \mathbf{f(x}_n) = \mathbf{f(s)},$$

and thus

$$\mathbf{s} = \lim_{n \rightarrow \infty} \mathbf{x}_n = \lim_{n \rightarrow \infty} \mathbf{x}_{n+1} = \lim_{n \rightarrow \infty} \mathbf{f(x}_n) = \mathbf{f(s)},$$

as desired.

Uniqueness of the solution $\mathbf{s}$ follows from the Lipschitz condition (*iii*), exactly as it does for one variable (see the end of §4.1).   Relation (5-9) finally follows by letting $m \to \infty$ in (5-11). This completes the proof of theorem 5.2. No use has been made of the fact that the number of equations and unknowns is two; both the theorem and its proof remain valid in a Euclidean space of arbitrary dimension, and even in some infinite-dimensional spaces.

Problems

1. Prove the relation (5-12) by means of (*iii*), without making use of the continuity of f and g.
2. Prove that condition (*iii*) (even when $L \geqq 1$) implies that both f and g are continuous at every point of R.

5.3 A Bound for the Lipschitz Constant†

We consider here the problem of verifying whether a given pair of functions (f, g) satisfies a Lipschitz condition of the form (5-8). The corresponding condition for a single function of one variable could be very easily checked by computing the derivative (see §4.1). If the absolute value of the derivative turned out to be bounded by a constant, then the Lipschitz condition was satisfied with that constant. A similar criterion is available for pairs of functions of two variables.

Theorem 5.3   Let the functions f and g have continuous partial derivatives in the region R defined in theorem 5.2. Then the inequality (5-8) holds with $L = J$, where

$$(5\text{-}13) \qquad J = \max_{(x,\, y)\, \in\, R} \sqrt{f_x^2 + f_y^2 + g_x^2 + g_y^2}$$

Proof. Let (x_0, y_0) and (x_1, y_1) be two points in R. We define for $0 \leqq t \leqq 1$

$$x_t = x_0 + th, \qquad y_t = y_0 + tk,$$

where

$$h = x_1 - x_0, \qquad k = y_1 - y_0,$$

and set

$$f_t = f(x_t, y_t), \qquad g_t = g(x_t, y_t).$$

It is to be proved that

$$(5\text{-}14) \qquad (f_1 - f_0)^2 + (g_1 - g_0)^2 \leqq (h^2 + k^2)J^2.$$

For the proof we shall require both the Schwarz inequality for sums,

$$(5\text{-}15) \qquad (ac + bd)^2 \leqq (a^2 + b^2)(c^2 + d^2),$$

---

† A reader who is willing to accept the statement of Theorem 5.3 may omit the remainder of the section without loss of continuity.

valid for any four real numbers a, b, c, d, and the Schwarz inequality for integrals,

$$(5\text{-}16) \qquad \left(\int_a^b p(x)q(x)\, dx \right)^2 \leq \int_a^b [p(x)]^2\, dx \int_a^b [q(x)]^2\, dx,$$

valid for any two functions p and q that are integrable on the interval $[a, b]$. Proofs of these well-known inequalities are sketched in the problems 3 and 4.

By the chain rule of differentiation (see Taylor [1959], p. 590) we have

$$\frac{df_t}{dt} = hf_x + kf_y,$$

where

$$f_x = f_x(x_t, y_t), \qquad f_y = f_y(x_t, y_t),$$

and hence

$$f_1 - f_0 = \int_0^1 (hf_x + kf_y)\, dt.$$

By (5-15),

$$(hf_x + kf_y)^2 \leq (h^2 + k^2)(f_x^2 + f_y^2).$$

Hence

$$(f_1 - f_0)^2 \leq (h^2 + k^2)\left[\int_0^1 \sqrt{f_x^2 + f_y^2}\, dt \right]^2.$$

We now use (5-16) where $a = 0$, $b = 1$, $p(t) = 1$, $q(t) = (f_x^2 + f_y^2)^{\frac{1}{2}}$. It follows that

$$(f_1 - f_0)^2 \leq (h^2 + k^2) \int_0^1 (f_x^2 + f_y^2)\, dt.$$

In exactly the same manner we find

$$(g_1 - g_0)^2 \leq (h^2 + k^2) \int_0^1 (g_x^2 + g_y^2)\, dt.$$

Adding the last two inequalities and using $f_x^2 + f_y^2 + g_x^2 + g_y^2 \leq J^2$, we obtain (5-14).

EXAMPLE

2. Let

$$f(x, y) = A \sin x + B \cos y, \qquad g(x, y) = A \cos x - B \sin y$$

where A and B are constants. We find

$$f_x^2 + f_y^2 + g_x^2 + g_y^2 = A^2 + B^2,$$

thus (5-8) holds with $L = \sqrt{A^2 + B^2}$. The conditions of theorem 5.2

thus are satisfied, for instance for $a = -2, b = 2$, whenever $A^2 + B^2 < 1$. Convergence of the process for $A = 0.7$, $B = 0.2$, $x_0 = y_0 = 0$ is illustrated by the values given in table 5.3.

Table 5.3

n	x_n	y_n
1	0.200000	0.700000
2	0.292037	0.557203
3	0.371280	0.564599
4	0.422927	0.545289
	...	...
28	0.526519	0.507921
29	0.526521	0.507921
30	0.526521	0.507920
31	0.526522	0.507920

Naturally, the condition of theorem 5.3 is not *necessary* for convergence of the iterative process. For instance for the problem considered in example 2 convergence also takes place for $A = B = 0.71$.

Problems

3. Prove (5-15) by observing that the equation for x,

$$(a + cx)^2 + (b + dx)^2 = 0$$

can have at most one real solution.

4. Prove (5-16) by noting that the equation in x,

$$\int_a^b (p(t) + xq(t))^2 = 0$$

can have at most one real solution. What are the conditions on p and q in order that there is a real solution?

5.4   Quadratic Convergence

Suppose the functions f and g satisfy the conditions of theorem 5.2 and have continuous derivatives up to order 2 in R. The sequence of points (x_n, y_n) defined by algorithm 5.2 then converges to a solution (s, t) of the system (5-1). What is the behavior of the errors

$$d_n = x_n - s,$$
$$e_n = y_n - t,$$

as $n \to \infty$?

An application of Taylor's theorem for functions of two variables (see Buck [1956], p. 200) shows that

$$
\begin{aligned}
d_{n+1} &= x_{n+1} - s \\
&= f(x_n, y_n) - f(s, t) \\
&= f(s + d_n, t + e_n) - f(s, t) \\
&= f_x(s, t)\, d_n + f_y(s, t)e_n + 0(\|\mathbf{d}_n\|^2),
\end{aligned}
$$

and similarly

$$
e_{n+1} = g_x(s, t)\, d_n + g_y(s, t)e_n + 0(\|\mathbf{d}_n\|^2).
$$

Here $\mathbf{d}_n$ denotes the vector of errors,

$$
\mathbf{d}_n = \begin{pmatrix} d_n \\ e_n \end{pmatrix},
$$

and $0(\|\mathbf{d}_n\|^2)$ denotes a quantity bounded by $C\|\mathbf{d}_n\|^2$. Introducing the *Jacobian matrix* of the functions f and g,

$$
\mathbf{J} = \begin{pmatrix} f_x & f_y \\ g_x & g_y \end{pmatrix},
$$

the above relations can be written in abbreviated form as follows:

(5-17) $$\mathbf{d}_{n+1} = \mathbf{J}(s, t)\, \mathbf{d}_n + 0(\|\mathbf{d}_n\|^2).$$

Relation (5-17) is the multidimensional generalization of (4-8). If $\mathbf{J}(s, t) \neq \mathbf{0}$ (that is, if the elements of the matrix $\mathbf{J}(s, t)$ are not all zero), it shows that at each step of the iteration the error vector is approximately multiplied by a constant matrix. In this sense we again may speak of *linear convergence*. If $\mathbf{J}(s, t) = \mathbf{0}$ (that is, if all four elements of $\mathbf{J}$ are zero), then we see that the norm of the error at the $(n + 1)$st step is of the order of the square of the norm of the error at the nth step. This is similar to what earlier has been called *quadratic convergence*.

The following analog of theorem 4.6 holds in the case where $\mathbf{J}(s, t) = \mathbf{0}$:

Theorem 5.4   Let the functions f and g be defined in a region R, and let them satisfy the following conditions:

(*i*) The first partial derivatives of f and g exist and are continuous in R.

(*ii*) The system (5-1) has a solution (s, t) in the interior of R such that $\mathbf{J}(s, t) = \mathbf{0}$.

Then there exists a number $d > 0$ such that algorithm 5.2 converges to (s, t) for any choice of the starting point within the distance d of the solution.

The conclusion can be expressed by saying that the algorithm always converges if the starting vector is "sufficiently close" to the solution vector. The proof of theorem 5.4 is based on the fact that by virtue of the continuity of the first partial derivatives

$$\sqrt{f_x^2 + f_y^2 + g_x^2 + g_y^2} = L < 1$$

in a certain neighborhood of $(s, t) \in R$. It then follows from theorem 5.3 that the conditions of theorem 5.2 are satisfied in that neighborhood. Further details are omitted.

5.5 Newton's Method for Systems of Equations

We now shall consider the problem of solving systems of two equations with two unknowns which are of the form

(5-18)
$$\begin{cases} F(x, y) = 0, \\ G(x, y) = 0, \end{cases}$$

where both functions F and G are defined and twice continuously differentiable on a certain rectangle R of the (x, y) plane. We suppose that the system (5-18) has a solution (s, t) in the interior of R, and that the Jacobian determinant

(5-19)
$$D(x, y) = \begin{vmatrix} F_x(x, y) & F_y(x, y) \\ G_x(x, y) & G_y(x, y) \end{vmatrix}$$

is different from zero when $(x, y) = (s, t)$. It then follows from a theorem of calculus (see Buck [1956], p. 216) that the system (5-18) has no solution other than (s, t) in a certain neighborhood of the point (s, t).

Newton's method for solving a single equation $F(x) = 0$ could be understood as arising from replacing $F(x + \delta)$ by $F(x) + F'(x)\delta$ and solving the equation

$$F(x) + F'(x)\delta = 0$$

for δ, thus obtaining a supposedly better approximation $x + \delta$ to s than x. We now apply the same principle to the system (5-18). Assuming that (x, y) is a point "near" the desired solution (s, t), we replace the function $F(x + \delta, y + \varepsilon)$ by its first degree Taylor polynomial at the point (x, y), that is, by $F(x, y) + F_x(x, y)\delta + F_y(x, y)\varepsilon$. A similar replacement is made for the function $G(x + \delta, y + \varepsilon)$. Setting the Taylor polynomials equal to zero, we obtain a system of two linear equations for δ and ε,

(5-20)
$$\begin{cases} F_x(x, y)\delta + F_y(x, y)\varepsilon = -F(x, y), \\ G_x(x, y)\delta + G_y(x, y)\varepsilon = -G(x, y). \end{cases}$$

The determinant of this system is just the quantity $D(x, y)$ defined by (5-19). Since D is continuous and $D(s, t) \neq 0$ by hypothesis, it follows that $D(x, y) \neq 0$ for all points (x, y) sufficiently close to (s, t). Thus for all these (x, y) the system (5-20) has a unique solution $\delta = \delta(x, y)$, $\varepsilon = \varepsilon(x, y)$. The point $(x + \delta(x, y), y + \varepsilon(x, y))$ is now chosen as the next approximation to (s, t). Algorithmically speaking, the procedure can be described as follows:

Algorithm 5.5 (Newton's method for two variables.) Choose (x_0, y_0), and determine the sequence of points (x_n, y_n) by

(5-21) $\qquad \begin{cases} x_{n+1} = f(x_n, y_n) \\ y_{n+1} = g(x_n, y_n), \end{cases} \quad n = 0, 1, 2, \ldots,$

where the functions f and g are defined by

$$f(x, y) = x + \delta(x, y), \qquad g(x, y) = y + \varepsilon(x, y),$$

and where δ and ε denote the solution of the linear system (5-20).

We assert that the sequence (x_n, y_n) converges quadratically to (s, t) for all (x_0, y_0) sufficiently close to (s, t). In order to verify this statement, it suffices by theorem 5.4 to show that all elements of the Jacobian matrix of the iteration functions f and g,

$$\mathbf{J}(x, y) = \begin{pmatrix} f_x(x, y) & f_y(x, y) \\ g_x(x, y) & g_y(x, y) \end{pmatrix}$$

are zero for $(x, y) = (s, t)$. Omitting arguments, we have

(5-22) $\qquad \begin{cases} f_x = 1 + \delta_x, & f_y = \delta_y, \\ g_x = \varepsilon_x, & g_y = 1 + \varepsilon_y. \end{cases}$

The values of the derivatives $\delta_x, \delta_y, \varepsilon_x, \varepsilon_y$ are best determined from (5-20) by implicit differentiation. Differentiating the first of these equations, we get

$$F_{xx}\delta + F_x\delta_x + F_{xy}\varepsilon + F_y\varepsilon_x = -F_x,$$
$$F_{xy}\delta + F_x\delta_y + F_{yy}\varepsilon + F_y\varepsilon_y = -F_y.$$

Two similar relations involving the function G are obtained by differentiating the second relation. We now set $(x, y) = (s, t)$ and observe that $\delta(s, t) = \varepsilon(s, t) = 0$. We thus obtain for $(x, y) = (s, t)$

$$F_x\delta_x + F_y\varepsilon_x = -F_x,$$
$$G_x\delta_x + G_y\varepsilon_x = -G_x,$$
$$F_x\delta_y + F_y\varepsilon_y = -F_y,$$
$$G_x\delta_y + G_y\varepsilon_y = -G_y.$$

The determinant of each of these two systems of linear equations is again the Jacobian determinant $D(s, t)$ and hence is different from zero.   We now easily find

$$\delta_x = -1, \qquad \delta_y = 0,$$
$$\varepsilon_x = 0, \qquad \varepsilon_y = -1.$$

According to equation (5-22), this implies that all elements of the matrix $\mathbf{J}$ are zero at the point (s, t), as desired.

For actual execution of algorithm 5.5 (although not for the above theoretical analysis) it is necessary at each step of the iteration to solve the system (5-20) for δ and ε.   An application of Cramer's rule† yields

$$\delta = \frac{\begin{vmatrix} -F & F_y \\ -G & G_y \end{vmatrix}}{\begin{vmatrix} F_x & F_y \\ G_x & G_y \end{vmatrix}} = \frac{GF_y - FG_y}{F_x G_y - F_y G_x},$$

$$\varepsilon = \frac{\begin{vmatrix} F_x & -F \\ G_x & -G \end{vmatrix}}{\begin{vmatrix} F_x & F_y \\ G_x & G_y \end{vmatrix}} = \frac{FG_x - GF_x}{F_x G_y - F_y G_x}.$$

EXAMPLE

3.   We solve the system considered in example 2 by writing it in the form

$$x - A \sin x - B \cos y = 0,$$
$$y - A \cos x + B \sin y = 0,$$

and applying Newton's method.   The following values are obtained for $A = 0.7$, $B = 0.2$, $x_0 = y_0 = 0$:

n	x_n	y_n
1	0.6666667	0.5833333
2	0.5362400	0.5088490
3	0.5265620	0.5079319
4	0.5265226	0.5079197
5	0.5265226	0.5079197

The much greater rapidity of convergence is clearly evident.

---

† It is well known that Cramer's rule should never be used for the solution of linear systems of any sizable order.   The solution of such systems is much more conveniently found by a process of elimination.   For systems of small order (such as 2 or 3) Cramer's rule is perfectly applicable, however.

Problems

5. Generalizing the approach outlined in §4.7, one might try to generalize Newton's method to systems by suitably choosing functions h and k so that iteration of the functions

$$f(x, y) = x + h(x, y)F(x, y)$$
$$g(x, y) = y + k(x, y)G(x, y)$$

yields a quadratically convergent process. Show that this is *not* possible in general.

6*. Determine the matrix $\mathbf{H} = \mathbf{H}(\mathbf{x})$ such that application of iteration to

$$\mathbf{f}(\mathbf{x}) = \mathbf{x} + \mathbf{H}(\mathbf{x})\mathbf{F}(\mathbf{x})$$

yields a quadratically convergent process. Show that the process obtained is identical with Newton's method.

7. Find a solution near $x = 0.8$, $y = 0.4$ of the system

$$x - x^2 - y^2 = 0,$$
$$y - x^2 + y^2 = 0,$$

using Newton's method.

8. Solve by Newton's method the pair of equations

$$x + 13 \log x - y^2 = 0,$$
$$2x^2 - xy - 5x + 1 = 0,$$

starting with $x_0 = 3.4$, $y_0 = 2.2$.

9. Solve the system

$$4x^3 - 27xy^2 + 25 = 0,$$
$$4x^2 - 3y^3 - 1 = 0,$$

by Newton's method, starting with $x_0 = y_0 = 1$.

5.6   The Determination of Quadratic Factors

In this section we shall prepare the ground for the application of Newton's method to the problem of finding pairs of complex conjugate zeros of polynomials with real coefficients. We first consider the following preliminary problem.

Given a polynomial p of degree $N \geq 2$ with real coefficients,

$$p(x) = a_0 x^N + a_1 x^{N-1} + \cdots + a_N,$$

and a quadratic polynomial $x^2 - ux - v$, to determine constants $b_0, b_1, \ldots, b_N$ such that the identity

(5-23) $$p(x) = (x^2 - ux - v)q(x) + b_{N-1}(x - u) + b_N$$

holds, where

$$q(x) = b_0 x^{N-2} + b_1 x^{N-3} + \cdots + b_{N-2}.$$

Multiplying the polynomials on the right side of (5-23) and comparing coefficients of like powers of x, we find the following conditions on $b_0, b_1, \ldots, b_N$:

$$b_0 = a_0,$$
$$b_1 = a_1 + ub_0,$$
$$b_2 = a_2 + ub_1 + vb_0,$$

and generally for $n = 2, \ldots, N - 2$

(5-24) $$b_n = a_n + ub_{n-1} + vb_{n-2}.$$

By comparing the coefficients of x^1 and x^0 it is seen that (5-24) also holds for $n = N - 1$ and $n = N$. Furthermore, if we agree to put $b_{-1} = b_{-2} = 0$, the relation holds also for $n = 0$ and $n = 1$. Thus we find the following algorithm for determining the coefficients $b_0, b_1, \ldots, b_N$ in the representation (5-23) of the polynomial p:

Algorithm 5.6 If $a_0, a_1, \ldots, a_N, u, v$ are given numbers, determine $b_0, b_1, \ldots, b_N$ from the recurrence relation $b_{-2} = b_{-1} = 0$,

$$b_n = a_n + ub_{n-1} + vb_{n-2}, \qquad n = 0, 1, \ldots, N.$$

Clearly, the coefficients $b_0, b_1, \ldots, b_N$ thus obtained are functions of the variables u and v. The reason for considering the b_n is contained in the following theorem:

Theorem 5.6 The polynomial $x^2 - ux - v$ is a quadratic factor of the real polynomial $p(x) = a_0x^N + a_1x^{N-1} + \cdots + a_N$ if and only if $b_{N-1} = b_N = 0$.

Proof. (a) If $b_{N-1} = b_N = 0$, then (5-23) reduces to the relation

$$p(x) = (x^2 - ux - v)q(x).$$

It shows that a zero of $x^2 - ux - v$ is also a zero of p. By considering the derivative we find that a double zero of $x^2 - ux - v$ is also a double zero of p. Thus $x^2 - ux - v$ is a quadratic factor of p.

(b) We now *assume* that $x^2 - ux - v$ is a quadratic factor. Denoting its zeros by z_1, z_2, we first assume that $z_1 \neq z_2$. We then have from (5-23), since $z_k^2 - uz_k - v = 0$ $(k = 1, 2)$

$$0 = p(z_k) = b_{N-1}(z_k - u) + b_N, \qquad k = 1, 2,$$

or, written more explicitly,

$$(z_1 - u)b_{N-1} + b_N = 0,$$
$$(z_2 - u)b_{N-1} + b_N = 0.$$

This homogeneous system of two linear equations has the determinant

$z_1 - z_2 \neq 0$, hence its only solution is $b_{N-1} = b_N = 0$. If the two zeros of the quadratic factor $x^2 - ux - v$ coincide, then $p(z_1) = p'(z_1) = 0$, hence from (5-23)

$$b_{N-1}(z_1 - u) + b_N = 0,$$
$$b_{N-1} \qquad\qquad = 0,$$

and it follows again that $b_{N-1} = b_N = 0$.

Theorem 5.6 shows that the problem of determining a quadratic factor of the polynomial p is equivalent to the problem of determining u and v such that

(5-25)
$$\begin{cases} b_{N-1}(u, v) = 0, \\ b_N(u, v) = 0. \end{cases}$$

The solution of this pair of simultaneous equations by Newton's method is known as Bairstow's method.

5.7 Bairstow's Method

In order to apply Newton's method to the system (5-25) we require the partial derivatives

$$\frac{\partial b_{N-1}}{\partial u}(u, v), \qquad \frac{\partial b_{N-1}}{\partial v}(u, v),$$
$$\frac{\partial b_N}{\partial u}(u, v), \qquad \frac{\partial b_N}{\partial v}(u, v).$$

We shall obtain recurrence relations for these derivatives by differentiating the recurrence relations (5-24).

We first differentiate with respect to u and observe that $\partial b_0/\partial u = 0$. Writing

(5-26)
$$c_n = \frac{\partial b_{n+1}}{\partial u}$$

for notational convenience, we obtain

$$c_0 = b_0,$$
$$c_1 = b_1 + uc_0,$$
$$c_2 = b_2 + uc_1 + vc_0,$$

and generally

$$c_n = b_n + uc_{n-1} + vc_{n-2},$$

where $n = 0, 1, 2, \ldots, N - 1$, $c_{-2} = c_{-1} = 0$. We see that the c_n are generated from the b_n exactly as the b_n were generated from the a_n. From (5-26) we have

$$\frac{\partial b_{N-1}}{\partial u} = c_{N-2}, \qquad \frac{\partial b_N}{\partial u} = c_{N-1}.$$

We next differentiate with respect to v. We observe that now $\partial b_0 / \partial v = \partial b_1 / \partial v = 0$. Thus, writing

(5-27) $$d_n = \frac{\partial b_{n+2}}{\partial v},$$

we obtain $d_{-2} = d_{-1} = 0$,

$$d_0 = b_0,$$
$$d_1 = b_1 + u d_0,$$
$$d_2 = b_2 + u d_1 + v d_0,$$

and generally

$$d_n = b_n + u d_{n-1} + v d_{n-2},$$

$n = 0, 1, 2, \ldots, N - 2$. These recurrence relations and initial conditions are exactly the same as those for the c_n, hence we have $d_n = c_n$, $n = 0, 1, 2, \ldots, N - 2$, and from (5-27) we obtain

$$\frac{\partial b_{N-1}}{\partial v} = c_{N-3}, \qquad \frac{\partial b_N}{\partial v} = c_{N-2}.$$

If the increments of u and v are denoted by δ and ε, respectively, their values as determined by Newton's method must satisfy

$$c_{N-2} \delta + c_{N-3} \varepsilon = -b_{N-1},$$
$$c_{N-1} \delta + c_{N-2} \varepsilon = -b_N$$

and hence are given by

(5-28) $$\delta = \frac{b_N c_{N-3} - b_{N-1} c_{N-2}}{c_{N-2}^2 - c_{N-1} c_{N-3}}, \qquad \varepsilon = \frac{b_{N-1} c_{N-1} - b_N c_{N-2}}{c_{N-2}^2 - c_{N-1} c_{N-3}}.$$

The whole procedure is summarized in

Algorithm 5.7 Given the polynomial

$$p(x) = a_0 x^N + a_1 x^{N-1} + a_2 x^{N-2} + \cdots + a_N$$

and an arbitrary tentative quadratic factor $x^2 - u_0 x - v_0$, determine a sequence $\{x^2 - u_k x - v_k\}$ of quadratic factors as follows: For each $k = 0, 1, 2, \ldots$ determine the sequence $\{b_n\} = \{b_n^{(k)}\}$ from $b_{-2} = b_{-1} = 0$,

$$b_n = a_n + u_k b_{n-1} + v_k b_{n-2}, \qquad n = 0, 1, \ldots, N$$

and the sequence $\{c_n\} = \{c_n^{(k)}\}$ from $c_{-2} = c_{-1} = 0$,

$$c_n = b_n + u_k c_{n-1} + v_k c_{n-2}, \qquad n = 0, 1, \ldots, N - 1.$$

Then set $u_{k+1} = u_k + \delta$, $v_{k+1} = v_k + \varepsilon$, where δ and ε are given by (5-28).

Schematically the algorithm can be described as follows:

$$
\begin{array}{ccc}
a_0 & b_0 & c_0 \\
a_1 & v \begin{pmatrix} b_1 \\ \end{pmatrix} u & v \begin{pmatrix} c_1 \\ \end{pmatrix} u \\
a_2 \longrightarrow b_2 \longrightarrow c_2 \\
\vdots & \vdots & \vdots \\
a_{N-3} & b_{N-3} & c_{N-3} \\
a_{N-2} & b_{N-2} & c_{N-2} \\
a_{N-1} & \mathbf{b_{N-1}} & c_{N-1} \\
a_N & \mathbf{b_N} \\
\end{array}
$$

Scheme 5.8

The coefficients that are needed in formula (5-28) are printed in bold face.

EXAMPLE

4. To determine a quadratic factor of

$$p(x) = x^3 - x^2 + 2x + 5$$

near $x^2 - 2x + 5$. Algorithm 5.7 yields for $u_0 = 2$, $v_0 = -5$ (watch the signs!)

n	a_n	b_n	c_n	
0	1	1	1	$3\delta + \varepsilon = 1$
1	−1	1	3	$0\delta + 3\varepsilon = 2$
2	2	−1	0	$\delta = 0.11111111$
3	5	−2		$\varepsilon = 0.66666667$

Continuing with $u_1 = u_0 + \delta = 2.11111111$, $v_1 = v_0 + \varepsilon = -4.33333333$, we find

n	a_n	b_n	c_n	
0	1	1	1	$3.22222222\delta + \varepsilon$ = −0.01234568
1	−1	1.11111111	3.22222222	$2.48148148\delta + 3.22222222\varepsilon$ = −0.2112483
2	2	0.01234568	2.48148148	$\delta = 0.02170139$
3	5	0.21124830		$\varepsilon = 0.08227238$

yielding $u_2 = 2.1328125$, $v_2 = -4.41560571$. Proceeding in a similar manner, we find

$$
\begin{array}{ll}
u_3 = 2.13249371, & v_3 = -4.41503560, \\
u_4 = 2.13249369, & v_4 = -4.41503564.
\end{array}
$$

Calculating the b_n with the last values, we obtain

n	a_n	b_n
0	1	1.00000000
1	-1	1.13249369
2	2	0.00000000
3	5	0.00000000

indicating that convergence has been achieved. The two complex zeros are thus found to be

$$z_{1,2} = \frac{u}{2} \pm i\sqrt{-\left(\frac{u}{2}\right)^2 - v}$$

$$= 1.06624685 \pm i1.81056712.$$

Problems

10. Determine all zeros of the polynomial

$$p(x) = x^4 - 8x^3 + 39x^2 - 62x + 51$$

using the fact that

$$p(x) = (x^2 - 2x + 2)(x^2 - 6x + 25) + 1.$$

11. Determine quadratic factors of the polynomial

$$p(x) = 3x^6 + 9x^5 + 9x^4 + 5x^3 + 3x^2 + 8x + 5$$

near $x^2 + 1.541451x + 1.487398$ and $x^2 - 1.127178x + 0.831492$. Also determine the real zeros near -1.86 and -0.72.

5.8   Convergence of Bairstow's Method

In order to justify the application of Newton's method to the problem of solving (5-25) we have to show, according to §5.5, that the Jacobian determinant

$$(5\text{-}29) \qquad D(u, v) = \begin{vmatrix} \dfrac{\partial b_{N-1}}{\partial u}(u, v) & \dfrac{\partial b_{N-1}}{\partial v}(u, v) \\ \dfrac{\partial b_N}{\partial u}(u, v) & \dfrac{\partial b_N}{\partial v}(u, v) \end{vmatrix}$$

is different from zero for $(u, v) = (s, t)$, where $x^2 - sx - t$ is a quadratic factor of the polynomial p. One condition under which this is the case is as follows.

Theorem 5.8   Let $x^2 - sx - t$ be a quadratic factor of the polynomial p, and let its zeros z_1, z_2 be two distinct, simple zeros of p.   Then $D(s, t) \neq 0$.

Proof.   We differentiate the identity (5-23) with respect to both u and v. Since p does not depend on u or v, the result is

$$0 = -xq(x) + (x^2 - ux - v)\frac{\partial q(x)}{\partial u} + \frac{\partial b_{N-1}}{\partial u}(x - u) - b_{N-1} + \frac{\partial b_N}{\partial u},$$

$$0 = -q(x) + (x^2 - ux - v)\frac{\partial q(x)}{\partial v} + \frac{\partial b_{N-1}}{\partial v}(x - u) + \frac{\partial b_N}{\partial v}.$$

Setting here $u = s$, $v = t$, $x = z_k$ $(k = 1, 2)$ we obtain by virtue of $z_k^2 - sz_k - t = 0$ $(k = 1, 2)$ the four relations

(5-30) $\dfrac{\partial b_{N-1}}{\partial u}(z_k - s) + \dfrac{\partial b_N}{\partial u} = z_k q_k$ $(k = 1, 2)$,

(5-31) $\dfrac{\partial b_{N-1}}{\partial v}(z_k - s) + \dfrac{\partial b_N}{\partial v} = q_k$ $(k = 1, 2)$,

where $q_k = q(z_k)$, and where the derivatives are taken at $(u, v) = (s, t)$. The two relations (5-30) can be regarded as a system of two linear equations for the two unknowns $\partial b_{N-1}/\partial u$ and $\partial b_N/\partial u$ with the nonvanishing determinant $z_1 - z_2$.   We thus have

$$\frac{\partial b_{N-1}}{\partial u} = \frac{z_1 q_1 - z_2 q_2}{z_1 - z_2}, \qquad \frac{\partial b_N}{\partial u} = \frac{z_1 z_2(q_2 - q_1) + s(z_1 q_1 - z_2 q_2)}{z_1 - z_2}.$$

Solving the system (5-31) in a similar manner, we find

$$\frac{\partial b_{N-1}}{\partial v} = \frac{q_1 - q_2}{z_1 - z_2}, \qquad \frac{\partial b_N}{\partial v} = \frac{z_1 q_2 - z_2 q_1 + s(q_1 - q_2)}{z_1 - z_2}.$$

Some algebraic manipulation now yields

$$D(s, t) = \begin{vmatrix} \dfrac{\partial b_{N-1}}{\partial u} & \dfrac{\partial b_{N-1}}{\partial v} \\[2mm] \dfrac{\partial b_N}{\partial u} & \dfrac{\partial b_N}{\partial v} \end{vmatrix} = q_1 q_2 = q(z_1)q(z_2).$$

Since z_1 and z_2 are both simple zeros of p, we have $q(z_k) \neq 0$, $k = 1, 2$, and the conclusion of the theorem follows.

The theory given in §5.5 now shows that under the hypotheses of theorem 5.8, algorithm 5.7 actually defines a sequence of quadratic polynomials $\{x^2 - u_k x - v_k\}$ which converges to $x^2 - sx - t$ whenever the initial quadratic $x^2 - u_0 x - v_0$ is sufficiently close to $x^2 - sx - t$.

For most polynomials even crude approximations to quadratic factors are hard to obtain by mere inspection.   In chapter 8 we shall discuss a method that will automatically produce good first approximations to quadratic factors (with complex zeros) of almost any real polynomial.

5.9   Steffensen's Iteration for Systems†

We now return to ordinary iteration as applied to systems of equations (algorithm 5.2).   Our goal is to extend the Aitken-Steffensen formula (4-11) to systems of equations.   As explained in §4.4, the rationale behind that formula consisted in neglecting the small term ε_n in (4-8).   Proceeding in the same vein, we now assume heuristically that the asymptotic error formula (5-17) is true without the term denoted by $0(\|d_n\|^2)$.

Denoting by $\{x_n\}$ a sequence of vectors generated by algorithm 5.2 and by s a solution of $x = f(x)$, our assumption implies that

$$(5\text{-}32) \qquad\qquad x_{n+1} - s = J(x_n - s),$$

for $n = 0, 1, 2, \ldots$, where $J = J(s)$ denotes the Jacobian matrix of the function f taken at the solution s.   The problem is to determine s from several consecutive iterates $x_n, x_{n+1}, x_{n+2}, \ldots$, notwithstanding the fact that J is unknown.

Subtracting two consecutive equations (5-32) from each other, we find, using the symbol Δ to denote forward differences,

$$(5\text{-}33) \qquad\qquad \Delta x_{n+1} = J\,\Delta x_n, \qquad n = 0, 1, 2, \ldots.$$

We define X_n to be the matrix‡ with the columns x_n and x_{n+1},

$$X_n = (x_n, x_{n+1}) = \begin{pmatrix} x_n & x_{n+1} \\ y_n & y_{n+1} \end{pmatrix}.$$

Defining ΔX_n in the obvious way, (5-33) shows that

$$J\,\Delta X_n = \Delta X_{n+1}, \qquad n = 0, 1, 2, \ldots.$$

If the matrix ΔX_n is nonsingular, we can solve for J, finding

$$(5\text{-}34) \qquad\qquad J = \Delta X_{n+1}(\Delta X_n)^{-1}.$$

We now solve (5-32) for s.   Assuming that $I - J$ is nonsingular (I = unit matrix), we get

$$(I - J)s = x_{n+1} - Jx_n$$
$$= (I - J)x_n + \Delta x_n$$

---

† This section may be omitted without loss of continuity.

‡ Here we use implicitly the fact that the vectors considered have two components. In the case of N components we would have to put

$$X_n = (x_n, x_{n+1}, \ldots, x_{n+N-1}).$$

and hence

$$\mathbf{s} = \mathbf{x}_n + (\mathbf{I} - \mathbf{J})^{-1}\varDelta\mathbf{x}_n.$$

Using equation (5-34) and applying the matrix identity $(\mathbf{AB})^{-1} = \mathbf{B}^{-1}\mathbf{A}^{-1}$ we have, always proceeding in a purely formal manner,

$$\begin{aligned}
(\mathbf{I} - \mathbf{J})^{-1} &= (\mathbf{I} - \varDelta\mathbf{X}_{n+1}(\varDelta\mathbf{X}_n)^{-1})^{-1} \\
&= ((\varDelta\mathbf{X}_n - \varDelta\mathbf{X}_{n+1})(\varDelta\mathbf{X}_n)^{-1})^{-1} \\
&= -\varDelta\mathbf{X}_n(\varDelta^2\mathbf{X}_n)^{-1}
\end{aligned}$$

and hence finally

$$\mathbf{s} = \mathbf{x}_n - \varDelta\mathbf{X}_n(\varDelta^2\mathbf{X}_n)^{-1}\varDelta\mathbf{x}_n.$$

This formula for the exact solution has been derived under the assumption that (5-17) is true without the $0(\|\mathbf{d}_n\|^2)$ term. If this term is present, we still may hope that the vector

$$(5\text{-}35) \qquad\qquad \mathbf{x}'_n = \mathbf{x}_n - \varDelta\mathbf{X}_n(\varDelta^2\mathbf{X}_n)^{-1}\varDelta\mathbf{x}_n$$

is closer to $\mathbf{s}$ than $\mathbf{x}_n$, provided that the matrix $\varDelta^2\mathbf{X}_n$ is nonsingular. It will be noted that the formula (5-35) is built like (4-11), with the difference that certain scalars are now replaced by appropriately defined vectors and matrices.

Formula (5-35) can be used in either of two ways. Either we can apply it to a sequence of vectors $\{\mathbf{x}_n\}$ already constructed to obtain a sequence $\{\mathbf{x}'_n\}$ which presumably converges to $\mathbf{s}$ faster. More effectively, the formula can be used as in algorithm 4.12, as follows.

Algorithm 5.9 Choose a vector $\mathbf{x}^{(0)}$, and construct the sequence of vectors $\{\mathbf{x}^{(k)}\}$ as follows: For each $k = 0, 1, 2, \ldots$, set $\mathbf{x}_0 = \mathbf{x}^{(k)}$, calculate $\mathbf{x}_1, \mathbf{x}_2, \mathbf{x}_3$, from

$$\mathbf{x}_{n+1} = \mathbf{f}(\mathbf{x}_n), \qquad n = 0, 1, 2,$$

and let $\mathbf{x}^{(k+1)} = \mathbf{x}'_0$, where $\mathbf{x}'_0$ is defined by (5-35).

This algorithm has not yet been fully investigated from the theoretical point of view. As it stands, it is not even fully defined, since there is no indication of what is to be done if the matrix $\varDelta^2\mathbf{X}_0$ is singular. It is therefore impossible to prove that the algorithm converges, let alone that it converges quadratically. Substantial experimental evidence, and also some theoretical considerations, seem to indicate, however, that the algorithm is indeed quadratically convergent in a large number of cases, even when ordinary iteration diverges.

EXAMPLE

5. To find the solution of the system

$$x = x^2 + y^2,$$
$$y = x^2 - y^2,$$

near (0.8, 0.4). Algorithm 5.9 yields the following values:

Table 5.9

k	$x^{(k)}$	$y^{(k)}$	n	x_n	y_n
0	0.8	0.4			
			1	0.8000000	0.4800000
			2	0.8070400	0.4096000
			3	0.9253683	0.5898240
1	0.7741243	0.4194303			
			1	0.7751902	0.4233468
			2	0.7801424	0.4216974
			3	0.7864508	0.4307934
2	0.7718671	0.4196500			
			1	0.7718850	0.4196728
			2	0.7719317	0.4196813
			3	0.7720109	0.4197462
3	0.7718445	0.4196434			
			1	0.7718445	0.4196434

The fact that $x_1 = x_0$ for $k = 3$ indicates that convergence has been accomplished. It is evident in this example that the sequences $\{x_n\}$ would not converge, as they begin to diverge for each k already for small values of n.

Problems

12. Assuming (5-32) to be exact, show that applying Aitken's formula (4-11) individually to each component of x does, in general, not produce the correct solution vector s.

13*. Assuming (5-32) to be exact, show that the matrix $\Delta^2 X_n$ is singular if and only if at least one of the following situations obtains:

   (*i*) $x_n = s$;
   (*ii*) $x_n - s$ is an eigenvector of J;
   (*iii*) the matrix J is similar to a matrix cI, where c is real.

   Show that in the cases (*ii*) and (*iii*) the procedure described in problem 12 is effective.

14. Find the solution of the system considered in example 2 by means of algorithm 5.9.

Recommended Reading

The abstract background of the method of iteration is discussed in most textbooks on functional analysis; see, for example, Liusternik and Sobolev [1961], p. 27. A beautiful discussion of Newton's method for systems of equations is given by Kantorovich [1948] (see also Henrici [1962], pp. 367–371).

Research Problems

1. Formulate Newton's method for systems of equations with an arbitrary number of unknowns.

2. Generalize Bairstow's method to extract from a polynomial of degree N a factor of arbitrary degree $n < N$.

3. Discuss Newton's method in the "singular" case when the determinant (5-19) is zero at the solution (s, t).

4. Discuss experimentally the stability of algorithm 5.9 when the matrix $\Delta^2 \mathbf{X}_n$ is nearly singular.

5. Develop a theory of iteration for systems if the vectors $\mathbf{x}_n$ are formed according to

$$x_{n+1} = f(x_n, y_n),$$
$$y_{n+1} = g(x_{n+1}, y_n)$$

(i.e., the most recent information is used at each step). What is the correct formulation of the Δ^2 procedure in this case?

chapter 6 linear difference equations

Linear difference equations were first encountered in §3.3, where the main topic was the study of linear difference equations of the first order. In the present chapter, we shall consider linear difference equations of arbitrary order. The theory of such difference equations is required for the understanding of some important algorithms to be discussed in the chapters 7 and 8. Furthermore, linear difference equations are a useful tool in the study of many other processes of numerical analysis; see in particular the chapters 14 and 16. Much unnecessary complication is avoided by considering difference equations with *complex* coefficients and solutions.

6.1 Notation

We recall that a linear difference equation of order N has the form

$$(6\text{-}1) \qquad a_{0,n}x_n + a_{1,n}x_{n-1} + \cdots + a_{N,n}x_{n-N} = b_n.$$

Here $\{a_{0,n}\}, \{a_{1,n}\}, \ldots, \{a_{N,n}\}$ and $\{b_n\}$ are given sequences, and $\{x_n\}$ is a sequence to be determined. The difference equation (6-1) is called *homogeneous* if $\{b_n\}$ is the zero sequence, i.e., if all its elements are zero.

Although much of the subsequent theory can easily be extended to the general case, we shall be concerned exclusively with the case of linear difference equations with constant coefficients. In this case, $a_{k,n} = a_k$ for all values of n and for certain (real or complex) constants $a_0, a_1, \ldots, a_N$. (It is not required that $b_n = b$, however.) Without loss of generality we may assume that $a_0 \neq 0$, $a_N \neq 0$, for otherwise the order of the difference equation could be reduced. Dividing through by a_0 and renaming the constants and the elements of the sequence $\{b_n\}$, the linear difference equation with constant coefficients appears in the form

$$(6\text{-}2) \qquad x_n + a_1 x_{n-1} + a_2 x_{n-2} + \cdots + a_N x_{n-N} = b_n,$$

where $a_N \neq 0$.

119

We can aid the understanding by two notational simplifications.   First, we shall no longer refer to a sequence by explicitly exhibiting its elements, as in the symbol $\{x_n\}$, but instead denote a sequence by a single capital letter.   The nth element of a sequence X will be denoted by x_n, or sometimes also by $(X)_n$.   Secondly, if $X = \{x_n\}$ is any sequence, we shall denote by $\mathscr{L}X$ the sequence whose nth element is given by

$$(\mathscr{L}X)_n = x_n + a_1 x_{n-1} + a_2 x_{n-2} + \cdots + a_N x_{n-N}.$$

With these notations, the problem of solving the difference equation (6-2) is the same as the problem of finding a sequence X such that

(6-3) $$\mathscr{L}X = B,$$

where B denotes the sequence of the nonhomogeneous terms b_n.

We note that the operator $\mathscr{L}$ defined above is a *linear* operator.   Defining the product aX of a scalar a and of a sequence X by

$$(aX)_n = a(X)_n$$

and the sum of two sequences X and Y by

$$(X + Y)_n = (X)_n + (Y)_n,$$

we have for arbitrary scalars a and b and for any two sequences X and Y

(6-4) $$\mathscr{L}(aX + bY) = a\mathscr{L}X + b\mathscr{L}Y.$$

The comprehension of the above notation may be helped by considering analogous simplifications in the theory of functions.   Writing X in place of $\{x_n\}$ is much like writing f in place of $f(x)$ to denote a function.   The operator $\mathscr{L}$ plays a role similar to that, say, of the differentiation operator D, which associates with a function f its derivative $Df = f'$.   Relation (6-4) is the analog of the familiar fact that differentiation is a linear operation, i.e., that $D(af + bg) = aDf + bDg$.

6.2   Particular Solutions of the Homogeneous Equation of Order Two

We shall consider in some detail the case $N = 2$.   Here we have, for some constants a_1 and $a_2 \neq 0$,

(6-5) $$(\mathscr{L}X)_n = x_n + a_1 x_{n-1} + a_2 x_{n-2}.$$

Our first task is to find solutions of the homogeneous equation $\mathscr{L}X = 0$.†

---

† Here the symbol 0 does not denote the scalar zero, but rather the sequence whose elements are all zero.   Since no misunderstanding is possible, we shall not attempt to make a graphical distinction between the two concepts.

Let us take a look at the corresponding problem in the theory of differential equations. The linear homogeneous differential equation of order 2 with constant coefficients,

$$x'' + a_1 x' + a_2 x = 0 \qquad \left(x' = \frac{dx}{dt} \right)$$

has, for suitably chosen r, solutions of the form e^{rt}. Is the same true also for the difference equation

(6-6) $$x_n + a_1 x_{n-1} + a_2 x_{n-2} = 0 ?$$

Replacing t by the discrete variable n, we are tempted to seek solutions of the form $x_n = e^{rn} = (e^r)^n$, or, putting $e^r = z$, of the form $x_n = z^n$. With this definition of $X = \{x_n\}$ we have

$$\begin{aligned}(\mathscr{L}X)_n &= z^n + a_1 z^{n-1} + a_2 z^{n-2} \\ &= z^{n-2}(z^2 + a_1 z + a_2).\end{aligned}$$

This expression is 0 not only if $z = 0$—which would yield the *trivial* solution of the difference equation—but also if z is a zero of the polynomial p defined by

(6-7) $$p(z) = z^2 + a_1 z + a_2.$$

This polynomial is called the *characteristic polynomial* of the difference equation (6-6). We know from the theory of equations that p has exactly two zeros, z_1 and z_2, which may be real or complex, and which may coincide. If $z_1 \neq z_2$, then the sequences with elements z_1^n and z_2^n represent two distinct, nonzero solutions of $\mathscr{L}X = 0$.

We can also find two distinct solutions if $z_1 = z_2$. We recall that if z_1 is a zero of multiplicity > 1 of p, then by theorem 2.6 not only $p(z_1) = 0$ but also $p'(z_1) = 0$. This suggests finding a second solution by differentiation. If $(X)_n = z^n$, we have identically in z

$$\begin{aligned}(\mathscr{L}X)_n &= z^n + a_1 z^{n-1} + a_2 z^{n-2} \\ &= z^{n-2} p(z).\end{aligned}$$

Differentiating with respect to z, we get

$$nz^{n-1} + a_1(n-1)z^{n-2} + a_2(n-2)z^{n-3}$$
$$= (n-2)z^{n-3}p(z) + z^{n-2}p'(z).$$

For $z = z_1$ the expression on the right is zero for all n, showing that the sequence with nth element nz_1^{n-1} is also a solution. We thus have obtained:

Theorem 6.2 Let a_1 and $a_2 \neq 0$ be constants, and consider the difference equation $\mathscr{L}X = 0$, where $\mathscr{L}$ is defined by (6-5). If z_1, z_2

are two distinct zeros of the characteristic polynomial, then the two sequences $X^{(1)}$, $X^{(2)}$ defined by

$$(X^{(1)})_n = z_1^n, \qquad (X^{(2)})_n = z_2^n$$

are solutions of $\mathscr{L} X = 0$. If z_1 is a zero of multiplicity 2, then the sequences defined by

$$(X^{(1)})_n = z_1^n, \qquad (X^{(2)})_n = n z_1^{n-1}$$

are solutions.

EXAMPLES

1. We consider the difference equation $x_n = x_{n-1} + x_{n-2}$. The characteristic polynomial $p(z) = z^2 - z - 1$ has the two zeros

$$z_1 = \frac{1 + \sqrt{5}}{2}, \qquad z_2 = \frac{1 - \sqrt{5}}{2},$$

thus two solutions are given by

$$(X^{(1)})_n = \left(\frac{1 + \sqrt{5}}{2}\right)^n, \qquad (X^{(2)})_n = \left(\frac{1 - \sqrt{5}}{2}\right)^n.$$

2. Consider $x_n - 2x_{n-1} + x_{n-2} = 0$. The characteristic polynomial is $p(z) = z^2 - 2z + 1$; $z_1 = 1$ is a double zero. Thus we find the two solutions

$$(X^{(1)})_n = 1^n = 1, \qquad (X^{(2)})_n = n1^{n-1} = n,$$

as can be verified directly.

6.3 The General Solution

Throughout this section, $\mathscr{L}$ will denote the difference operator defined by (6-5), where $a_2 \neq 0$. The first tool is the following:

Lemma 6.3a If $X^{(1)}$ and $X^{(2)}$ are any two solutions of $\mathscr{L} X = 0$, and if c_1 and c_2 are any two constants, then the sequence

$$X = c_1 X^{(1)} + c_2 X^{(2)}$$

is also a solution.

Proof. By the linearity of the operator $\mathscr{L}$,

$$\mathscr{L} X = \mathscr{L}(c_1 X^{(1)} + c_2 X^{(2)})$$
$$= c_1 \mathscr{L} X^{(1)} + c_2 \mathscr{L} X^{(2)} = 0.$$

As an exercise in the application of lemma 6.3a, we consider the problem of finding real solutions of real difference equations.   Let $\mathscr{L}$ be defined as above, where a_1 and a_2 are real.   The characteristic polynomial

$$p(z) = z^2 + a_1 z + a_2,$$

although a polynomial with real coefficients, may still have nonreal zeros z_1, z_2.   However, these zeros according to theorem 2.5d then are complex conjugate, i.e.,

$$z_2 = \bar{z}_1,$$

or, denoting by Re a and Im a the real and imaginary parts of a complex number a,

$$\text{Re } z_1 = \text{Re } z_2, \qquad \text{Im } z_2 = -\text{Im } z_1.$$

If the two zeros are nonreal, two distinct solutions are given by

$$(X^{(1)})_n = z_1^n, \qquad (X^{(2)})_n = (\bar{z}_1)^n.$$

Since the product of complex conjugate numbers is equal to the complex conjugate number of the product, we can also write

$$(X^{(2)})_n = (\overline{z_1^n}).$$

It now follows from lemma 6.3a that the two real sequences $Y^{(1)}$, $Y^{(2)}$ defined by

$$(Y^{(1)})_n = \tfrac{1}{2}[z_1^n + \overline{z_1^n}] = \text{Re } z_1^n,$$

$$(Y^{(2)})_n = \frac{1}{2i} [z_1^n - \overline{z_1^n}] = \text{Im } z_1^n,$$

are likewise solutions of the difference equation.

EXAMPLE
3.   Let $-1 < t < 1$, and consider the difference equation

(6-8) $$x_n - 2tx_{n-1} + x_{n-2} = 0.$$

The characteristic polynomial $p(z) = z^2 - 2tz + 1$ has the zeros

$$z_1 = t + i\sqrt{1 - t^2}, \qquad z_2 = t - i\sqrt{1 - t^2}.$$

If we define the angle φ by the condition

$$\cos \varphi = t \qquad (0 < \varphi < \pi)$$

then the zeros appear in the form

$$z_1 = \cos \varphi + i \sin \varphi = e^{i\varphi},$$
$$z_2 = \cos \varphi - i \sin \varphi = e^{-i\varphi}.$$

Consequently, the two solutions given by theorem 6.2 appear in the form

$$(X^{(1)})_n = (e^{i\varphi})^n = e^{in\varphi},$$
$$(X^{(2)})_n = (e^{-i\varphi})^n = e^{-in\varphi},$$

and are complex. However, in view of $e^{in\varphi} = \cos n\varphi + i \sin n\varphi$, also the sequences defined by

$$(Y^{(1)})_n = \cos n\varphi, \qquad (Y^{(2)})_n = \sin n\varphi$$

are solutions. This may be verified directly by using the addition theorems of the trigonometric functions.

It is readily seen that the elements of the sequence $Y^{(1)}$ are polynomials of degree n in the variable $t = \cos \varphi$. In fact, $(Y^{(1)})_0 = 1$, $(Y^{(1)})_1 = t$, and (6-8) shows that if this property holds for the integers $n - 2$ and $n - 1$, it holds for the integer n. Conventionally, these polynomials are denoted by $T_n(t)$ and are called Chebyshev polynomials of degree n. By the above we can also write

$$T_n(t) = \cos (n \text{ arc cos } t).$$

The Chebyshev polynomials are important in many branches of numerical analysis (see §9.4).

We return to the problem of finding *all* solutions of $\mathscr{L}X = 0$. Our second tool is the following simple observation.

Lemma 6.3b Let I be a set of consecutive integers, and let the integers m and $m - 1$ be in I. Let the sequence $B = \{b_n\}$ be defined on I. Then the difference equation $\mathscr{L}X = B$ has precisely one solution which assumes given values for $n = m$ and $n = m - 1$.

Proof. Let $X^{(1)}$ and $X^{(2)}$ be two solutions having the same values at $n = m$ and $n = m - 1$. The sequence $D = X^{(1)} - X^{(2)} = \{d_n\}$ then is a solution of $\mathscr{L}X = 0$ assuming the values 0 for $n = m$ and $n = m - 1$. If $X^{(1)}$ and $X^{(2)}$ are not identical, then $d_n \neq 0$ for some $n > m$ or some $n < m - 1$. To fix ideas, let $d_n \neq 0$ for some $n > m$. If we denote by n the smallest integer for which this is the case, then, because D is a solution,

$$d_n + a_1 d_{n-1} + a_2 d_{n-2} = 0.$$

However, since $d_{n-1} = d_{n-2} = 0$, $d_n \neq 0$, this equation reveals a contradiction. The case where some $d_n \neq 0$ with $n < m - 1$ is dealt with similarly, making use of the fact that $a_2 \neq 0$. It follows that D is the zero sequence, and hence that the sequences $X^{(1)}$ and $X^{(2)}$ are identical.

The zero sequence is a solution of $\mathscr{L}X = 0$ with the property that its elements are zero for any two consecutive values of n. Lemma 6.3b shows that this property characterizes the zero solution. In other words: *If a solution of $\mathscr{L}X = 0$ vanishes at two consecutive integers, it vanishes identically.*

Let now $X^{(1)} = \{x_n^{(1)}\}$ and $X^{(2)} = \{x_n^{(2)}\}$ be two known solutions of $\mathscr{L}X = 0$, defined for $-\infty < n < \infty$, and let $X = \{x_n\}$ be an arbitrary third solution. Is it possible to find constants c_1 and c_2 such that

$$X = c_1 X^{(1)} + c_2 X^{(2)}?$$

If that is the case, then we are entitled to call the sequence $c_1 X^{(1)} + c_2 X^{(2)}$ the *general solution* of $\mathscr{L}X = 0$, for any special solution can be obtained by assigning special values to the constants c_1 and c_2.

By lemma 6.3a, the sequence $c_1 X^{(1)} + c_2 X^{(2)}$ is in any case a solution of $\mathscr{L}X = 0$. By lemma 6.3b, this solution is identical with X if it agrees with X at two consecutive integers, n and $n - 1$, say. In order to obtain this agreement, we must be able to determine the constants c_1 and c_2 such that the two equations

(6-9)
$$\begin{cases} c_1 x_n^{(1)} + c_2 x_n^{(2)} = x_n, \\ c_1 x_{n-1}^{(1)} + c_2 x_{n-1}^{(2)} = x_{n-1}, \end{cases}$$

are satisfied. By the theory of linear equations, this system has a solution for arbitrary x_n and x_{n-1} if its determinant

$$w_n = \begin{vmatrix} x_n^{(1)} & x_n^{(2)} \\ x_{n-1}^{(1)} & x_{n-1}^{(2)} \end{vmatrix}$$

is different from zero.

The determinant w_n is called the Wronskian determinant of the sequences $X^{(1)}$ and $X^{(2)}$ at the point n. We have obtained

Theorem 6.3 Let $X^{(1)}$ and $X^{(2)}$ be two particular solutions of $\mathscr{L}X = 0$. Then every solution of $\mathscr{L}X = 0$ can be expressed in the form $c_1 X^{(1)} + c_2 X^{(2)}$ if and only if the Wronskian determinant w_n of $X^{(1)}$ and $X^{(2)}$ is different from zero for all values of n.

The requirement that $w_n \neq 0$ for *all* n is less stringent than it appears, as will be seen presently.

EXAMPLES

4. Let the two zeros z_1 and z_2 of the characteristic polynomial p of $\mathscr{L}X = 0$ be different. We calculate the Wronskian of the two solutions

$$x_n^{(1)} = z_1^n, \qquad x_n^{(2)} = z_2^n$$

given by theorem 6.2. We find

$$w_n = \begin{vmatrix} z_1^n & z_2^n \\ z_1^{n-1} & z_2^{n-1} \end{vmatrix} = (z_1 z_2)^{n-1}(z_1 - z_2).$$

Evidently $w_n \neq 0$ for all n.   Hence the general solution of the difference equation is given by $c_1 z_1^n + c_2 z_2^n$.

5. If z_1 is a zero of multiplicity 2 of the characteristic polynomial, two solutions of the difference equation are given by

$$x_n^{(1)} = z_1^n, \qquad x_n^{(2)} = n z_1^{n-1}.$$

We find

$$w_n = \begin{vmatrix} z_1^n & n z_1^{n-1} \\ z_1^{n-1} & (n-1) z_1^{n-2} \end{vmatrix} = -z_1^{2n-2} \neq 0;$$

hence

$$x_n = c_1 z_1^n + c_2 n z_1^{n-1}$$

is the general solution.

6. As a somewhat more special example, we consider the real solutions of the difference equation

$$x_n - 2t x_{n-1} + x_{n-2} = 0 \qquad (-1 < t < 1)$$

found in example 3.   If $t = \cos \varphi$, we were able to express these solutions in the form $x_n^{(1)} = \cos n\varphi$, $x_n^{(2)} = \sin n\varphi$.   For these solutions,

$$w_n = \begin{vmatrix} \cos n\varphi & \sin n\varphi \\ \cos (n-1)\varphi & \sin (n-1)\varphi \end{vmatrix}$$
$$= \cos n\varphi \sin (n-1)\varphi - \cos (n-1)\varphi \sin n\varphi$$
$$= -\sin \varphi \neq 0.$$

Thus the general solution can be expressed in the form

$$c_1 \cos n\varphi + c_2 \sin n\varphi.$$

(Note, however, the restriction on t.)

We are now in a position to solve arbitrary initial value problems for the linear difference equation $\mathscr{L} X = 0$.   All we have to do is to find two particular solutions $X^{(1)}$, $X^{(2)}$ with nonvanishing Wronskian and to determine the constants c_1 and c_2 such that the sequence $X = c_1 X^{(1)} + c_2 X^{(2)}$ satisfies the given initial conditions.

EXAMPLE

7. Let us find the solution of

$$x_n = x_{n-1} + x_{n-2}$$

satisfying the conditions $x_{-1} = 0$, $x_0 = 1$.   By the examples 1 and 4, the general solution of the difference equation is

$$x_n = c_1 \left(\frac{1 + \sqrt{5}}{2} \right)^n + c_2 \left(\frac{1 - \sqrt{5}}{2} \right)^n.$$

The initial conditions yield the following conditions on c_1 and c_2:

$$c_1 + c_2 = 1,$$

$$c_1\left(\frac{1 + \sqrt{5}}{2}\right)^{-1} + c_2\left(\frac{1 - \sqrt{5}}{2}\right)^{-1} = 0.$$

We easily find the solution

$$c_1 = \frac{1 + \sqrt{5}}{2\sqrt{5}}, \qquad c_2 = -\frac{1 - \sqrt{5}}{2\sqrt{5}},$$

thus the solution of our initial value problem is given by the formula

$$x_n = \frac{1}{\sqrt{5}}\left[\left(\frac{1 + \sqrt{5}}{2}\right)^{n+1} - \left(\frac{1 - \sqrt{5}}{2}\right)^{n+1}\right].$$

It is not immediately evident from this representation that the numbers x_n all are integers. The sequence thus defined is called the Fibonacci sequence. It has many interesting number-theoretical properties.

Problems

1. Express in real terms a general solution of the following difference equations:

    (a) $\qquad\qquad\qquad x_n - x_{n-2} = 0$;
    (b) $\qquad\qquad\qquad x_n + x_{n-2} = 0$;
    (c) $\qquad\qquad x_n + 2x_{n-1} + x_{n-2} = 0.$

2. Find a general solution of the difference equation

$$x_n + 2ix_{n-1} + x_{n-2} = 0.$$

3. Let a, b be real, $a^2 - 4b < 0$. Show that the general solution of the difference equation

$$x_n + ax_{n-1} + bx_{n-2} = 0$$

can be expressed in the form

$$x_n = r^n(c_1 \cos n\varphi + c_2 \sin n\varphi),$$

where r and φ are real.

4. Let $X = \{x_n\}$ denote the Fibonacci sequence. Show that

$$\lim_{n \to \infty} \frac{x_{n+1}}{x_n} = \frac{1 + \sqrt{5}}{2}.$$

5*. Let b, c be real. Show that a necessary and sufficient condition in order that *all* solutions of the difference equation

$$x_n + 2bx_{n-1} + cx_{n-2} = 0$$

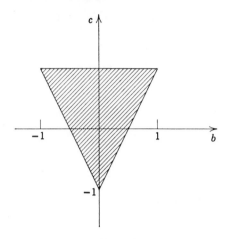

Figure 6.3

tend to zero for $n \to \infty$ is as follows: The point (b, c) lies in the interior of the triangular region of the (b, c) plane bounded by the straight lines

$$c = 1, \qquad 2b - 1 = c, \qquad -2b - 1 = c$$

(see Fig. 6.3). (Hint: Treat separately the two cases where the zeros of the characteristic polynomial are real or imaginary.)

6*. Formulate and prove results analogous to those of §6.3 for linear difference equations with variable coefficients.

7. Solve the following initial value problems for difference equations:

(a) $\qquad x_n = 3x_{n-1} - x_{n-2}, \qquad x_0 = 1, \qquad x_1 = 2;$
(b) $\qquad x_n - 2x_{n-1} + 2x_{n-2} = 0, \qquad x_0 = x_1 = 1.$

6.4   Linear Dependence

The results of §6.3 can be formulated somewhat more elegantly by introducing the concepts of linear dependence and linear independence of sequences. Two sequences $X^{(1)}$ and $X^{(2)}$ defined on a set I of integers (but not necessarily solutions of a difference equation such as $\mathscr{L} X = 0$) are called *linearly dependent* if there exist two constants c_1, c_2, not both zero, such that the sequence

$$X = c_1 X^{(1)} + c_2 X^{(2)}$$

is the zero sequence on I. If no such constants exist, the sequences are called *linearly independent*.

EXAMPLES

8. Let I be the set of integers from 1 to 20 and let

$$X^{(1)} = \{\underbrace{1, 1, \ldots, 1,}_{\text{10 elements}} \ \underbrace{0, 0, \ldots, 0\}}_{\text{10 elements}}$$

$$X^{(2)} = \{0, 0, \ldots, 0, \quad 1, 1, \ldots, 1\}$$

The sequences $X^{(1)}$ and $X^{(2)}$ are independent.   For assume that

$$c_1 x_n^{(1)} + c_2 x_n^{(2)} = 0$$

for $n = 1, \ldots, 20$.   For $1 \leq n \leq 10$ this implies $c_1 = 0$, for $11 \leq n \leq 20$ it implies $c_2 = 0$.   Thus $c_1 X^{(1)} + c_2 X^{(2)} = 0$ is possible only for $c_1 = c_2 = 0$.

9.   If one of the sequences $X^{(1)}$ and $X^{(2)}$ is the zero sequence, the two sequences are always linearly dependent.

To find out whether two given sequences are independent may be difficult in general.,   However, if the two sequences are both solutions of the same linear difference equation $\mathscr{L}X = 0$, their linear dependence or independence is closely connected with their Wronskian determinants.

Theorem 6.4   Let $X^{(1)}$ and $X^{(2)}$ be two solutions of $\mathscr{L}X = 0$, and let $W = \{w_n\}$ be the sequence of their Wronskian determinants.

(i) If $X^{(1)}$ and $X^{(2)}$ are linearly dependent, then W is the zero sequence.

(ii) If $w_m = 0$ for some integer m, then the two solutions are linearly dependent.

An immediate consequence of theorem 6.4 is the following

Corollary 6.4   If $w_m \neq 0$ for some integer m, then $w_n \neq 0$ for all n, and the solutions $X^{(1)}$ and $X^{(2)}$ are linearly independent.

Proof of theorem 6.4.   (i) If $X^{(1)}$ and $X^{(2)}$ are linearly dependent, then there exist, by definition, two constants c_1 and c_2, not both zero, so that

$$c_1 X^{(1)} + c_2 X^{(2)} = 0.$$

In particular, for an arbitrary integer n,

$$c_1 x_n^{(1)} + c_2 x_n^{(2)} = 0,$$
$$c_1 x_{n-1}^{(1)} + c_2 x_{n-1}^{(2)} = 0.$$

This is a homogeneous linear system of two equations with the nontrivial solution (c_1, c_2).   Its determinant must therefore vanish, showing that $w_n = 0$, as required.

(ii) Let m be an integer such that $w_m = 0$.   The homogeneous system of two linear equations with two unknowns,

$$c_1 x_m^{(1)} + c_2 x_m^{(2)} = 0,$$
$$c_1 x_{m-1}^{(1)} + c_2 x_{m-1}^{(2)} = 0,$$

then has a nontrivial solution (c_1, c_2).   We define the sequence

$$X = c_1 X^{(1)} + c_2 X^{(2)}.$$

By lemma 6.3a, X is a solution of $\mathscr{L}X = 0$. By construction, this solution is zero at the two consecutive points $n = m$ and $n = m - 1$. Hence, by lemma 6.3b, X is the zero sequence. It follows that $X^{(1)}$ and $X^{(2)}$ are linearly dependent.

If A and B are any two sequences, and if c_1, c_2 are any two constants, the sequence $c_1A + c_2B$ is called a *linear combination* of the sequences A and B. Using the concept of linear dependence, the content of theorem 6.3 can now be phrased more elegantly as follows: *Every solution of $\mathscr{L}X = 0$ can be expressed as a linear combination of two fixed, linearly independent solutions.*

Problems

8. Show that if two sequences are linearly dependent, one is a constant multiple of the other.

9. Let $W = \{w_n\}$ be the sequence of Wronskian determinants of two solutions of $\mathscr{L}X = 0$. Show that W satisfies the first order difference equation

$$w_n = a_2 w_{n-1}.$$

Hence give an independent proof of the fact that either W is the zero sequence, or $w_n \neq 0$ for all n.

10. Let $X = \{x_n\}$ be a solution of $\mathscr{L}X = 0$ such that $x_n \neq 0$ for all n. Show that any solution $Y = \{y_n\}$ of the linear difference equation of order 1 with variable coefficients,

$$y_n = \frac{x_n}{x_{n-1}} y_{n-1} + a_2^n$$

satisfies $\mathscr{L}Y = 0$, and that the solutions X and Y are linearly independent.

11*. Extend the results of §6.4 to linear difference equations with variable coefficients.

6.5   The Non-homogeneous Equation of Order Two

We now shall discuss the solution of the equation $\mathscr{L}X = B$ or, more explicitly,

$$(6\text{-}10) \qquad x_n + a_1 x_{n-1} + a_2 x_{n-2} = b_n,$$

where the sequence $B = \{b_n\}$ is defined on some set I of consecutive integers. Much of the corresponding theory for differential equations again carries over. For instance we have the following result.

Theorem 6.5   Let Y be a special solution of $\mathscr{L}Y = B$, and let $X^{(1)}$ and $X^{(2)}$ be two linearly independent solutions of the corresponding homogeneous equation $\mathscr{L} X = 0$.   Then every solution X of $\mathscr{L} X = B$ can be expressed in the form

$$X = c_1 X^{(1)} + c_2 X^{(2)} + Y,$$

where c_1, c_2 are suitable constants.

Proof. By hypothesis, if $Y = \{y_n\}$

(6-11) $y_n + a_1 y_{n-1} + a_2 y_{n-2} = b_n$

for all $n \in I$. If $X = \{x_n\}$ is any solution of (6-10), we obtain by subtracting (6-11) from (6-10) and setting $D = \{d_n\} = X - Y$,

$$d_n + a_1 d_{n-1} + a_2 d_{n-2} = 0.$$

Thus, the sequence D is a solution of the homogeneous equation.   By theorem 6.3 it can be written in the form

$$D = c_1 X^{(1)} + c_2 X^{(2)}$$

for suitable c_1, c_2.   Since $X = Y + D$, the desired result follows.

Theorem 6.5 reduces the problem of finding the general solution of the nonhomogeneous equation to the problem of finding one particular solution of it.   As in the case of differential equations, such a particular solution can frequently be found by an inspired guess.   For instance, if the elements of B are bq^n, where b and $q \neq 0$ are constants, a special solution of $\mathscr{L} X = B$ can frequently be found in the form $x_n = aq^n$, where a is a constant to be determined.   Substituting into (6-10) yields

$$a(q^n + a_1 q^{n-1} + a_2 q^{n-2}) = bq^n,$$

or, denoting by p the characteristic polynomial of the homogeneous equation and dividing by q^n,

$$aq^{-2} p(q) = b.$$

If $p(q) \neq 0$, we find

$$a = \frac{q^2 b}{p(q)};$$

the method breaks down, however, if q is a zero of the characteristic polynomial.

Similarly, if the elements of B are polynomials in n, there often exist solutions X whose elements are polynomials of the same degree in n. These solutions can be found by the method of undetermined coefficients. Again the method may break down in exceptional cases.

EXAMPLE

10.   To find a particular solution of

$$x_n - x_{n-1} - x_{n-2} = -n^2.$$

We set $x_n = an^2 + bn + c$, with constants a, b, c to be determined. Substitution yields

$$a[n^2 - (n - 1)^2 - (n - 2)^2] + b[n - (n - 1) - (n - 2)]$$
$$+ c[1 - 1 - 1] = -n^2$$

or, after simplification,

$$a(-n^2 + 6n - 5) + b(-n + 3) + c(-1) = -n^2.$$

Comparing coefficients of like powers of n, we get

$$\begin{aligned} -a &= -1, \\ 6a - b &= 0, \\ -5a + 3b - c &= 0, \end{aligned}$$

yielding $a = 1$, $b = 6$, $c = 13$. Thus we have found the solution

$$x_n = n^2 + 6n + 13.$$

Problems

12. Find general solutions of the difference equations

(a) $x_n - 5x_{n-1} + 6x_{n-2} = 2$;
(b) $8x_n - 6x_{n-1} + x_{n-2} = 2^n$;
(c) $x_n - x_{n-1} - 2x_{n-2} = n^2$.

13. Find the solution of equation (a) above that satisfies the initial condition $x_0 = 1$, $x_1 = -1$.

14. Determine a general solution of the difference equation

$$2x_n - 3x_{n-1} - 2x_{n-2} + 3 = 0.$$

What relation must hold between the values x_0 and x_1 of a particular solution $\{x_n\}$ of the above equation in order that $|x_n| \leqq C$ for some constant C and for all $n > 0$?

15. Difference equations have remarkable applications in the theory of economics. Let y_n be the national income in the year n, a the marginal propensity to consume, and b the ratio of private investment to increase in consumption. Assuming that government expenditure is constant and equal to 1, a certain economic theory states that the following difference equation for y_n holds (see S. Goldberg [1958], p. 6):

$$y_n = ay_{n-1} + ab(y_{n-1} - y_{n-2}) + 1.$$

(a) Solve this equation with $a = 0.5$, $b = 1$ under the initial condition

$$y_{1946} = 2, \qquad y_{1947} = 3.$$

(b) How frequent are depressions in this economy?
(c) For what values of a, b is the economy thus described noninflationary?

6.6 Variation of Constants

The methods discussed in §6.5 for finding a particular solution of the nonhomogeneous equation are, at best, heuristic and furnish a solution only in special cases. With differential equations, the method of variation of constants permits us to find a solution of any nonhomogeneous equation if the general solution of the homogeneous equation is known (see Coddington [1962], p. 67). A similar formula exists for difference equations. For simplicity we shall consider the case in which the sequence B is defined on the set of nonnegative integers.

Theorem 6.6 Let $X^{(1)} = \{x_n^{(1)}\}$ and $X^{(2)} = \{x_n^{(2)}\}$ be two linearly independent solutions of the homogeneous equation $\mathscr{L}X = 0$, and let $W = \{w_n\}$ be the sequence of their Wronskian determinants. If $B = \{b_n\}$ is defined on the set of nonnegative integers, a solution X of $\mathscr{L}X = B$ is given by

$$(6\text{-}12) \qquad x_n = \sum_{m=0}^{n} \frac{\begin{vmatrix} x_n^{(1)} & x_n^{(2)} \\ x_{m-1}^{(1)} & x_{m-1}^{(2)} \end{vmatrix}}{w_m} b_m, \qquad n \geq 0.$$

Proof. Formula (6-12) can be verified by direct computation. A different proof, which shows how the formula is obtained, is as follows.

It is clear that, for a given difference operator $\mathscr{L}$, the elements x_n of the required solution are *linear* functions of the preceding elements of the sequence B. We may thus write

$$(6\text{-}13) \qquad x_n = \sum_{m=0}^{n} d_{n,m} b_m,$$

where the coefficients $d_{n,m}$ depend only on $\mathscr{L}$ and not on the particular sequence B. The requirement that the sequence $X = \{x_n\}$ defined by (6-13) satisfies $\mathscr{L}X = B$ leads to

$$x_n + a_1 x_{n-1} + a_2 x_{n-2} = \sum_{m=0}^{n} d_{n,m} b_m + a_1 \sum_{m=0}^{n-1} d_{n-1,m} b_m + a_2 \sum_{m=0}^{n-2} d_{n-2,m} b_m$$
$$= b_n,$$

or, collecting factors of b_m,

$$\sum_{m=0}^{n-2} (d_{n,m} + a_1 d_{n-1,m} + a_2 d_{n-2,m}) b_m$$
$$+ (d_{n,n-1} + a_1 d_{n-1,n-1}) b_{n-1} + d_{n,n} b_n = b_n.$$

This identity must hold for all $n \geq 0$, no matter how the sequence B is chosen. It follows that for each m, the coefficients of b_m on both sides must be equal. On the right, there is a nonzero coefficient only for $m = n$. This leads to the relations

(6-14a) $d_{n,m} + a_1 d_{n-1,m} + a_2 d_{n-2,m} = 0,$ $n > m + 1,$

(6-14b) $d_{n,n-1} + a_1 d_{n-1,n-1} = 0,$

(6-14c) $d_{n,n} = 1.$

Formula (6-14a) shows that for a fixed value of m, the sequence $D = \{d_{n,m}\}$ is a solution of $\mathscr{L}D = 0$ for $n > m + 1$. Equation (6-14c) yields a first initial condition $d_{m,m} = 1$. If we impose the further initial condition $d_{m-1,m} = 0$, then (6-14b) can be replaced by (6-14a), where $n \geq m + 1$. The sequence D is thus characterized as solving $\mathscr{L}D = 0$ for $n > m$ and satisfying the initial conditions

(6-15) $d_{m-1,m} = 0,$ $d_{m,m} = 1.$

By theorem 6.3 there must exist constants $c_m^{(1)}$, $c_m^{(2)}$, depending only on m, such that

$$d_{n,m} = c_m^{(1)} x_n^{(1)} + c_m^{(2)} x_n^{(2)}.$$

The initial conditions (6-15) are satisfied if

$$c_m^{(1)} x_m^{(1)} + c_m^{(2)} x_m^{(2)} = 1,$$
$$c_m^{(1)} x_{m-1}^{(1)} + c_m^{(2)} x_{m-1}^{(2)} = 0.$$

The solution of this system of linear equations is readily found to be

$$c_m^{(1)} = \frac{x_{m-1}^{(2)}}{w_m}, \qquad c_m^{(2)} = -\frac{x_{m-1}^{(1)}}{w_m}.$$

We thus find

(6-16) $d_{n,m} = \dfrac{x_n^{(1)} x_{m-1}^{(2)} - x_n^{(2)} x_{m-1}^{(1)}}{w_m}.$

Substituting this into (6-13), we obtain (6-12).

Although theorem 6.6 does provide a solution to the nonhomogeneous equation that works in all cases, there is of course no guarantee that the sum appearing in (6-13) can be expressed in any simple form. (Neither can the integrals appearing in the variation-of-constants formula for differential equations always be expressed in closed form.) But formula (6-13) can be effective in cases where heuristic methods fail.

EXAMPLE

11. To find a particular solution of

(6-17) $x_n - 2x_{n-1} + x_{n-2} = 1.$

By example 2, two linearly independent solutions of the homogeneous equation are given by

$$x_n^{(1)} = 1, \qquad x_n^{(2)} = n.$$

Their Wronskian determinants are, by example 5, $w_n = -1$. Formula (6-13) thus yields in view of $b_n = 1$ $(n = 0, 1, 2, \ldots)$

$$x_n = -\sum_{m=0}^{n} \begin{vmatrix} 1 & n \\ 1 & m-1 \end{vmatrix} = \sum_{m=0}^{n} (n + 1 - m).$$

Changing the index of summation from m to p, where $p = n + 1 - m$, we have

$$x_n = \sum_{p=1}^{n+1} p = 1 + 2 + 3 + \cdots + (n + 1).$$

By a well-known summation formula we obtain

$$x_n = \frac{(n + 1)(n + 2)}{2}.$$

We leave it to the reader to verify that this indeed is a solution of (6-17).

Problems

16. Let a, b be any two constants, $a \neq 0$, $b \neq 0$. Find a particular solution of the difference equation

$$x_n - (a + b)x_{n-1} + abx_{n-2} = a^n.$$

(Distinguish the cases $a = b$ and $a \neq b$.)
17*. Show that formula (6-13) is valid for linear difference equations with variable coefficients.

6.7 The Linear Difference Equation of Order N

The theory developed in the sections §6.3 through §6.6 for difference equations of order two carries over, without essential change, to equations of arbitrary order. We now consider the difference operator $\mathscr{L}$ of order N defined by

(6-18) $\qquad (\mathscr{L}X)_n = x_n + a_1 x_{n-1} + \cdots + a_N x_{n-N},$

where $a_1, a_2, \ldots, a_N$ are arbitrary (real or complex) constants, $a_N \neq 0$. We are interested in both the homogeneous equation

(6-19) $\qquad \mathscr{L}X = 0,$

where 0 denotes the null sequence, and the nonhomogeneous equation

(6-20) $\qquad \mathscr{L}X = B,$

where B denotes an arbitrary given sequence.

The following analogs of the results given above hold and are proved similarly.

Lemma 6.3a′   Any linear combination of two solutions of $\mathscr{L}X = 0$ is again a solution.

Lemma 6.3b′   Let I be a set of consecutive integers, let the integers $m, m - 1, \ldots, m - N + 1$ be in I, and let the sequence B be defined on I. Then the difference equation $\mathscr{L}X = B$ has precisely one solution which assumes given values for $n = m, m - 1, \ldots, m - N + 1$.

Again we have the corollary that if a solution of $\mathscr{L}X = 0$ vanishes at N consecutive integers, it vanishes identically.

Let now the sequences $X^{(1)} = \{x_n^{(1)}\}$, $X^{(2)} = \{x_n^{(2)}\}, \ldots, X^{(N)} = \{x_n^{(N)}\}$ be N solutions of $\mathscr{L}X = 0$. Their Wronskian determinant at the point n is defined by

$$W_n = \begin{vmatrix} x_n^{(1)} & x_n^{(2)} & \cdots & x_n^{(N)} \\ x_{n-1}^{(1)} & x_{n-1}^{(2)} & \cdots & x_{n-1}^{(N)} \\ \cdot & \cdot & \cdots & \cdot \\ x_{n-N+1}^{(1)} & x_{n-N+1}^{(2)} & \cdots & x_{n-N+1}^{(N)} \end{vmatrix}.$$

The following result is obtained as in §6.3:

Theorem 6.3′   Let $X^{(1)}, \ldots, X^{(N)}$ be N solutions of $\mathscr{L}X = 0$. Then every solution of $\mathscr{L}X = 0$ can be expressed as a linear combination of $X^{(1)}, \ldots, X^{(N)}$ if and only if the Wronskian determinant of these solutions is different from zero for all values of n.

N sequences $A^{(1)}, A^{(2)}, \ldots, A^{(N)}$ are called *linearly dependent* if there exist constants $c_1, c_2, \ldots, c_N$, not all zero, such that

$$c_1 A^{(1)} + c_2 A^{(2)} + \cdots + c_N A^{(N)}$$

is the zero sequence. If no such constants exist, the sequences are called linearly independent. As in §6.4 we can show:

Theorem 6.4′   Let $X^{(1)}, \ldots, X^{(N)}$ be N solutions of $\mathscr{L}X = 0$, and let $W = \{w_n\}$ be the sequence of their Wronskian determinants.
(*i*) If the solutions $X^{(1)}, \ldots, X^{(N)}$ are linearly dependent, then W is the zero sequence.
(*ii*) If W contains a zero element, then the solutions $X^{(1)}, \ldots, X^{(N)}$ are linearly dependent.

It follows that if $w_m \neq 0$ for some integer m, then $w_n \neq 0$ for all n. Thus the condition of theorem 6.3 is satisfied if the solutions $X^{(1)}, \ldots, X^{(N)}$ are linearly independent, and it follows that every solution of $\mathscr{L}X = 0$ can be expressed as a linear combination of a system of linearly independent solutions.

We now turn to the nonhomogeneous equation $\mathscr{L}X = B$.   As above, we have

Theorem 6.5′   Let Y be a special solution of $\mathscr{L}Y = B$, and let $X^{(1)}, \ldots, X^{(N)}$ be a system of linearly independent solutions of the corresponding homogeneous equation $\mathscr{L}X = 0$.   Then every solution X of $\mathscr{L}X = B$ can be expressed in the form

$$X = c_1 X^{(1)} + \cdots + c_N X^{(N)} + Y,$$

where $c_1, \ldots, c_N$ are suitable constants.

Particular solutions of the nonhomogeneous equation can frequently be found by guessing the general form of the solution and determining the parameters such that the equation is satisfied.   A more generally applicable method is given by

Theorem 6.6′   Let $X^{(1)} = \{x_n^{(1)}\}, \ldots, X^{(N)} = \{x^{(N)}\}$ be N linearly independent solutions of the homogeneous equation $\mathscr{L}X = 0$, and let $W = \{w_n\}$ be the sequence of their Wronskian determinants.   If $B = \{b_n\}$ is defined on the set of nonnegative integers, then a solution X of $\mathscr{L}X = B$ is given by

$$(6\text{-}21) \quad x_n = \sum_{m=0}^{n} \frac{\begin{vmatrix} x_n^{(1)} & x_n^{(2)} & \cdots & x_n^{(N)} \\ x_{m-1}^{(1)} & x_{m-1}^{(2)} & \cdots & x_{m-1}^{(N)} \\ \cdot & \cdot & \cdot & \cdot \\ x_{m-N+1}^{(1)} & x_{m-N+1}^{(2)} & \cdots & x_{m-N+1}^{(N)} \end{vmatrix}}{w_m} b_m.$$

Problem

18*. Let W denote the sequence of the Wronskian determinants of N solutions of $\mathscr{L}X = 0$.   Find a linear difference equation of order 1 satisfied by W.

6.8   A System of Linearly Independent Solutions

To complete the discussion of the linear difference equation of order N with constant coefficients, we need to determine a system of N linearly independent solutions of the homogeneous equation $\mathscr{L}X = 0$.   Special solutions of $\mathscr{L}X = 0$ can again be found by means of the characteristic polynomial

$$p(z) = z^N + a_1 z^{N-1} + \cdots + a_N$$

associated with the operator $\mathscr{L}$. If z is a zero of the characteristic polynomial, it is easy to see that the sequence $X = \{x_n\}$ defined by $x_n = z^n$ is a
solution of $\mathscr{L}X = 0$, for we have

$$
\begin{aligned}
(\mathscr{L}X)_n &= x_n + a_1 x_{n-1} + \cdots + a_N x_{n-N} \\
&= z^n + a_1 z^{n-1} + \cdots + a_N z^{n-N} \\
&= z^{n-N} p(z) \\
&= 0.
\end{aligned}
$$

We now distinguish two cases.

(*i*) The polynomial p has N distinct zeros $z_1, z_2, \ldots, z_N$. Then the
sequences $X^{(k)} = \{z_k^n\}$ $(k = 1, 2, \ldots, N)$ represent N solutions of $\mathscr{L}X = 0$.
Their Wronskian at $n = N - 1$ is

$$
\begin{vmatrix}
z_1^{N-1} & z_2^{N-1} & \cdots & z_N^{N-1} \\
z_1^{N-2} & z_2^{N-2} & \cdots & z_N^{N-2} \\
\cdot & \cdot & \cdots & \cdot \\
1 & 1 & \cdots & 1
\end{vmatrix}.
$$

This determinant is known to have the value

$$
\prod_{m<n} (z_m - z_n) \neq 0.
$$

It follows that the solutions found are linearly independent.

(*ii*) General case: The polynomial p has zeros of multiplicity > 1.
The above method does not furnish sufficiently many solutions. However, further solutions can be found by differentiation. Let z be a zero of
multiplicity $k + 1$ where $k > 0$. We then have

(6-22) $$p(z) = p'(z) = \cdots = p^{(k)}(z) = 0.$$

It was shown above that the sequence $X^{(0)} = \{z^n\}$ is a solution. We now
assert that also the sequences $X^{(1)}, X^{(2)}, \ldots, X^{(k)}$ defined by

$$
\begin{aligned}
x_n^{(1)} &= nz^{n-1}, \\
x_n^{(2)} &= n(n-1)z^{n-2}, \\
\cdot\ \cdot\ &\cdot\ \cdot\ \cdot\ \cdot \\
x_n^{(k)} &= n(n-1)\ldots(n-k+1)z^{n-k},
\end{aligned}
$$

are solutions. Indeed, if $0 \leq m \leq k$, then

$$
\begin{aligned}
(\mathscr{L}X^{(m)})_n &= x_n^{(m)} + a_1 x_{n-1}^{(m)} + \cdots + a_N x_{n-N}^{(m)} \\
&= n(n-1)\ldots(n-m+1)z^{n-m} \\
&\quad + a_1(n-1)(n-2)\ldots(n-m)z^{n-m-1} \\
&\quad + \cdots \\
&\quad + a_N(n-N)(n-N-1)\ldots(n-N-m+1)z^{n-N-m} \\
&= (z^{n-N}p(z))^{(m)}.
\end{aligned}
$$

By the Leibnitz rule for differentiating a product (see Kaplan [1953], p. 19),

$$(z^{n-N}p(z))^{(m)} = \binom{m}{0}(n - N)(n - N - 1)\ldots(n - N - m + 1)z^{n-N-m}p(z)$$

$$+ \binom{m}{1}(n - N)\ldots(n - N - m + 2)z^{n-N-m+1}p'(z)$$

$$+ \cdots$$

$$+ \binom{m}{m}z^{n-N}p^{(m)}(z),$$

which vanishes for all n, by (6-22).

We are thus led to the following analog of theorem 6.2:

Theorem 6.8 Let the characteristic polynomial of the Nth order equation $\mathscr{L}X = 0$ have the distinct zeros $z_1, z_2, \ldots, z_k$ ($k \leq N$), and let the multiplicity of z_i be $m_i + 1$ ($\sum m_i = N - k$). Then the sequences defined by

(6-23) $$x_n = n(n - 1)\ldots(n - m + 1)z_i^{n-m},$$
$$m = 0, 1, \ldots, m_i; \qquad i = 1, 2, \ldots, k$$

form a system of N linearly independent solutions of $\mathscr{L}X = 0$.

We omit the proof of the fact that the N solutions given by (6-23) are linearly independent.

EXAMPLES

12. Let

$$(\mathscr{L}X)_n = x_n - 2x_{n-2} + x_{n-4}.$$

The characteristic polynomial

$$p(z) = z^4 - 2z^2 + 1 = (z + 1)^2(z - 1)^2$$

has the two distinct zeros $z_1 = 1$, $z_2 = -1$, each with multiplicity 2. Theorem 6.8 thus yields the four solutions

$$x_n^{(1)} = 1, \qquad x_n^{(2)} = n, \qquad x_n^{(3)} = (-1)^n, \qquad x_n^{(4)} = (-1)^n n.$$

13. We shall show: If z is a zero of multiplicity 4 of the characteristic polynomial, then the sequence $X = \{n^3 z^n\}$ is a solution. Proof: We seek to represent X as a linear combination of the solutions given by theorem 6.8. We have

$$n(n - 1)(n - 2) = n^3 - 3n^2 + 2n$$
$$n(n - 1) = n^2 - n$$
$$n = n$$

and hence

$$n^3 = n(n - 1)(n - 2) + 3n(n - 1) + n.$$

Thus

$$n^3 z^n = z^3[n(n - 1)(n - 2)z^{n-3}] + 3z^2[n(n - 1)z^{n-2}] + z[nz^{n-1}],$$

which is a combination of the desired form.

Example 13 leads us to conjecture the following fact:

Corollary 6.8   Under the hypotheses of theorem 6.8 a set of N independent solutions of $\mathscr{L}X = 0$ is also given by the sequences

$$(6\text{-}24) \quad x_n = n^m z_i^n, \qquad m = 0, 1, \ldots, m_i; \qquad i = 1, 2, \ldots, k.$$

The proof that the sequences X defined by (6-24) are solutions boils down to showing that n^m can be written as a linear combination of

$$n, n(n - 1), \ldots, n(n - 1)\ldots(n - m + 1),$$

with coefficients that are independent of n. This is readily verified by an induction argument. The proof that the N sequences given by (6-24) are linearly independent is again omitted.

EXAMPLE

14.   To find the general solution of the difference equation

$$x_n - \binom{N}{1}x_{n-1} + \binom{N}{2}x_{n-2} - \cdots + (-1)^N\binom{N}{N}x_{n-N} = 0.$$

The characteristic polynomial is

$$p(z) = z^N - \binom{N}{1}z^{N-1} + \binom{N}{2}z^{N-2} - \cdots + (-1)^N.$$

As is known from the binomial theorem,

$$p(z) = (z - 1)^N.$$

It follows that $z = 1$ is a zero of multiplicity N. Corollary 6.8 yields the solutions $1, n, n^2, \ldots, n^{N-1}$, and the general solution can be written in the form

$$x_n = c_0 + c_1 n + c_2 n^2 + \cdots + c_{N-1} n^{N-1}.$$

Problems

19. Let $N > 0$ be an integer. Find a general solution of the difference equation

$$x_n + x_{n-1} + x_{n-2} + \cdots + x_{n-N} = 0.$$

Deduce that for any real angle α and any integer m, $1 \leq m \leq N$, if $\varphi = 2\pi m/(N + 1)$,

$$\cos \alpha + \cos (\alpha + \varphi) + \cdots + \cos (\alpha + N\varphi) = 0,$$
$$\sin \alpha + \sin (\alpha + \varphi) + \cdots + \sin (\alpha + N\varphi) = 0.$$

20. Construct a difference equation that has the solutions

(a) $x_n = 1,$ $x_n = 2^n,$ $x_n = 3^n;$

(b) $x_n = n,$ $x_n = n^2,$ $x_n = n^3.$

Also find a difference equation that has both the solutions given under (a) and (b).

21*. Let $0 \leq \varphi_k < 2\pi$, $k = 1, 2, \ldots, N$, $\varphi_k \neq \varphi_l$ for $k \neq l$, and let the numbers α_k be arbitrary. Show that the N sequences

$$X^{(k)} = \{e^{i(\alpha_k + n\varphi_k)}\}, \qquad k = 1, 2, \ldots, N$$

are linearly independent.

6.9   The Backward Difference Operator

An important special difference operator of the kind defined by (6-18) is the *backward difference operator*, traditionally denoted by ∇ (read "nabla" from the arabic word for harp) and defined by

$$(6\text{-}25) \qquad (\nabla X)_n = x_n - x_{n-1}.$$

This difference operator is of order 1. It is closely related to the forward difference operator Δ which was introduced in §4.4. In fact, we have

$$(\Delta X)_n = (\nabla X)_{n+1}.$$

Thus every relation valid for the forward difference operator can also be expressed in terms of the backward operator, and conversely. In the framework of our present notations it is a little more convenient to work with the backward operator.

Integral powers of the operator ∇ are defined inductively by the relation

$$(6\text{-}26) \qquad \nabla^k X = \nabla(\nabla^{k-1} X), \qquad k = 2, 3, \ldots.$$

EXAMPLE

15. $\nabla^2 X = \nabla(\nabla X)$, hence

$$\begin{aligned}
(\nabla^2 X)_n &= (\nabla X)_n - (\nabla X)_{n-1} \\
&= x_n - x_{n-1} - (x_{n-1} - x_{n-2}) \\
&= x_n - 2x_{n-1} + x_{n-2}.
\end{aligned}$$

The example suggests an explicit expression for $(\nabla^k X)_n$ in terms of binomial coefficients. In fact,

$$(6\text{-}27) \quad (\nabla^k X)_n = x_n - \binom{k}{1} x_{n-1} + \binom{k}{2} x_{n-2} - \cdots + (-1)^k \binom{k}{k} x_{n-k}.$$

The proof of (6-27) is by induction with respect to k.   By definition, the formula is true for $k = 1$.   Assuming its truth for $k = m$, we have

$$(\nabla^{m+1}X)_n = (\nabla^m X)_n - (\nabla^m X)_{n-1}$$

$$= \left[x_n - \binom{m}{1}x_{n-1} + \binom{m}{2}x_{n-2} - \cdots + (-1)^m \binom{m}{m}x_{n-m} \right]$$

$$- \left[x_{n-1} - \binom{m}{1}x_{n-2} + \cdots + (-1)^{m-1}\binom{m}{m-1}x_{n-m} \right.$$

$$\left. + (-1)^m x_{n-m-1} \right]$$

By virtue of the identity (3-13) this expression simplifies to

$$x_n - \binom{m+1}{1}x_{n-1} + \binom{m+1}{2}x_{n-2} - \cdots + (-1)^{m+1}\binom{m+1}{m+1}x_{n-m-1},$$

which is equal to the term on the right of (6-27) when k is replaced by $m + 1$.   Thus, (6-27) must be true for all positive integers k.

   In practice, the operator ∇ is mainly used in connection with sequences F defined by

$$f_n = f(x_n),$$

where $x_n = nh$, h being a positive constant.   One ordinarily writes ∇f_n in place of the logically more consistent, but cumbersome notation $(\nabla F)_n$.   The values of successive powers of ∇ are conveniently arranged in a two-dimensional array, as in table 6.9a.

Table 6.9a

f_n			
	∇f_{n+1}		
f_{n+1}		$\nabla^2 f_{n+2}$	
	∇f_{n+2}		$\nabla^3 f_{n+3}$
f_{n+2}		$\nabla^2 f_{n+3}$	
	∇f_{n+3}		$\vdots$
f_{n+3}		$\vdots$	
	$\vdots$		
$\vdots$			

   Table 6.9a is called the *difference table* of the function f, constructed with the step h.   Each entry in the table is the difference of the two entries immediately to the left of it.

EXAMPLE

16.   From Comrie's table of the exponential function $f(x) = e^x$ (Comrie [1961]), which has the step $h = 10^{-3}$, we can construct the following difference table (beginning at $x_n = 1.35$):

Table 6.9b

x_n	f_n	∇f_n	$\nabla^2 f_n$
1.350	3.857420		
		0.003859	
1.351	3.861285		0.000004
		0.003863	
1.352	3.865148		0.000004
		0.003867	
1.353	3.869015		

The study of properties and applications of the difference table, in particular the relation between differences and derivatives of a function, will be one of our chief concerns in part II.   Here we are content with stating the following fundamental fact.

Theorem 6.9   Let the function f be defined on the whole real line, let k be a positive integer, and let $h > 0$.   If $f_n = f(nh)$, a necessary and sufficient condition in order that

(6-28) $\nabla^k f_n = 0$

for all integers n is that $f(nh) = P(nh)$ for all n, where P is a polynomial of degree not exceeding $k - 1$.

Briefly, the theorem states that the kth differences of a function are identically zero if and only if, at the points where the differences are taken, the function agrees with a polynomial of degree $< k$.

Proof.   We have to show: (a) If the kth differences are zero, then the values of f at $x = nh$ are those of a polynomial of degree $< k$; (b) If the values of f agree with those of a polynomial P of degree $< k$ at all points $x = nh$, then the kth differences are zero.

To prove (a), we note that (6-28) means that the sequence $F = \{f_n\}$ is a solution of the difference equation

(6-29) $\nabla^k F = 0$.

By (6-27), the characteristic polynomial of this equation is

$$p(z) = z^k - \binom{k}{1} z^{k-1} + \binom{k}{2} z^{k-2} - \cdots + (-1)^k \binom{k}{k}$$

$$= (z - 1)^k.$$

This polynomial has a single zero, of multiplicity k, at $z = 1$. By corollary 6.8, any solution of (6-29) can thus be represented in the form

$$f_n = c_1 + c_2 n + c_3 n^2 + \cdots + c_k n^{k-1}.$$

Clearly, $f_n = P(nh)$, where P is the polynomial defined by

$$P(x) = c_1 + \frac{c_2}{h} x + \frac{c_3}{h^2} x^2 + \cdots + \frac{c_k}{h^{k-1}} x^{k-1}.$$

To prove (b), let $f(x) = P(x)$ for $x = nh$, where P is a polynomial of degree $< k$,

$$P(x) = a_0 x^{k-1} + a_1 x^{k-2} + \cdots + a_{k-1}.$$

We then have

$$f_n = P(nh) = a_0 (nh)^{k-1} + a_1 (nh)^{k-2} + \cdots + a_{k-1}.$$

This relation shows that the sequence $F = \{f_n\}$ is a linear combination of the sequences $X^{(m)}$ defined by

$$X^{(m)} = n^m, \qquad m = 0, 1, 2, \ldots, k - 1.$$

These sequences are solutions of the difference equation $\nabla^k X = 0$, by corollary 6.8. It follows that the sequence F is a solution of the same difference equation, that is, (6-28) holds.

EXAMPLE

17. Let $P(x) = x^3 - 2x + 1$. The difference table with step $h = 1$ (begun at $x = 0$) is shown in table 6.9c.

Table 6.9c

1				
0	−1			
5	5	6		
22	17	12	6	0
57	35	18	6	0
116	59	24	6	0
205	89	30	6	0
330	125	36	6	:
:	:	:	:	:

Problems

22. If $T = T(t)$ denotes the sequence of Chebyshev polynomials introduced in example 3, calculate

(a) $(\nabla^2 T(t))_{n+1}$,   (b) $(\nabla^4 T(t))_{n+2}$.

23. Give an explicit formula for the differences of the exponential function, $f(x) = e^x$, and show that these differences are never zero. Explain the apparent contradiction with the numerical values given in table 6.9b.

24. The first differences given in table 6.9b are very nearly equal to 10^{-3} times the average of the adjacent function values. Is this an accident?

25. Formula (6-27) expresses $(\nabla^k X)_n$ in terms of $x_n, x_{n-1}, \ldots, x_{n-k}$. Show conversely how to express x_{n-k} in terms of $(\nabla^0 X)_n, (\nabla X)_n, \ldots, (\nabla^k X)_n$.

Recommended Reading

A more thorough treatment of difference equations will be found in Milne-Thomson [1933]. Goldberg [1958] gives an enjoyable elementary account with many interesting applications.

Research Problem

Try to generalize as many results of this chapter as possible to linear difference equations with variable coefficients. Do not attempt to find solutions in explicit form.

chapter 7 Bernoulli's method

With the present chapter we return to the problem of solving nonlinear, and in particular polynomial, equations. The methods discussed in the chapters devoted to iteration are very effective, but only if a reasonable first approximation to the desired solution is known. How to obtain such a first approximation is a problem which, for equations without special properties, is of such generality that it cannot be solved by generally applicable rules or algorithms. For polynomials, however, there do exist algorithms that furnish the desired first approximation using no other information than the coefficients of the polynomial. Two such algorithms —a classical one due to D. Bernoulli and one of its modern extensions due to Rutishauser—form the subject of this and the next chapter. Bernoulli's method, in particular, is one which yields all *dominant* zeros of a polynomial. By a dominant zero we mean a zero whose modulus is not exceeded by the modulus of any other zero.

7.1 Single Dominant Zero

In chapter 6 we have seen how the general linear difference equation with constant coefficients can be solved analytically by determining the zeros of the associated characteristic polynomial. Bernoulli's method consists in reversing this procedure. The polynomial whose zeros are sought is considered the characteristic polynomial of some difference equation, and this associated difference equation is solved *numerically* by solving the recurrence relation implied by it. From this solution it is easy to extract information about the zeros of the polynomial, as we shall see.

To begin with the simplest case, let us assume that the polynomial of degree N,

$$(7\text{-}1) \qquad p(z) = a_0 z^N + a_1 z^{N-1} + \cdots + a_N,$$

whose coefficients may be complex, has N *distinct* zeros $z_1, z_2, \ldots, z_N$. What happens if we solve the difference equation

$$(7\text{-}2) \qquad a_0 x_n + a_1 x_{n-1} + \cdots + a_N x_{n-N} = 0$$

which has (7-1) as its characteristic polynomial? According to §6.7, the solution $X = \{x_n\}$ (whatever its starting values are) must be representable in the form

$$(7\text{-}3) \qquad x_n = c_1 z_1^n + c_2 z_2^n + \cdots + c_N z_N^n,$$

where $c_1, c_2, \ldots, c_N$ are suitable constants. To proceed further we make two assumptions:

(*i*) The polynomial p has a single dominant zero, i.e., one of the zeros— we may call it z_1—has a larger modulus than all the others:

$$(7\text{-}4) \qquad |z_1| > |z_k|, \qquad k = 2, 3, \ldots, N.$$

(*ii*) The starting values are such that the dominant zero is represented in the solution (7-3), i.e., we have

$$(7\text{-}5) \qquad c_1 \neq 0.$$

We now consider the ratio of two consecutive values of the solution sequence X. Using (7-3) we find

$$\frac{x_{n+1}}{x_n} = \frac{c_1 z_1^{n+1} + c_2 z_2^{n+1} + \cdots + c_N z_N^{n+1}}{c_1 z_1^n + c_2 z_2^n + \cdots + c_N z_N^n}.$$

By virtue of (7-5) this may be written

$$(7\text{-}6) \qquad \frac{x_{n+1}}{x_n} = z_1 \frac{1 + \frac{c_2}{c_1}\left(\frac{z_2}{z_1}\right)^{n+1} + \cdots + \frac{c_N}{c_1}\left(\frac{z_N}{z_1}\right)^{n+1}}{1 + \frac{c_2}{c_1}\left(\frac{z_2}{z_1}\right)^{n} + \cdots + \frac{c_N}{c_1}\left(\frac{z_N}{z_1}\right)^{n}}.$$

By (7-4), $|z_k/z_1| < 1$ for $k = 2, 3, \ldots, N$. It follows that

$$(7\text{-}7) \qquad \left(\frac{z_k}{z_1}\right)^n \to 0 \quad \text{as} \quad n \to \infty$$

for $k = 2, 3, \ldots, N$. The fraction multiplying z_1 thus tends to 1 as $n \to \infty$, and we find

$$\lim_{n \to \infty} \frac{x_{n+1}}{x_n} = z_1.$$

We thus have the following tentative formulation of Bernoulli's method:

Algorithm 7.1 Choose arbitrary values $x_0, x_{-1}, \ldots, x_{-N+1}$, and determine the sequence $\{x_n\}$ from the recurrence relation

$$x_n = -\frac{a_1 x_{n-1} + a_2 x_{n-2} + \cdots + a_N x_{n-N}}{a_0}, \qquad n = 1, 2, \ldots.$$

Then form the sequence of quotients

(7-8)
$$q_n = \frac{x_{n+1}}{x_n}.$$

We have proved:

Theorem 7.1   If the polynomial p given by (7-1) has a single dominant zero, and if the starting values are such that (7-5) holds, then the quotients q_n are ultimately defined and converge to the dominant zero of p.

EXAMPLES

1. Applying Bernoulli's method to the polynomial

$$p(z) = z^2 - z - 1$$

with the starting values $x_0 = 1$, $x_{-1} = 0$ yields the *Fibonacci* sequence considered in example 7 of chapter 6.   Conditions (*i*) and (*ii*) are clearly satisfied here.   The ratios of consecutive elements thus converge to the dominant zero $z_1 = (1 + \sqrt{5})/2$.

2. Let

$$p(z) = 70z^4 - 140z^3 + 90z^2 - 20z + 1.$$

The difference equation (7-2) (solved for x_n) here takes the form

$$x_n = \frac{140x_{n-1} - 90x_{n-2} + 20x_{n-3} - x_{n-4}}{70}.$$

The first three columns of table 7.1 give the values of n, of the sequence x_n (started with $x_0 = 1$, $x_n = 0$, $n < 0$) and of the sequence $\{q_n\}$.   (The other columns will be explained in §7.2.)

The preceding is a mere outline of Bernoulli's method in the simplest possible case.   A number of complications have yet to be dealt with, such as the following:

1. Slow convergence.
2. Zeros of multiplicity > 1.
3. Unfortunate choice of initial values.
4. Several dominant zeros.
5. Calculation of nondominant zeros.

These questions will be dealt with in the subsequent sections.

Problems

1. Find, by Bernoulli's method, the dominant zero of the polynomial
$$p(z) = 32z^3 - 48z^2 + 18z - 1$$
to three significant digits.

Table 7.1

n	x_n	q_n	Δq_n	$\Delta^2 q_n$	$-\dfrac{(\Delta q_n)^2}{\Delta^2 q_n}$	q'_n
0	1					
1	2					
2	2.7142857					
3	3.1428571	1.1578947	−0.0910116		−0.1982597	0.9596350
4	3.3530611	1.0668831	−0.0492325	0.0417791	−0.1214233	0.9454595
5	3.4122447	1.0176506	−0.0292706	0.0199619	−0.0792746	0.9383760
6	3.3725945	0.9883800	−0.0184630	0.0108076	−0.0536856	0.9346444
7	3.2711367	0.9699170	−0.0121134	0.0063496	−0.0371658	0.9327512
8	3.1331066	0.9578036	−0.0081653	0.0039481		
9	2.9753180	0.9496383	−0.0056093	0.0025560		
10	2.8087866	0.9440290	−0.0039052	0.0017041		
11	2.6406071	0.9401238	−0.0027449	0.0011603		
12	2.4752493	0.9373789	−0.0019425	0.0008024		
13	2.3154382	0.9354364	−0.0013814	0.0005611		
14	2.1627466	0.9340550	−0.0009858	0.0003956	−0.0034633	0.9305917
15	2.0179923	0.9330692	−0.0007052	0.0002806	−0.0024890	0.9305802
16	1.8815034	0.9323640	−0.0005054	0.0001998		
17	1.7532951	0.9318586				
		. . .				. . .
		0.93057				0.93057

2. Use the value obtained in problem 1 to obtain the dominant zero to six significant digits, using either Newton's or Steffensen's method.

3. Look up the closed formula for finding the zeros of a cubic polynomial and use it to determine the dominant zero of the polynomial of problem 1.

4. What is the meaning of the phrase "the quotients are ultimately defined" in theorem 7.1? Give an example of a polynomial violating condition (*i*) where infinitely many of the quotients q_n are undefined.

7.2   Accelerating Convergence

Even if the conditions of convergence of Bernoulli's method are satisfied, the speed of convergence may be slow.   By this we mean that the error of the approximation x_{n+1}/x_n to the zero z_1,

$$d_n = \frac{x_{n+1}}{x_n} - z_1$$

tends to zero only slowly.   As in the case of iteration, it may be possible to speed up convergence by making judicious use of information about the *manner* in which d_n approaches zero.   In order to discover this manner of convergence, we shall analyze the errors d_n more closely.   We shall continue to make the assumptions (*i*) and (*ii*) of §7.1.   In addition, we assume that

(7-9) $$|z_1| > |z_2| > |z_k|, \qquad k = 3, 4, \ldots, N$$

(the next-to-dominant zero is the only zero of its modulus), and that

(7-10) $$c_2 \neq 0$$

in (7-3) (the next-to-dominant zero is represented in the solution $\{x_n\}$).
   Under these hypotheses, the error

$$d_n = \frac{c_1 z_1^{n+1} + c_2 z_2^{n+1} + \cdots + c_N z_N^{n+1} - z_1(c_1 z_1^n + \cdots + c_N z_N^n)}{c_1 z_1^n + c_2 z_2^n + \cdots + c_N z_N^n}$$

$$= \frac{c_2(z_2 - z_1)z_2^n + \cdots + c_N(z_N - z_1)z_N^n}{c_1 z_1^n + c_2 z_2^n + \cdots + c_N z_N^n}$$

can be written in the form

(7-11) $$d_n = At^n(1 + \varepsilon_n),$$

where

$$A = \frac{c_2(z_2 - z_1)}{c_1}, \qquad t = \frac{z_2}{z_1},$$

and

$$1 + \varepsilon_n = \frac{1 + \dfrac{c_3(z_3 - z_1)}{c_2(z_2 - z_1)}\left(\dfrac{z_3}{z_2}\right)^n + \cdots + \dfrac{c_N(z_N - z_1)}{c_2(z_2 - z_1)}\left(\dfrac{z_N}{z_2}\right)^n}{1 + \dfrac{c_2}{c_1}\left(\dfrac{z_2}{z_1}\right)^n + \cdots + \dfrac{c_N}{c_1}\left(\dfrac{z_N}{z_1}\right)^N}.$$

By virtue of condition (7-9), we have, in addition to (7-7),

$$\left(\frac{z_k}{z_2}\right)^n \to 0 \quad \text{as} \quad n \to \infty$$

for $k = 3, 4, \ldots, N$, and the ratio on the right has the limit 1 as $n \to \infty$. It follows that

(7-12) $$\lim_{n \to \infty} \varepsilon_n = 0.$$

As a consequence,

$$\frac{d_{n+1}}{d_n} = t\frac{1 + \varepsilon_{n+1}}{1 + \varepsilon_n} = t(1 + \delta_n),$$

where

$$\delta_n = \frac{\varepsilon_{n+1} - \varepsilon_n}{1 + \varepsilon_n} \to 0 \quad \text{as} \quad n \to \infty.$$

The errors d_n thus satisfy precisely the condition (4-12) of theorem 4.5 which makes Aitken's Δ^2 process effective. We thus have as an immediate consequence of theorem 4.5:

Theorem 7.2 Under the hypotheses stated at the beginning of the present section, the sequence $\{q_n'\}$ derived from the sequence $\{q_n\}$ by means of Aitken's Δ^2 formula,

$$q_n' = q_n - \frac{(\Delta q_n)^2}{\Delta^2 q_n},$$

converges faster to the dominant zero z_1 than the sequence $\{q_n\}$ in the sense that

$$\lim_{n \to \infty} \frac{q_n' - z_1}{q_n - z_1} = 0.$$

EXAMPLE

3. In table 7.1 the values q_n' are shown in the last column. Also given are the intermediate values required to calculate q_n'. The faster convergence of the sequence $\{q_n'\}$ to the exact zero $z_1 = 0.93057$ is evident.

Problems

5. Find, by Bernoulli's method speeded up by the Δ^2 process, the dominant zero of the polynomial

$$p(z) = z^4 - 7z^3 + 13z^2 - 8z + 12.$$

6. Apply the Δ^2 method to the sequence $\{q_n\}$ obtained in problem 1.
7. Explain why the Δ^2 method does not speed up appreciably the Bernoulli sequence for the polynomial

$$p(z) = z^3 - 4z^2 + 6z - 4.$$

($z = 2$ is the sole dominant zero.)

8. If suitable conditions are satisfied, the error of the sequence $\{q_n\}$ generated by Bernoulli's method tends to zero like $(z_2/z_1)^n$. How does the error of the accelerated sequence $\{q_n'\}$ tend to zero? What do you conclude about the number of steps necessary to achieve a given accuracy
(a) if $|z_1| \sim |z_2| \gg |z_3|$,
(b) if $|z_1| \gg |z_2| \sim |z_3|$?

7.3 Zeros of Higher Multiplicity

If the polynomial p has repeated nondominant zeros $z_2, \ldots, z_N$, then formula (7-3) for the general solution of the difference equation (7-2) also contains (by corollary 6.8) terms like $n^k z_2^n$. Thus the expression (7-6) contains terms like

$$n^k \left(\frac{z_2}{z_1}\right)^n \quad \text{in addition to} \quad \left(\frac{z_2}{z_1}\right)^n.$$

However, these terms cannot disturb the convergence of the method, since if $|q| < 1$, then we have not only $q^n \to 0$ as $n \to \infty$ but also $n^k q^n \to 0$ for any fixed value of k.

The situation is different if the *dominant* zero has multiplicity >1. (We still assume that there is only one dominant zero.) To fix ideas, let the multiplicity be 2. Relation (7-3) then takes the form

$$x_n = c_1 n z_1^n + c_2 z_1^n + c_3 z_3^n + \cdots,$$

where $|z_3| < |z_1|$. Thus the ratio (7-6) now becomes (still assuming $c_1 \neq 0$)

$$(7\text{-}13) \quad \frac{x_{n+1}}{x_n} = \frac{c_1(n+1)z_1^{n+1} + c_2 z_1^{n+1} + c_3 z_3^{n+1} + \cdots}{c_1 n z_1^n + c_2 z_1^n + c_3 z_3^n + \cdots}$$

$$= z_1 \frac{(n+1)c_1 + c_2}{nc_1 + c_2} \cdot \frac{1 + \dfrac{c_3}{(n+1)c_1 + c_2}\left(\dfrac{z_3}{z_1}\right)^{n+1} + \cdots}{1 + \dfrac{c_3}{nc_1 + c_2}\left(\dfrac{z_3}{z_1}\right)^{n} + \cdots}.$$

Convergence still takes place, but, due to the factor

$$\frac{(n+1)c_1 + c_2}{nc_1 + c_2} = 1 + \frac{1}{n + c_2/c_1}$$

at a much slower rate; the error after n steps is now of the order of $1/n$ as against $(z_1/z_2)^n$ if the dominant zero has multiplicity one. An example for this slowdown of convergence, as well as a possible remedy, is given in the following section.

Problems

9. Devise an Aitken-like acceleration scheme for the case of a single dominant zero of multiplicity 2. [Sketch of a possible solution: From (7-13) we find for the error $d_n = q_n - z_1$

$$d_n = \frac{c_1 z_1 + 0(t^n)}{c_1 n + c_2 + 0(t^n)},$$

where $t = z_3/z_1$. Neglecting $0(t^n)$ and setting $c = c_2/c_1$, we have

$$(7\text{-}14) \qquad\qquad q_n - z_1 = \frac{z_1}{n + c}.$$

The unknowns c and, if desired, n can be eliminated from consecutive relations (7-14) as in the derivation of Aitken's formula, yielding a formula for z_1.]

10. Apply the procedure devised in problem 9 to the calculation of the dominant zero of the polynomial

$$p(z) = z^4 - 4z^3 - 2z^2 + 12z + 9.$$

7.4   Choice of Starting Values

One of the conditions for convergence of algorithm 7.1 was that $c_1 \neq 0$. It can be shown by means of complex variable theory that this condition is always satisfied if the starting values are chosen as follows:

$$(7\text{-}15) \qquad x_{-N+1} = x_{-N+2} = \cdots = x_{-1} = 0, \qquad x_0 = 1.$$

A different, more sophisticated choice of starting values is defined by the following algorithm:

Algorithm 7.4   If the coefficients $a_0, a_1, \ldots, a_N$ are given, calculate $x_0, x_1, \ldots, x_{N-1}$ by the formulas

$$x_0 = -\frac{a_1}{a_0},$$

$$x_1 = -\frac{1}{a_0}(2a_2 + a_1x_0),$$

$$x_2 = -\frac{1}{a_0}(3a_3 + a_2x_0 + a_1x_1),$$

and generally

$$(7\text{-}16) \quad x_k = -\frac{1}{a_0}[(k+1)a_{k+1} + a_kx_0 + a_{k-1}x_1 + \cdots + a_1x_{k-1}],$$

$$k = 1, 2, \ldots, N-1.$$

The starting values generated by algorithm 7.4 have the following very desirable property:

Theorem 7.4   Let the polynomial $p(z) = a_0z^N + a_1z^{N-1} + \cdots + a_N$ have the distinct zeros $z_1, z_2, \ldots, z_M$ $(M \leq N)$, and let the multiplicity of z_i be m_i $(i = 1, 2, \ldots, M)$. If the starting values for Bernoulli's method are determined by algorithm 7.4, then relation (7-3) takes the form

$$(7\text{-}17) \qquad x_n = m_1z_1^{n+1} + m_2z_2^{n+1} + \cdots + m_Mz_M^{n+1}, \qquad n = 0, 1, \ldots.$$

The proof is again most easily accomplished by complex variable theory, and is omitted here.   Relation (7-17) is remarkable for the fact that no powers of n appear, notwithstanding the possible presence of zeros of multiplicity higher than 1.   The difficulty mentioned in §7.3 thus can always be avoided by a proper choice of the starting values.   The ratio x_{n+1}/x_n then converges at a rate determined only by the magnitude of the two largest zeros.

EXAMPLE

4. We compare the sequences $\{q_n\}$ for the polynomial

$$p(z) = (z - 3)^2(z + 1)^2 = z^4 - 4z^3 - 2z^2 + 12z + 9,$$

generated by starting Bernoulli's method by (7-15) and by algorithm 7.4. The recurrence relation is in both cases

$$x_n = 4x_{n-1} + 2x_{n-2} - 12x_{n-3} - 9x_{n-4}.$$

Table 7.4	Method (7.15)		Algorithm 7.4	
n	x_n	q_n	x_n	q_n
-3	0			
-2	0			
-1	0			
0	1		4	
1	4	4	20	5
2	18	4.5	52	2.6
3	68	3.78	164	3.15385
4	251	3.69	484	2.95122
5	888	3.54	1 460	3.01653
6	3 076	3.46	4 372	2.99452
7	10 456	3.40	13 124	3.00183
8	35 061	3.35	39 364	2.99939
9	116 252	3.32	118 100	3.00020
10	381 974	3.29	354 292	2.99993
11	1 245 564	3.26086	1 062 884	3.00002
12	4 035 631	3.24000	3 188 644	2.99999
13	13 003 696	3.22222	9 565 940	3.0000025
14	41 701 512	3.20690	28 697 812	2.9999992
		$\cdots$		$\cdots$
		3.00000		3.0000000

Problems

11. Use algorithm 7.4 to obtain starting values for the application of Bernoulli's method to the polynomial of problem 10. Verify relation (7-17).

12. Using Vieta's formulas, verify (7-17) for $n = 0$ and $n = 1$ in the case of an arbitrary polynomial.

13. Show that if the rational function

$$r(z) = -\frac{a_1 + 2a_2z + \cdots + Na_Nz^{N-1}}{a_0 + a_1z + a_2z^2 + \cdots + a_Nz^N}$$

is expanded in powers of z, then the coefficients of $1, z, \ldots, z^{N-1}$ are identical with the numbers $x_0, x_1, \ldots, x_{N-1}$ defined by algorithm 7.4.

7.5   Two Conjugate Complex Dominant Zeros

The theory developed so far holds equally well whether the coefficients of the polynomial p are real or complex. However, we have always assumed that z_1 is the sole dominant zero of p. Let us now consider the case where p is a polynomial with real coefficients which has a pair of complex conjugate dominant zeros z_1 and $z_2 = \bar{z}_1$, both of multiplicity one. The remaining zeros shall satisfy

(7-18) $$|z_k| < |z_1|, \qquad k = 3, 4, \ldots, N;$$

for the purpose of our analysis we assume that the nondominant zeros, too, have multiplicity one, although this is not essential for the result.

If the starting values for the sequence $\{x_n\}$ are real, equation (7-3) takes the form

$$x_n = c_1 z_1^n + \bar{c}_1 \bar{z}_1^n + c_3 z_3^n + \cdots + c_N z_N^n.$$

Representing the complex numbers c_1 and z_1 in polar form, we write $z_1 = re^{i\varphi}$, $c_1 = ae^{i\delta}$, where $r > 0$ and $a > 0$. We may assume, furthermore, that z_1 is the zero in the upper half-plane, and consequently $0 < \varphi < \pi$. The expression for x_n now becomes

$$x_n = 2ar^n \cos(n\varphi + \delta) + c_3 z_3^n + \cdots + c_N z_N^n,$$

This may be written

(7-19) $$x_n = 2ar^n[\cos(n\varphi + \delta) + \theta_n],$$

where

$$\theta_n = \frac{c_3}{2a}\left(\frac{z_3}{r}\right)^n + \cdots + \frac{c_N}{2a}\left(\frac{z_N}{r}\right)^n.$$

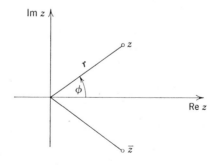

Figure 7.5

and hence, for some suitable constant C,

$$|\theta_n| \le Ct^n,$$

where t denotes the largest of the ratios $|z_k/r|$, $k = 3, \ldots, N$. In view of (7-18), the number t is less than 1, and hence

(7-20) $$\lim_{n \to \infty} \theta_n = 0.$$

Our problem is to recover the quantities r and φ from the sequence $\{x_n\}$. To find its solution, let us begin by assuming that $\theta_n = 0$. We recall that the sequence $\{x_n\}$ with the elements

$$x_n = 2ar^n \cos (n\varphi + \delta)$$

is a solution of the difference equation

(7-21) $$x_n + Ax_{n-1} + Bx_{n-2} = 0,$$

where $B = r^2$, $A = -2r \cos \varphi$ (see problem 3, chapter 6). To determine the coefficients A and B from the known solution $\{x_n\}$, we observe that equation (7-21) together with the corresponding equation with n increased by 1,

$$x_{n+1} + Ax_n + Bx_{n-1} = 0$$

represents a system of two linear equations for the two unknowns A and B. The determinant

(7-22) $$D_n = \begin{vmatrix} x_{n-1} & x_{n-2} \\ x_n & x_{n-1} \end{vmatrix}$$

equals, by a trigonometric identity,

$$4a^2r^{2n-2}\{\cos^2 [(n-1)\varphi + \delta] - \cos (n\varphi + \delta) \cos [(n-2)\varphi + \delta]\}$$
$$= 4a^2r^{2n-2} \sin^2 \varphi,$$

and hence is different from zero, as $0 < \varphi < \pi$. We may thus solve for A and B, finding

$$A = -\frac{E_n}{D_n}, \qquad B = \frac{D_{n+1}}{D_n},$$

where

(7-23) $$E_n = \begin{vmatrix} x_n & x_{n-2} \\ x_{n+1} & x_{n-1} \end{vmatrix}.$$

The desired quantities r and φ can now be found by

(7-24) $$r = \sqrt{B} = \sqrt{\frac{D_{n+1}}{D_n}}, \qquad \cos \varphi = -\frac{A}{2r} = \frac{E_n}{2\sqrt{D_n D_{n+1}}}.$$

The relations (7-24) solve our problem if $\{x_n\}$ is a solution of the difference equation (7-2) such that $\theta_n = 0$ in (7-19).   If $\{x_n\}$ is any solution of (7-2), $\theta_n \neq 0$ in general; however, using (7-20) it is not too hard to show that

$$(7\text{-}25) \qquad \frac{D_{n+1}}{D_n} \to r^2, \qquad \frac{E_n}{2D_n} \to r \cos \varphi$$

as $n \to \infty$.   We thus have obtained the convergence of the following algorithm for the determination of a pair of complex conjugate dominant zeros.

Algorithm 7.5   From the solution $\{x_n\}$ of the difference equation (7-2), calculate the determinants D_n and E_n defined by (7-22) and (7-23).   With these, form the sequence of ratios D_{n+1}/D_n and $E_n/2D_n$.

The corresponding convergence theorem is as follows:

Theorem 7.5   Let the polynomial (7-1) have exactly two dominant zeros $z_{1,2} = re^{\pm i\varphi}$, both of multiplicity one, where $0 < \varphi < \pi$.   If the solution of the difference equation (7-2) is such that $c_1 \neq 0$ in (7-3), then the limit relations (7-25) hold.

EXAMPLE

5.   For the polynomial

$$p(z) = 81z^4 - 108z^3 + 24z + 20$$

(exact dominant zeros: $z_{1,2} = 1 \pm i\frac{1}{3}$) the computation proceeds as in table 7.5.   It is evidently unnecessary to calculate the determinants D_n and E_n from the beginning.

Algorithm 7.5 has the disadvantage of not being very accurate when φ, the argument of the dominant zero, is small.   Let us assume that we have determined r accurately and $x = r \cos \varphi = \lim E_n/2D_n$ with an error dx. We then have to calculate φ from the relation

$$\cos \varphi = \frac{x}{r}$$

and find for the differential of φ

$$d\varphi = -\frac{1}{r \sin \varphi} dx.$$

For small values of φ this may be quite large even for a small value of dx.

Problems

14. Determine the dominant zeros of the polynomial

$$p(z) = z^3 - z^2 + 2$$

to three decimal places.

Table 7.5

x_n	D_n	D_{n+1}/D_n	E_n	$E_n/2D_n$
1				
1.333333				
1.777778				
2.074074				
2.123457				
1.975309				
1.580247				
0.965707				
0.178022				
−0.718589				
−1.634438				
−2.470444				
−3.124966				
−3.504914	1.229324	1.111187	2.458740	1.000037
−3.537670	1.366009	1.111125	2.732037	1.000006
−3.180991	1.517807	1.111095	3.035585	0.999990
−2.431232	1.686428	1.111115	3.372868	1.000003
−1.328033	1.873816	1.111111	3.747630	0.999999
0.045304	2.082018	1.111113	4.164035	1.000000
1.566200	2.313358		4.626705	
		. . .		. . .
		1.111111		1.000000

15. Starting from the exact representation (7-19) of x_n, prove the relations (7-25).

16. Devise a modification of Bernoulli's method for the case that the polynomial p has two real dominant zeros, both of multiplicity one, of opposite signs, and apply your algorithm to the polynomial

$$p(z) = z^4 - 2z^3 - 2z^2 + 8z - 8.$$

7.6 Sign Waves†

How can we detect a pair of conjugate complex zeros? Formula (7-19) shows that in the presence of a pair of conjugate complex dominant zeros the elements of the sequence $\{x_n\}$ behave like

(7-26) $$x_n = 2ar^n \cos(n\varphi + \delta).$$

This equation implies, in particular, that the signs of the x_n oscillate. Moreover, by looking at the frequency with which these oscillations occur,

---

† This section may be omitted at first reading.

we can hope to be able to make a statement about the angle φ. Clearly, when φ is small, the "sign waves" are long, and when φ is close to π, they are short. (In the extreme case, where $\varphi = \pi$, plus and minus signs alternate in strict order.)

In order to establish a more precise result, let us assume that the polynomial p has degree 2. Equation (7-26) then is exact; we now write it in the form

$$(7\text{-}27) \qquad x_n = 2ar^n \sin \left(n\varphi + \delta + \frac{\pi}{2} \right).$$

If n were a continuous variable, the signs of x_n would change from minus to plus whenever

$$n\varphi + \delta + \frac{\pi}{2} = 2k\pi,$$

where k is any integer. Let now, for any positive integer k, n_k be the first integer following the kth change of sign from minus to plus of the sine-function in (7-27), i.e., let

$$(7\text{-}28) \qquad n_k = \frac{2k\pi - \delta - \dfrac{\pi}{2}}{\varphi} + \theta_k,$$

where $0 \leq \theta_k < 1$. This means that for each positive integer k,

$$\text{sign } x_{n_k - 1} = -1, \qquad \text{sign } x_{n_k} = 0 \quad \text{or} \quad 1.$$

Now subtract from (7-28) the corresponding relation for $k = 0$. We find, after dividing by k,

$$\frac{n_k - n_0}{k} = \frac{2\pi}{\varphi} + \frac{\theta_k - \theta_0}{k}.$$

We now consider the limit of this expression as $k \to \infty$. Since the θ_k are numbers between zero and one,

$$\lim_{k \to \infty} \frac{\theta_k - \theta_0}{k} = 0.$$

It follows that

$$T = \lim_{k \to \infty} \frac{n_k - n_0}{k}$$

exists and has the value $2\pi/\varphi$. Solving for φ, we find

$$(7\text{-}29) \qquad \varphi = \frac{2\pi}{T}.$$

The limit T appearing here has a very simple interpretation. The integer $n_k - n_{k-1}$ represents the length or period of the kth sign wave in the sequence of signs of the x_n. By virtue of the identity

$$n_k - n_0 = (n_k - n_{k-1}) + (n_{k-1} - n_{k-2}) + \cdots + (n_1 - n_0),$$

the expression $n_k - n_0$ is the sum of the periods of all sign waves from the first to the kth.   Hence T is just the average period of all sign waves.

We thus have obtained the following algorithm for the determination of the arguments of a pair of complex conjugate dominant zeros.

Algorithm 7.6 In the sequence $\{x_n\}$ generated by algorithm 7.1, delete everything but the signs† of the x_n. For $k = 0, 1, 2, \ldots$, let n_k be the index of the kth $+$ which is preceded by a $-$, and calculate the average value of all sign wave periods $n_k - n_{k-1}$.

Although our analysis applies only to polynomials of degree 2, the following result can be proved for polynomials of arbitrary degree (Brown [1962]).

Theorem 7.6   Under the hypotheses of theorem 7.5, the argument φ is given by (7-29), where T denotes the average sign wave period.

In electrical engineering, the quantity $2\pi/T$ is called the circular frequency of an alternating current with period T.   In this sense we can say that φ is equal to the circular frequency of the sequence of signs of the x_n.

EXAMPLE

6.   For the polynomial

$$\begin{aligned} p(z) &= z^4 - 8z^3 + 39z^2 - 62z + 50 \\ &= (z^2 - 2z + 2)(z^2 - 6z + 25) \end{aligned}$$

the sequence of signs of the x_n turned out as follows:

```
+ + − − − − + + + − − − + + + + − − − + + + + − − − + + + − − − −
        |___|  |___|        |___|            |___|        |___|
          6      7              7                7
```

```
+ + + − − − + + + − − − − + + + + − − − + + + − − − − + + + · · ·
|_____|  |___|        |___|        |___|        |___|
   6        7              7          7              7
```

The average length of the periods of the full sign waves recorded above is 6.75.   We thus find the approximate value

$$\varphi = \frac{2\pi}{6.75} \sim 53{,}3°.$$

The exact value is

$$\varphi = \text{arctg } \tfrac{4}{3} \sim 53{,}1°.$$

---

† A zero may be considered positive or negative.

It is only fair to point out that, in spite of the favorable showing in example 6, the convergence of algorithm 7.6 is rather slow in general. The algorithm is not useful for purposes of exact numerical computation, but it does give very quickly, with almost no computation, an idea about the location of a pair of complex conjugate zeros.

Problems

17. Bernoulli's method is applied unsuspectedly to the polynomial

$$p(z) = z^4 + 2z^3 + 4z^2 - 2z - 5.$$

Show that the polynomial has a pair of complex conjugate dominant zeros.   In which quadrants of the plane do they lie?

18. Algorithm 7.6 fails for the polynomial

$$p(z) = z^3 - 2z^2 + 2z - 1..$$

Explain!   [Hint: How many dominant zeros are there?]

19. Devise an algorithm for removing a quadratic factor $z^2 - uz - v$ from a given polynomial.   [Hint: Write $p(z) = (z^2 - uz - v)q(z)$ and compare coefficients of like powers of z, as in §4.10.]

20. Verify that $z^2 - 3.711245z + 3.728699$ is (approximately) a quadratic factor of the polynomial

$$p(z) = z^5 - 3z^4 - 20z^3 + 60z^2 - z - 78$$

and remove it from p by the algorithm devised in problem 19.

Recommended Reading

Bernoulli's method, and some extensions of it, are dealt with by Aitken [1926]; other accounts are given in Householder [1953] and Hildebrand [1956], pp. 458–462.   An entirely different procedure for determining the dominant zeros of a polynomial is known as Graeffe's method.   It is dealt with in the above standard numerical analysis texts; see also Ostrowski [1940] and Bareiss [1960].   A complete account of modern automatic procedures for polynomials is given by Wilkinson [1959].

Research Problems

1.   How does the presence of more than two dominant zeros manifest itself in the sequence of signs of the x_n, and how can the arguments be recovered?

2.   Suppose a polynomial has exactly three dominant zeros, one real, two complex conjugate.   Devise a modification of Bernoulli's method that deals with this situation.

chapter 8 the quotient-difference algorithm

Bernoulli's method has the disadvantage of furnishing only the dominant zeros of a polynomial. If it is desired to compute a non-dominant zero by Bernoulli's method, it is necessary first to compute all larger zeros, and then to remove them from the polynomial by the method of §4.10. Only rarely these zeros will be known exactly. Thus the successive deflations will have a tendency to falsify the remaining zeros. We now shall discuss a modern extension of Bernoulli's method, due to Rutishauser, which has the advantage of providing simultaneous approximations to *all* zeros. Since the prerequisites for this volume do not include complex function theory, we are unable to provide the proofs for the convergence theorems in this chapter. But even though its theoretical background cannot be fully exposed, we feel that Rutishauser's algorithm is of sufficient interest to warrant its presentation at this point.

8.1 The Quotient-Difference Scheme

The Quotient-Difference (QD) algorithm can be looked at as a generalization of Bernoulli's method. As in chapter 7, we are given a polynomial

$$(8\text{-}1) \qquad p(z) = a_0 z^N + a_1 z^{N-1} + \cdots + a_N$$

and form a solution of the associated difference equation

$$(8\text{-}2) \qquad a_0 x_n + a_1 x_{n-1} + \cdots + a_N x_{n-N} = 0.$$

The sequence $\{x_n\}$ may for instance be started by setting

$$(8\text{-}3) \qquad x_{-N+1} = x_{-N+2} = \cdots = x_{-1} = 0; \qquad x_0 = 1.$$

In chapter 7 we now formed the quotients

$$(8\text{-}4) \qquad q_n = \frac{x_{n+1}}{x_n} = q_n^{(1)}.$$

If the polynomial p has a single dominant zero, then, as was shown in chapter 7, the sequence $\{q_n\}$ converges to it.

The elements of the sequence $\{q_n\}$ will now be denoted by $q_n^{(1)}$; they form the first column of the two-dimensional scheme called the Quotient-Difference (QD) scheme. The elements of the remaining columns are conventionally denoted by $e_n^{(1)}, q_n^{(2)}, e_n^{(2)}, q_n^{(3)}, \ldots, e_n^{(N-1)}, q_n^{(N)}$ and are generated by alternately forming differences and quotients, as follows:

(8-5a)
$$e_n^{(k)} = (q_{n+1}^{(k)} - q_n^{(k)}) + e_{n+1}^{(k-1)},$$

(8-5b)
$$q_n^{(k+1)} = \frac{e_{n+1}^{(k)}}{e_n^{(k)}} q_{n+1}^{(k)},$$

where $k = 1, 2, \ldots, N - 1$, $n = 0, 1, 2, \ldots$. In (8-5a) we set $e_n^{(0)} = 0$ when $k = 1$. The number of q columns formed is equal to the degree of the given polynomial.

EXAMPLE

1. For $N = 4$, the general QD scheme looks as follows:

$$
\begin{array}{ccccccc}
& q_0^{(1)} & & & & & \\
0 & & e_0^{(1)} & & & & \\
& q_1^{(1)} & & q_0^{(2)} & & & \\
0 & & e_1^{(1)} & & e_0^{(2)} & & \\
& q_2^{(1)} & & q_1^{(2)} & & q_0^{(3)} & \\
0 & & e_2^{(1)} & & e_1^{(2)} & & e_0^{(3)} \\
& q_3^{(1)} & & q_2^{(2)} & & q_1^{(3)} & & q_0^{(4)} \\
0 & & e_3^{(1)} & & e_2^{(2)} & & e_1^{(3)} \\
& q_4^{(1)} & & q_3^{(2)} & & q_2^{(3)} & & q_1^{(4)} \\
& \vdots & e_4^{(1)} & \vdots & e_3^{(2)} & \vdots & e_2^{(3)} & \vdots \\
& & \vdots & & \vdots & & \vdots &
\end{array}
$$

Scheme 8.1

In each column of the scheme the superscripts are constant, and in each diagonal the subscripts. The rules (8-5) can be memorized by observing that in each of the rhombus-like configurations shown in the scheme either the sums or the products of the SW and of the NE pair of elements are equal. If a rhombus is centered in a q column, sums are equal; if it is centered in an e column, products are equal. In view of this interpretation the formulas (8-5) are occasionally referred to as the *rhombus rules*.

The QD scheme can be described in yet another way if we introduce, in addition to the forward difference operator already introduced in §4.4, the quotient operator Q defined by

$$Qx_n = \frac{x_{n+1}}{x_n}.$$

The relations (8-5) then can be written more compactly thus:

$$(8\text{-}6) \qquad e_n^{(k)} = e_{n+1}^{(k-1)} + \Delta q_n^{(k)}, \qquad q_n^{(k+1)} = q_{n+1}^{(k)} Q e_n^{(k)}.$$

Here it must be understood that the operators Δ and Q act on the *subscript*.

EXAMPLE

2. We conclude this section with the numerical QD scheme for the polynomial

$$p(z) = z^2 - z - 1.$$

The sequence $\{x_n\}$ is the Fibonacci sequence (see example 7, §6.3).

Table 8.1

x_n	$e_n^{(0)}$	$q_n^{(1)}$	$e_n^{(1)}$	$q_n^{(2)}$	$e_n^{(2)}$
1	0				
		1.000000			
1	0		1.000000		
		2.000000		-1.000000	
2	0		-0.500000		-0.000001
		1.500000		-0.500001	
3	0		0.166667		-0.000001
		1.666667		-0.666669	
5	0		-0.066667		0.000002
		1.600000		-0.600000	
8	0		0.025000		0.000025
		1.625000		-0.624975	
13	0		-0.009615		-0.000049
		1.615385		-0.615409	
21	0		0.003663		-0.000171
		1.619048		-0.619243	
34	0		-0.001401		
		1.617647			
55	0				
		$\cdots$		$\cdots$	
		$\downarrow$		$\downarrow$	
		$\dfrac{1 + \sqrt{5}}{2}$		?	

Problems

1. Generate the QD scheme for the polynomial $p(z) = z^3 + 5x^2 + 9z + 5$.
2. Let $x_n = n!$, $n = 0, 1, 2, \ldots$. Determine the QD scheme corresponding to the sequence $\{x_n\}$ (a) numerically, (b) analytically. (The sequence $\{x_n\}$ does not arise as solution of a difference equation in this case.)
3. Give analytical formulas for the entries of the QD scheme, if $x_n = 1 + q^n$, where $0 < |q| < 1$. Show that

$$\lim_{n \to \infty} q_n^{(1)} = 1, \qquad \lim_{n \to \infty} e_n^{(1)} = 0, \qquad \lim_{n \to \infty} q_n^{(2)} = q,$$

and that $e_n^{(2)} = 0$ for all n.

8.2   Existence of the QD Scheme

Evidently the QD scheme fails to exist if a coefficient $e_n^{(k)}$ with $0 < k < N$ becomes zero, and it is easy to construct examples for which this actually occurs.   Another, trivial, case of nonexistence of the scheme arises when a x_n becomes accidentally zero.   It appears to be difficult to state explicit necessary and sufficient conditions for the existence of the scheme in terms of the polynomial p.   In terms of the sequence $\{x_n\}$, a necessary and sufficient condition is that the determinants

$$(8\text{-}7) \qquad H_n^{(k)} = \begin{vmatrix} x_n & x_{n+1} & \cdots & x_{n+k-1} \\ x_{n+1} & x_{n+2} & \cdots & x_{n+k} \\ \cdot & \cdot & \cdot \cdot \cdot \cdot & \cdot \\ x_{n+k-1} & x_{n+k} & \cdots & x_{n+2k-2} \end{vmatrix}$$

should be different from zero for $k = 1, 2, \ldots, N$ and for $n = 0, 1, 2, \ldots$. It is possible to state simple *sufficient* (but not necessary) conditions for this to be the case.   Among them are the following:

(*i*) The zeros $z_1, z_2, \ldots, z_N$ of p are positive, and the sequence $\{x_n\}$ is started by algorithm 7.4.

(*ii*) The zeros $z_1, z_2, \ldots, z_N$ of p are simple (but not necessarily real) and have distinct absolute values:

$$(8\text{-}8) \qquad |z_1| > |z_2| > \cdots > |z_N| > 0.$$

In case (*ii*) we can assert only that $H_n^{(k)} \neq 0$, and consequently $e_n^{(k)} \neq 0$, for all sufficiently large values of n.

There is a good deal of numerical evidence that the QD scheme exists in many cases even if neither of the above sufficient conditions is satisfied, for instance if p is a polynomial with real coefficients having pairs of complex conjugate zeros.

Problems

4. Show that

$$q_n^{(1)} = \frac{H_{n+1}^{(1)}}{H_n^{(1)}}, \qquad e_n^{(1)} = \frac{H_n^{(2)}}{H_n^{(1)} H_{n+1}^{(1)}}, \qquad q_n^{(2)} = \frac{H_{n+1}^{(2)} H_n^{(1)}}{H_n^{(2)} H_{n+1}^{(1)}}.$$

5. Show that the determinants $H_n^{(1)}$ and $H_n^{(2)}$ formed with the elements $x_n = 1 + q^n$, where $0 < |q| < 1$, are always different from zero.

6. Let $x_n = r^n \cos(n\varphi + \delta)$, where $0 < \varphi < \pi$. Show that the corresponding determinants $H_n^{(2)}$ are always different from zero. What are the conditions on φ and δ in order that $H_n^{(1)} \neq 0$ for all n?

8.3   Convergence Theorems

If the QD scheme exists, some remarkable statements are possible about the limits of its elements as $n \to \infty$. The simplest situation arises if the zeros of the polynomial p satisfy (8-8). We then have

Theorem 8.3a   Under the conditions just stated,

$$(8\text{-}9) \qquad\qquad \lim_{n \to \infty} q_n^{(k)} = z_k, \qquad k = 1, 2, \ldots, N,$$

i.e., the kth q-column of the QD scheme converges to the kth zero of the polynomial.

It follows from (8-9) by virtue of (8-5a) that

$$\lim_{n \to \infty} e_n^{(1)} = 0,$$

and from this we get easily by induction

$$(8\text{-}10) \qquad\qquad \lim_{n \to \infty} e_n^{(k)} = 0, \qquad k = 1, 2, \ldots, N - 1.$$

Thus, under the condition (8-8), all e-columns of the QD scheme tend to zero.

EXAMPLE

3.   In table 8.1 the column headed $q_n^{(2)}$ tends to $(1 - \sqrt{5})/2$, the smaller zero of $p(z) = z^2 - z - 1$. (Concerning the column $e_n^{(2)}$, see §8.4.)

If several zeros of p have the same absolute value (this happens, for instance, every time when a polynomial with real coefficients has a pair of complex conjugate zeros), the convergence properties of the scheme are more complicated. We still assume that the zeros are numbered such that

$$(8\text{-}11) \qquad\qquad |z_1| \geq |z_2| \geq |z_3| \geq \cdots \geq |z_N| > 0.$$

For convenience in formulating some of the conditions below, we shall put

$$|z_0| = \infty, \qquad |z_{N+1}| = 0.$$

Always assuming that the scheme exists, we then have

Theorem 8.3b   For every k such that $|z_{k+1}| < |z_k| < |z_{k-1}|$,

$$(8\text{-}12) \qquad\qquad \lim_{n \to \infty} q_n^{(k)} = z_k.$$

For every k such that $|z_k| > |z_{k+1}|$,

$$(8\text{-}13) \qquad\qquad \lim_{n \to \infty} e_n^{(k)} = 0.$$

These facts can be used in the following manner: The e columns which tend to zero (a behavior which is numerically conspicuous) divide the QD

table into subtables. All zeros z_k whose subscripts agree with the super-scripts of the q's in one subtable have the same modulus. Thus, if a z_k is the only zero of its modulus, this will be evident from the fact that the corresponding subtable contains one q column only, and the value of z_k can be obtained as the limit of that q column.

It is not yet clear how to deal with several zeros having the same absolute value. (Most frequently this situation occurs in connection with complex conjugate zeros of real polynomials.) Such zeros, too, can be obtained from the QD table. We first consider the general case where m zeros $z_{k+1}, z_{k+2}, \ldots, z_{k+m}$ have the same modulus:

$$(8\text{-}14) \qquad |z_k| > |z_{k+1}| = |z_{k+2}| = \cdots = |z_{k+m}| > |z_{k+m+1}|.$$

Here it is necessary to construct polynomials $p_n^{(l)}, l = k, k+1, \ldots, k+m$, by means of the recurrence relations

$$(8\text{-}15a) \qquad p_n^{(k)}(z) = 1, \qquad n = 0, 1, 2, \ldots,$$

$$(8\text{-}15b) \qquad p_n^{(l)}(z) = z p_{n+1}^{(l-1)}(z) - q_n^{(l)} p_n^{(l-1)}(z),$$
$$l = 1, 2, \ldots, m; \quad n = 0, 1, 2, \ldots.$$

These polynomials can again be thought as being arranged in a two-dimensional array. Scheme 8.3 shows a segment of this array for $m = 2$.

$$
\begin{array}{ccccc}
1 & -q_n^{(k+1)} & & & \\
 & z & p_n^{(k+1)}(z) & & \\
1 & & & -q_n^{(k+2)} & \\
 & & z & & p_n^{(k+2)}(z) \\
 & p_{n+1}^{(k+1)}(z) & & & \\
1 & & & p_{n+1}^{(k+2)}(z) & \\
 & p_{n+2}^{(k+1)} & & & \\
1 & & & &
\end{array}
$$

Scheme 8.3

The zeros $z_{k+1}, z_{k+2}, \ldots, z_{k+m}$ can now be obtained from the polynomials $p_n^{(k+m)}$ by virtue of the following theorem:

Theorem 8.3c If the zeros satisfy (8-14), then for each fixed z

$$(8\text{-}16) \qquad \lim_{n \to \infty} p_n^{(k+m)}(z) = (z - z_{k+1})(z - z_{k+2}) \cdots (z - z_{k+m}),$$

i.e., the coefficients of the polynomials $p_n^{(k+m)}$ tend for $n \to \infty$ to the coefficients of the polynomial with zeros $z_{k+1}, \ldots, z_{k+m}$ and leading coefficient 1.

For $m = 1$ theorem 8.3c reduces to relation (8-3a). For $m = 2$ (the practically most frequent case) the converging polynomials are given by

$$p_n^{(k+2)}(z) = z[z - q_{n+1}^{(k+1)}] - q_n^{(k+2)}[z - q_n^{(k+1)}]$$
$$= z^2 - (q_{n+1}^{(k+1)} + q_n^{(k+2)})z + q_n^{(k+1)}q_n^{(k+2)}.$$

Relation (8-16) here means that the limits

(8-17a) $$\lim_{n \to \infty} (q_{n+1}^{(k+1)} + q_n^{(k+2)}) = A_k,$$

(8-17b) $$\lim_{n \to \infty} q_n^{(k+1)}q_n^{(k+2)} = B_k,$$

exist, and that the polynomial $z^2 - A_k z + B_k$ has the zeros z_{k+1} and z_{k+2}.

We finally mention the following fact, which could (and will) play the role of a computational check.

Theorem 8.3d   The quantities $e_n^{(N)}$ calculated from (8-5a) with $k = N$ are identically zero.

Examples illustrating the above theorems will be given in §8.5.

Problems

7. Prove theorem 8.3a in the case of a polynomial of degree $N = 2$ whose zeros satisfy $|z_1| > |z_2| > 0$.
8. Show that

$$q_n^{(1)}q_n^{(2)} = \frac{D_{n+3}}{D_{n+2}}, \qquad q_{n+1}^{(1)} + q_n^{(2)} = -\frac{E_{n+2}}{D_{n+2}},$$

where D_n and E_n denote the determinants defined by (7-22) and (7-23). Thus conclude that in the case $z_1 = \bar{z}_2$ theorem 8.3c is equivalent to theorem 7.5.
9. Obtain the two dominant complex conjugate zeros of the polynomial

$$p(z) = 81z^4 - 108z^3 + 24z + 20$$

from the QD scheme by using the relations (8-17) with $k = 0$.
10. Assuming that the quantities x_n are given by (7-17), prove theorem 8.3d for polynomials of degree $N = 1$ and $N = 2$.

8.4   Numerical Instability

As described above, the QD scheme is built up proceeding from the left to the right. The sequence $\{x_n\}$ determines the first q column $\{q_n^{(1)}\}$; from it we obtain in succession the columns $\{e_n^{(1)}\}, \{q_n^{(2)}\}, \ldots$, by means of the relations (8-5). The reader will be shocked to learn that this method of generating the QD scheme is not feasible in practice, because it suffers from *numerical instability* due to severe loss of *significant digits*.

The concept of numerical instability was already briefly alluded to in §1.5. In mathematical analysis, real (or complex) numbers are always conceived as being determined with infinite accuracy. Mathematically, this infinite accuracy is expressed by the Dedekind cut property (see Taylor [1959], p. 447; Buck [1956], p. 388). Numerically, the real numbers of mathematical analysis can be thought of as infinite decimal fractions. In computation a real number z is—with rare exceptions— never represented exactly, but rather approximated by some rational number z^*, e.g., by a decimal fraction with eight decimals.† The quantity $z^* - z$ is called *rounding error*, or sometimes also *absolute rounding error*, in order to distinguish it from the *relative rounding error* $|z^* - z|/|z|$. The following simple rules concerning the propagation of rounding errors follow easily from first principles:

(*i*) A sum or difference of two rounded numbers a^* and b^* has an *absolute* rounding error of the order of the sum of the *absolute* errors of a^* and b^*.

(*ii*) A product or quotient of two rounded numbers a^* and b^* has a *relative* error of the order of the sum of the *relative* errors of a^* and b^*.

The QD scheme offers an interesting illustration of these rules. To simplify matters, let us assume that the sequence $\{x_n\}$ is generated without rounding error (this is possible if p is a polynomial with integer coefficients and leading coefficient 1), and that the situation covered by theorem 8.3a obtains. If we carry t decimals, the numbers $q_n^{(1)}$ then will have (absolute) rounding errors of the order of 10^{-t}. The elements $e_n^{(1)}$, formed by differencing the $q_n^{(1)}$, will by rule (*i*) have absolute rounding errors of the same magnitude. However, since by (8-10) the $e_n^{(1)}$ tend to zero, their *relative* errors become larger and larger. Thus, by rule (*ii*), the ratios $e_{n+1}^{(1)}/e_n^{(1)}$ are formed less and less accurately, and, again by rule (*ii*), the relative error of the quantity $q_n^{(2)}$, as determined by (8-5b), increases without bound. Since $q_n^{(2)}$ tends to a nonzero limit as $n \to \infty$, the same is true for the absolute error of $q_n^{(2)}$. It is clear that this numerical instability becomes even more pronounced as k increases.

EXAMPLE

4. The loss of significant digits is illustrated already in example 2. By theorem 8.3d, the column $e_n^{(2)}$ should theoretically consist of zeros. The fact that these elements are not zero, and even increase with increasing n, shows the growing influence of rounding errors. The reader is asked to

---

† For more details the reader is referred to chapter 15. At the present stage we aim at a qualitative rather than quantitative understanding of rounding errors.

calculate the same example using numbers with ten decimals.   This will delay, but not ultimately prevent, the phenomenon of numerical instability.

The fact that the method of generating the QD scheme described in §8.1 is unstable does not, of course, prevent the theorems stated in §8.3 from being true.   These theorems concern the mathematically exact, unrounded QD scheme.   They can be used numerically as soon as we succeed in generating the QD scheme in a numerically stable manner.   One method of avoiding rounding errors, applicable to polynomials with rational coefficients, would be to perform all operations in exact rational arithmetic (see Henrici [1956]).   Fortunately, as will be seen in the following section, there is a simple way of generating a stable QD scheme also in conventional arithmetic.

8.5   Progressive Form of the Algorithm

The QD scheme can be generated in a stable manner if it is built up row by row instead of column by column, as in §8.1.   To see this, we solve each of the recurrence relations (8-5) for the south element of the rhombus involved:

$$(8\text{-}18a) \qquad q_{n+1}^{(k)} = (e_n^{(k)} - e_{n+1}^{(k-1)}) + q_n^{(k)},$$

$$(8\text{-}18b) \qquad e_{n+1}^{(k)} = \frac{q_n^{(k+1)}}{q_{n+1}^{(k)}} e_n^{(k)}.$$

Let us assume that a row of q's and a row of e's is known, each affected by "normal" rounding errors of several digits in the last place.   The new row of q's, as calculated from (8-18a), will then, by rule (i), have absolute errors of the same magnitude.   The fact that the relative errors in the e's are large (due to the smallness of the e's) is not important now.   Furthermore, the relative errors in the new row of e's determined by (8-18b) are of the same order as in the old row of e's, due to rule (ii).   While a normal amount of error propagation must be expected also in the present mode of generating the scheme, it is much less serious than when the scheme is generated column by column.

If the scheme is to be generated row by row, a first couple of rows must somehow be obtained.   The following algorithm shows how this is accomplished.

Algorithm 8.5   Let $a_0, a_1, \ldots, a_N$ be constants, all different from zero. Set

$$(8\text{-}19a) \qquad q_0^{(1)} = -\frac{a_1}{a_0}, \qquad q_{1-k}^{(k)} = 0, \qquad k = 2, 3, \ldots, N;$$

$$(8\text{-}19b) \qquad e^{(k)}_{(1-k)} = \frac{a_{k+1}}{a_k}, \qquad k = 1, 2, \ldots, N - 1.$$

Consider the elements thus generated as the first two rows of a QD scheme, and generate further rows by means of (8-18), using the side conditions

$$(8\text{-}20) \qquad e^{(0)}_n = e^{(N)}_n = 0, \qquad n = 1, 2, \ldots.$$

As before, there is the theoretical possibility of breakdown of the scheme due to the fact that a denominator is zero.   However, we have

Theorem 8.5   If the scheme of the elements $q^{(k)}_n$ and $e^{(k)}_n$ defined by algorithm 8.5 exists, it is (for $n \geqq 0$) identical with the QD scheme of the polynomial

$$p(z) = a_0 z^N + a_1 z^{N-1} + \cdots + a_N,$$

where the sequence $\{x_n\}$ is started in the manner (8-3).

The proof of theorem 8.5 requires some involved algebra and it is omitted here.   A necessary and sufficient condition for the existence of the scheme defined by algorithm 8.5 is that the determinants (8-7) be different from zero also for $n > -k$.   (Here we have to interpret $x_n = 0$ for $n < 0$.)

EXAMPLES

5.   The top rows of the scheme for $N = 4$ look as follows:

	$-\dfrac{a_1}{a_0}$		0		0		0
0		$\dfrac{a_2}{a_1}$		$\dfrac{a_3}{a_2}$		$\dfrac{a_4}{a_3}$	
	$q_1^{(1)}$		$q_0^{(2)}$		$q_{-1}^{(3)}$		$q_{-2}^{(4)}$
0		$e_1^{(1)}$		$e_0^{(2)}$		$e_{-1}^{(3)}$	
	$q_2^{(1)}$		$q_1^{(2)}$		$q_0^{(3)}$		$q_{-1}^{(4)}$
0		$e_2^{(1)}$		$e_1^{(2)}$		$e_0^{(3)}$	

Scheme 8.5

6.   For the polynomial

$$p(z) = 128z^4 - 256z^3 + 160z^2 - 32z + 1$$

we obtain the scheme shown in table 8.5a.

Table 8.5a

$q_n^{(1)}$	$e_n^{(1)}$	$q_n^{(2)}$	$e_n^{(2)}$	$q_n^{(3)}$	$e_n^{(3)}$	$q_n^{(4)}$
2.000000		.000000		.000000		.000000
	−.625000		−.200000		−.031250	
1.375000		.425000		.168750		.031250
	−.193182		−.079412		−.005787	
1.181818		.538770		.242375		.037037
	−.088068		−.035725		−.000884	
1.093750		.591114		.277215		.037921
	−.047596		−.016754		−.000121	
1.046154		.621956		.293848		.038042
	−.028297		−.007915		−.000016	
1.017857		.642337		.301748		.038058
	−.017857		−.003718		−.000002	
1.000000		.656476		.305464		.038060
	−.011723		−.001730		−.000000	
.988277		.666468		.307194		.038060
	−.007906		−.000798		−.000000	
.980372		.673576		.307992		.038060
	−.005432		−.000365		−.000000	
.974940		.678643		.308356		.038060
	−.003781		−.000166		−.000000	

All e columns tend to zero, thus by theorem 8.3a the polynomial has four zeros whose absolute values are different. The q columns converge to the zeros, whose exact values are as follows:

$$z_1 = 0.96194, \quad z_2 = 0.69134, \quad z_3 = 0.30866, \quad z_4 = 0.03806.$$

7. For the polynomial

$$p(z) = z^4 - 8z^3 + 39z^2 - 62z + 50$$

algorithm 8.5 yields the following QD scheme:

Table 8.5b

$q_n^{(1)}$	$e_n^{(1)}$	$q_n^{(2)}$	$e_n^{(2)}$	$q_n^{(3)}$	$e_n^{(3)}$	$q_n^{(4)}$
8.000000		.000000		.000000		.000000
	−4.875000		−1.589744		−.806452	
3.125000		3.285256		.783292		.806452
	−5.125000		−.379037		−.830296	
−2.000000		8.031220		.332033		1.636748
	20.580000		−.015670		−4.092923	
18.580000		−12.564451		−3.745220		5.729671
	−13.916921		−.004671		6.261609	
4.663079		1.347799		2.521060		−.531938
	−4.022497		−.008737		−1.321186	
.640581		5.361559		1.208612		.789248
	−33.667611		−.001970		−.862761	
−33.027030		39.027201		.347820		1.652009

We now have $e_n^{(2)} \to 0$, but $e_n^{(1)}$ and $e_n^{(3)}$ do not tend to zero. This is an indication that there are two pairs of complex conjugate zeros.   To make use of theorem 8.3c, we form the quantities

$$A_n^{(k)} = q_{n+1}^{(k+1)} + q_n^{(k+2)}, \qquad B_n^{(k)} = q_n^{(k+1)}q_n^{(k+2)}$$

for $k = 0$ and $k = 2$, obtaining the following values:

Table 8.5c

$A_n^{(0)}$	$B_n^{(0)}$	$A_n^{(2)}$	$B_n^{(2)}$
6.410256	26.282051	1.589744	0
6.031220	25.097561	1.968780	1.282051
6.015549	25.128902	1.984451	1.902439
6.010878	25.042114	1.989122	1.992225
6.002141	25.001372	1.997859	1.989741
6.000171	25.000110	1.999829	1.996637

The limits are 6, 25, 2, 2 respectively, indicating that the polynomials

$$z^2 - 6z + 25 \quad \text{and} \quad z^2 - 2z + 2$$

are quadratic factors of the given polynomial.   In fact,

$$(z^2 - 6z + 25)(z^2 - 2z + 2) = z^4 - 8z^3 + 39z^2 - 62z + 50.$$

We have yet to deal with the complication that arises if some of the coefficients $a_0, a_1, \ldots, a_N$ are zero.   In that case the extended QD scheme defined by algorithm 8.5 clearly does not exist, since some of the relations (8-19) may fail to make sense.   A possible remedy is to introduce a new variable

$$z^* = z - a$$

and to consider the polynomial

$$p^*(z^*) = p(a + z^*)$$

$$= p(a) + \frac{1}{1!}p'(a)z^* + \frac{1}{2!}p''(a)z^{*2} + \cdots + \frac{1}{N!}p^{(N)}(a)z^{*N}.$$

Here a denotes a suitably chosen parameter.   The coefficients of the polynomial p^* can easily be calculated by means of algorithm 3.6.   It can be shown that if p has some zero coefficients, then all coefficients of p^* are different from zero for sufficiently small values of $a \neq 0$.   If the zeros z_k^* of p^* have been computed, those of p are given by the formula

$$z_k = z_k^* + a.$$

EXAMPLE

8.   Let $p(z) = 81z^4 - 108z^3 + 24z + 20$.   Here $a_2 = 0$, and algorithm

8.5 cannot be started.   We form p^* with $a = 1$.   Scheme 3.6 turns out thus:

81	-108	0	24	20
81	-27	-27	-3	17
81	54	27	24	
81	135	162		
81	216			
81				

The new polynomial

$$p^*(z) = 81z^{*4} + 216z^{*3} + 162z^{*2} + 24z^* + 17$$

has all coefficients different from zero, and the first rows of its QD scheme are as follows:

	$-\dfrac{216}{81}$		0		0		0	
0		$\dfrac{162}{216}$		$\dfrac{24}{162}$		$\dfrac{17}{24}$		0

Problems

11. Construct the QD scheme for the polynomial

$$p(z) = 32z^3 - 48z^2 + 18z - 1$$

and determine its zeros to four significant digits.   Check your result by observing that $z = \frac{1}{2}$ is a zero.   (The convergence of the q columns may be sped up by Aitken's $\varDelta^2$-process.)

12. Determine approximate values for the zeros of the polynomial

$$p(z) = 70z^4 - 140z^3 + 90z^2 - 20z + 1.$$

Then find more exact values by Newton's method.

13. The polynomial

$$p(z) = z^5 - 3z^4 - 20z^3 + 60z^2 - z - 78$$

has two large real zeros of opposite sign, a pair of complex conjugate zeros, and a small real zero.   Find approximate values for the quadratic factors belonging to the two large real zeros and to the pair of complex zeros.

14. Prove theorem 8.5 for polynomials of degree $N = 2$.   [Hint: It suffices to show that algorithm 8.5 generates the correct values of $q_0^{(1)}$, $e_0^{(1)}$, and $q_0^{(2)}$.]

8.6   Computational Checks

Even if the QD scheme is generated by algorithm 8.5, excessively large (or small) elements may cause some loss of accuracy.   The mathematical results given below may be used for checking purposes; failure of any of these checks indicates excessive rounding error.

(*i*) Relation (8-5a) implies that the sum of the q values in any row of the scheme is constant.   For the top row this sum is $-a_1/a_0$, which by Vieta's formula (see §2.5) equals the algebraic sum of the zeros of the polynomial. Thus we have

$$(8\text{-}21) \qquad q_n^{(1)} + q_{n-1}^{(2)} + \cdots + q_{n-N+1}^{(N)} = -\frac{a_1}{a_0}.$$

Note that by theorem 8.3a this relation confirms Vieta's rule for $n \to \infty$!

(*ii*) It can be shown that the product of all q elements in any diagonal sloping downward is likewise independent of n and equals the product of all zeros of the polynomials p.   Thus again by Vieta,

$$(8\text{-}22) \qquad q_n^{(1)} q_n^{(2)} \ldots q_n^{(N)} = (-1)^N \frac{a_N}{a_0}.$$

It should be noted that while (8-21) checks only the additions and subtractions performed in constructing the scheme, equation (8-22) checks all operations.

(*iii*) If the QD scheme is generated by algorithm 8.5, the quantities x_n are not needed.   However, we may calculate the x_n from (8-2) and should find

$$(8\text{-}23) \qquad \frac{x_{n+1}}{x_n} = q_n^{(1)}, \qquad n = 0, 1, 2, \ldots.$$

Problem

15. Prove (8-22) for the QD scheme arising from a polynomial of degree $N = 2$.

8.7   QD versus Newton

In comparison with other methods for determining the zeros of a polynomial, the QD algorithm enjoys the tremendous advantage of furnishing simultaneously approximations to all zeros of a polynomial. No information about the polynomial other than the values of its coefficients is required.

These advantages have to be paid for by the rather slow convergence of the algorithm.   Since the QD method contains the Bernoulli method as a special case, the convergence can be no better than that of Bernoulli's method. In fact it can be shown that under the hypotheses of theorem 8.3b the errors $q_n^{(k)} - z_k$ tend to zero like the larger of the ratios

$$\left(\frac{z_k}{z_{k-1}}\right)^n \quad \text{and} \quad \left(\frac{z_{k+1}}{z_k}\right)^n.$$

Even if the figures in the q columns eventually settle down, the accuracy of the zeros thus obtained is somewhat uncertain, because the large number of arithmetic operations may have contaminated the scheme with rounding error.

For the above reasons, the QD algorithm is not recommended for the purpose of determining the zeros of a polynomial with final accuracy. Instead, the following two-stage procedure is advocated:

Stage 1: Use the QD algorithm to obtain crude first approximations to the zeros, respectively to the quadratic factors containing complex conjugate zeros.

Stage 2: Using these approximations as starting values, obtain the zeros accurately by Newton's or Bairstow's method.

This combination of several methods has the advantage that the final values of the zeros are obtained from the original, undisturbed polynomial, and thus are practically free of rounding error.

The choice of the point at which to make the change-over from QD to Newton-Bairstow is, to some extent, arbitrary. It is probably best to carry the QD scheme to a point where the division of the scheme into subschemes in the manner described after theorem 8.3b is clearly evident. On the other hand, if QD is pushed too far, a lot of computational effort may be wasted, since Newton-Bairstow usually takes only two or three steps to obtain the zeros very accurately even from mediocre first approximations. To fuse the three algorithms into one working program constitutes a challenging but rewarding problem in machine programming which is highly recommended to the reader. One such program is described by Watkins [1964], who also presents the results of extensive machine tests.

Problem

16. By combining QD and Newton-Bairstow, compute all zeros of the following polynomials with an error of less than 10^{-7}:

(a) $p(z) = z^4 - 8z^3 + 39z^2 - 62z + 51$;
(b) $p(z) = z^5 - 15z^4 + 85z^3 - 225z^2 + 264z - 120$;
(c) $p(z) = 4z^6 - 5z^5 + 4z^4 - 3z^3 + 7z^2 - 7z + 1$.

8.8   Other Applications

In addition to the calculation of the zeros of a polynomial, the QD

algorithm has many other applications, notably in the theory of continued fractions, in matrix computation, and in the summation of divergent series (see the references given below). It can also be used to furnish exact bounds (and not only approximations) for the location of the zeros of a polynomial. We shall mention only one further application which is related to the one discussed above.

Suppose the function f is defined by the power series

$$(8\text{-}24) \qquad\qquad f(z) = \sum_{n=0}^{\infty} a_n z^n,$$

where $a_n \neq 0$, $n = 0, 1, 2, \ldots$, and let it be known that the zeros z_k of f are real and positive, $0 < z_1 < z_2 < \cdots$.

EXAMPLE

9. The Bessel function of order zero can be defined by

$$(8\text{-}25) \qquad\qquad J_0(2\sqrt{x}) = \sum_{n=0}^{\infty} \frac{(-x)^n}{(n!)^2}$$

and has the required properties.

Obviously the QD scheme of such a function cannot be found in the ordinary way, because the horizontal rows are now infinite. However, the scheme may be generated in the following manner, as indicated by the arrows:

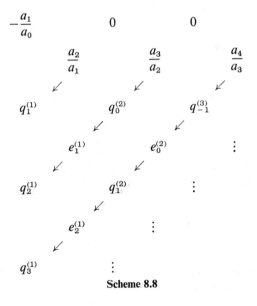

Scheme 8.8

Although the scheme does not terminate on the right, more and more diagonals sloping upward can be found. It can be shown that if the scheme exists,

$$\lim_{n \to \infty} q_n^{(k)} = \frac{1}{z_k}, \qquad k = 1, 2, \ldots.$$

Under certain conditions even complex conjugate zeros of transcendental functions can be found in this manner, using the method described in theorem 8.3c and in example 7.

EXAMPLE

10. For the function defined by (8-25) the coefficients a_n evidently satisfy

$$\frac{a_{n+1}}{a_n} = -\frac{1}{(n+1)^2}.$$

The first few diagonals of the QD scheme thus appear as follows:

```
1.000000        .000000         .000000         .000000         .000000
    -.250000        -.111111        -.062500        -.040000
.750000         .138889         .048611         .022500         .012222
    -.046296        -.038889        -.028929        -.021728
.703704         .146296         .058571         .029700
    -.009625        -.015570        -.014669
.694079         .140351         .059472
    -.001946        -.006597
.692133         .135700
    -.000382
.691751            ⋮
    ⋮              ⋮
.691660         .131271
```

Problems

17. Obtain an approximate value of $\sqrt{2/\pi}$ by applying the algorithm described above to the function

$$\cos \sqrt{z} = \sum_{n=0}^{\infty} \frac{(-z)^n}{(2n)!}.$$

18. Apply the above version of the QD algorithm to the problem of finding the small solutions of the transcendental equation

$$\tan z = cz$$

by setting

$$f(z^2) = \frac{\sin z}{z} - c \cos z.$$

Find the smallest positive solution for $c = 1.2$, and a pair of purely imaginary solutions for $c = 0.8$.

Recommended Reading

The QD algorithm was introduced by Rutishauser in a series of classical papers which are collected in the volume Rutishauser [1956].   A somewhat more elementary treatment is given by Henrici [1958].   A multitude of applications are discussed in Henrici [1963].

Research Problem

How can the argument of a complex zero of a real polynomial be determined from the signs of the elements of the corresponding q column? Consider the case $N = 2$ first.

PART TWO

INTERPOLATION

AND

APPROXIMATION

chapter 9 the interpolating polynomial

So far we have been concerned mainly with the problem of approximating *numbers* (such as the zeros of a polynomial or the solutions of systems of nonlinear equations). We now turn to the problem of approximating *functions* and, more generally, numbers, such as derivatives and integrals, that depend on an infinity of values of a function.

The most common method of approximating functions is the approximation by *polynomials*. Among the various types of polynomial approximation that are in use the one that is most flexible and most easily constructed (although not always the most effective) is the approximation by the *interpolating polynomial*.

9.1 Existence of the Interpolating Polynomial

Let the real function f be defined on an interval I, and let $x_0, x_1, \ldots, x_n$ be $n + 1$ *distinct* points of I. It is not assumed that these points are equidistant, nor even that they are in their natural order.

We shall write for brevity

$$f(x_k) = f_k, \qquad k = 0, 1, \ldots, n.$$

Theorem 9.1 There exists a unique polynomial P of degree not exceeding n (the so-called Lagrangian interpolating polynomial) such that

(9-1) $$P(x_k) = f_k, \qquad k = 0, 1, \ldots, n.$$

Proof. As usual, the proofs of existence and of uniqueness require separate arguments. The *existence* of the polynomial P is proved if we

183

can establish the existence of polynomials L_k ($k = 0, 1, \ldots, n$) with the following properties:

(*i*) Each L_k is a polynomial of degree $\leq n$;
(*ii*) For $x = x_k$, L_k has the special value

(9-2a) $$L_k(x_k) = 1;$$

however, if $m \neq k$, then

(9-2b) $$L_k(x_m) = 0.$$

Assuming the existence of these L_k, we can set

(9-3) $$P(x) = \sum_{k=0}^{n} f_k L_k(x).$$

The function P is a sum of polynomials of degree $\leq n$ with constant factors and thus itself a polynomial of degree $\leq n$. Furthermore, if we set $x = x_m$, then by (*ii*) all L_k are zero except for the one with $k = m$, and this has the value 1. Thus we find

$$P(x_m) = f_m,$$

as required.

The polynomials L_k are called the Lagrangian interpolation coefficients. To prove the existence of L_k, observe that the product

$$\prod_{\substack{m=0 \\ m \neq k}}^{n} \frac{x - x_m}{x_k - x_m} = \frac{(x - x_0)\ldots(x - x_{k-1})(x - x_{k+1})\ldots(x - x_n)}{(x_k - x_0)\ldots(x_k - x_{k-1})(x_k - x_{k+1})\ldots(x_k - x_n)}$$

has the required properties. Indeed, as a product of $n + 1 - 1 = n$ linear factors it represents a polynomial of degree n; furthermore, if $x = x_k$, then all factors have the value 1, thus the product has the value 1 also. On the other hand, if $x = x_m$, where $m \neq k$, then the factor containing $x - x_m$ is zero, and the product vanishes. Thus, the polynomials

(9-4) $$L_k(x) = \prod_{\substack{m=0 \\ m \neq k}}^{n} \frac{x - x_m}{x_k - x_m}$$

have the required properties, and the existence of the polynomial P is proved.

In order to show the uniqueness of the interpolating polynomial, assume there exist two interpolating polynomials, P and Q, say. Then their difference $D = P - Q$, being the difference of two polynomials of degree not exceeding n, is again a polynomial of degree not exceeding n. Moreover,

$$D(x_k) = P(x_k) - Q(x_k) = f_k - f_k = 0$$

for $k = 0, 1, \ldots, n$. The polynomial D thus has $n + 1$ zeros and hence, being a polynomial of degree $\leq n$, must vanish identically. It follows that $P = Q$. This completes the proof of theorem 9.1.

EXAMPLE

1. To find the interpolating polynomial for the following x_k and f_k:

x_k	2	3	-1	4
f_k	1	2	3	4

We first calculate the Lagrangian interpolation coefficients. Formula (9-4) yields

$$L_0(x) = \frac{(x - 3)(x + 1)(x - 4)}{(-1)3(-2)} = \tfrac{1}{6}(x - 3)(x + 1)(x - 4),$$

$$L_1(x) = \frac{(x - 2)(x + 1)(x - 4)}{1\cdot 4\cdot(-1)} = -\tfrac{1}{4}(x - 2)(x + 1)(x - 4),$$

$$L_2(x) = \frac{(x - 2)(x - 3)(x - 4)}{(-3)(-4)(-5)} = -\tfrac{1}{60}(x - 2)(x - 3)(x - 4),$$

$$L_3(x) = \frac{(x - 2)(x - 3)(x + 1)}{2\cdot 1\cdot 5} = \tfrac{1}{10}(x - 2)(x - 3)(x + 1).$$

Formula (9-3) thus yields

$$P(x) = \tfrac{1}{6}(x - 3)(x + 1)(x - 4) - \tfrac{1}{2}(x - 2)(x + 1)(x - 4)$$
$$- \tfrac{1}{20}(x - 2)(x - 3)(x - 4) + \tfrac{2}{5}(x - 2)(x - 3)(x + 1).$$

It can be verified that this polynomial has the required properties.

It will be noted that the representation of the interpolating polynomial given in the proof of theorem 9.1 does not give the polynomial in the customary form

$$P(x) = a_0 x^n + a_1 x^{n-1} + \cdots + a_n.$$

Of course, the polynomial *could* be put into the above standard form, but there usually is no particular reason for doing so. It is well to distinguish at this point between the *function P* and the various *representations* of *P*. As a function (i.e., as a set of ordered pairs $(x, P(x))$), P is unique. However, there may be many ways of representing P by an explicit formula. Each formula suggests a certain algorithm for calculating P. It is not claimed that the algorithm suggested by (9-3) is the most effective from the numerical point of view. Many other algorithms for constructing the polynomial will be discussed in the chapters 10 and 11.

Problems

1. Verify that the case $n = 1$ of (9-3) yields the familiar formula for linear interpolation. What is the meaning of theorem 9.1 when $n = 0$?

2. Is the interpolating polynomial constructed above always of the exact degree n?

3. Show that for $x \neq x_k$, $k = 0, 1, \ldots, n$, the interpolating polynomial can be represented in the form

$$P(x) = L(x) \sum_{k=0}^{n} \frac{f_k}{(x - x_k)L'(x_k)},$$

where

$$L(x) = (x - x_0)(x - x_1) \ldots (x - x_n).$$

Verify by L'Hopital's rule (see Taylor [1959], p. 456) that the limit of the expression on the right as $x \to x_m$ is f_m.

4. Prove: If f is a polynomial of degree n or less, then $P = f$.

9.2   The Error of the Interpolating Polynomial

Since we wish to use the interpolating polynomial to approximate the function f at points which do not belong to the set of *interpolating points* x_k, we are interested in estimating the difference $P(x) - f(x)$ for $x \in I$. It is clear that without further hypotheses nothing whatever can be said about this quantity.   For we can change the function f at will at points which are not interpolating points without changing the polynomial P at all (see Fig. 9.2).

A definite statement can be made, however, if we assume a qualitative knowledge of the derivatives of the function f.

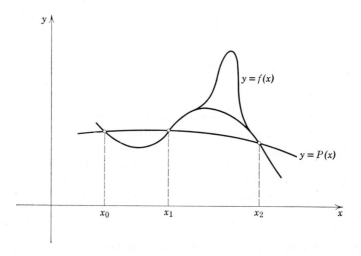

Figure 9.2

Theorem 9.2   In addition to the hypotheses of theorem 9.1, let f be $n + 1$ times continuously differentiable on the interval I.   Then to each $x \in I$ there exists a point ξ_x located in the smallest interval containing the points $x, x_0, x_1, \ldots, x_n$ such that

(9-5) $$f(x) - P(x) = \frac{1}{(n + 1)!} L(x) f^{(n+1)}(\xi_x)$$

where

$$L(x) = (x - x_0)(x - x_1)\ldots(x - x_n).$$

Proof.   If x is one of the points x_k, there is nothing to prove, since both sides of (9-5) vanish for arbitrary ξ.   If x has a fixed value different from any of the points x_k, consider the auxiliary function $F = F(t)$ defined by

(9-6) $$F(t) = f(t) - P(t) - cL(t),$$

where

$$c = \frac{f(x) - P(x)}{L(x)}.$$

We have

$$\begin{aligned}
F(x_k) &= f(x_k) - P(x_k) - cL(x_k) \\
&= f_k - f_k - 0 \\
&= 0, \quad k = 0, 1, 2, \ldots, n,
\end{aligned}$$

and also

$$F(x) = f(x) - P(x) - cL(x) = 0,$$

by the definition of c.   The function F thus has at least $n + 2$ distinct zeros in the interval I.   By Rolle's theorem, the derivative F' must have at least $n + 1$ zeros in the smallest interval containing x and the x_k, the second derivative must have no less than n zeros, and finally the $(n + 1)$st derivative must have at least one zero.   Let ξ_x be one such zero.   We now differentiate (9-6) $n + 1$ times and set $t = \xi_x$.   The $(n + 1)$st derivative of P is zero.   Since L is a polynomial with leading term x^{n+1}, the $(n + 1)$st derivative of cL is $c(n + 1)!$.   We thus have

$$0 = F^{(n+1)}(\xi_x) = f^{(n+1)}(\xi_x) - c(n + 1)!$$

or, remembering the definition of c and rearranging,

$$cL(x) = f(x) - P(x) = \frac{1}{(n + 1)!} L(x) f^{(n+1)}(\xi_x),$$

as was to be shown.

Equation (9-5) cannot be used, of course, to calculate the exact value of the error $f - P$, since ξ_x as a function of x is, in general, not known. (An exception occurs when the $(n + 1)$st derivative of f is constant; see below.)   However, as is shown in the examples below the formula can be used in many cases to find a *bound* for the error of the interpolating polynomial.

We also shall require the following fact.

Corollary 9.2   Under the hypotheses of theorem 9.2, the quantity $f^{(n+1)}(\xi_x)$ in (9-5) can be defined as a continuous function of x for $x \in I$.

Proof.   Define the function $g = g(x)$ by

$$g(x) = \begin{cases} \dfrac{f(x) - P(x)}{L(x)}, & x \neq x_k, \quad k = 0, 1, \ldots, n, \\ \dfrac{f'(x_k) - P'(x_k)}{L'(x_k)}, & x = x_k, \quad k = 0, 1, \ldots, n. \end{cases}$$

This function is continuous for $x \neq x_k$ and, by L'Hopital's rule (see Taylor [1959], p. 456) also at the points x_k. For $x \neq x_k$,

$$f^{(n+1)}(\xi_x) = (n+1)! g(x),$$

establishing the corollary. An application of this corollary will be made in the chapters 12 and 13.

EXAMPLES

2.   *One interpolating point.*   If there is only one interpolating point x_0, the interpolating polynomial reduces to the constant f_0. Formula (9-5) yields

$$f(x) - f(x_0) = (x - x_0) f'(\xi_x),$$

where ξ_x lies between x_0 and x. This is the familiar *mean value theorem* of the differential calculus.

3.   *Two interpolating points.*   The linear interpolating polynomial is given by

$$P(x) = \frac{(x_1 - x) f_0 + (x - x_0) f_1}{x_1 - x_0}.$$

Equation (9-5) yields the error formula

$$f(x) - P(x) = \frac{(x - x_0)(x - x_1)}{2!} f''(\xi_x).$$

What is the maximum error that can occur if we know that $|f''(x)| \leq M_2$ and x is between x_0 and x_1? The maximum of the function

$$|\tfrac{1}{2}(x - x_0)(x - x_1)|$$

between x_0 and x_1 occurs at $x = \tfrac{1}{2}(x_0 + x_1)$ and has the value $\tfrac{1}{8}(x_1 - x_0)^2$. Thus we find

$$|f(x) - P(x)| \leqq \frac{(x_1 - x_0)^2}{8} M_2$$

in this case.   Application: If we calculate the value of sin x from a sine table with step h, using linear interpolation, the error is bounded by $\frac{1}{8}h^2$, since $M_2 = 1$ in this case.

4.   Error in cubic interpolation.   We assume that the four interpolating points are equidistant, $x_k = x_0 + kh$, $k = 1, 2, 3$, and that x (the point where the value of the function f is sought) always lies between x_1 and x_2. We set

$$M_4 = \max_{x_0 \leq x \leq x_3} |f^{(4)}(x)|.$$

The interpolation error is then bounded by $(1/4!)M_4$ times the maximum of the absolute value of the function

$$L(x) = (x - x_0)(x - x_1)(x - x_2)(x - x_3)$$

in the interval $x_1 \leq x \leq x_2$.   For reasons of symmetry this maximum occurs at $x = (x_1 + x_2)/2$ and has the value

$$[(\tfrac{3}{2}h)(\tfrac{1}{2}h)]^2 = \tfrac{9}{16}h^4.$$

It follows that the interpolation error is bounded by

$$\frac{3}{128} h^4 M_4.$$

In a sine table, for instance, using cubic interpolation and a step as large as $h = 0.1$ we get a maximum error of less than 2.5×10^{-6}.

The basic error formula (9-5) requires the knowledge of a bound of the derivatives of the function f.   Such bounds can often be obtained very easily even for non-elementary functions by exploiting known functional relations.

EXAMPLE

5.   The Bessel function of order zero can be defined by

$$J_0(x) = \frac{1}{\pi} \int_0^\pi \cos (x \sin t)\, dt.$$

By differentiating under the integral sign,

$$J_0'(x) = -\frac{1}{\pi} \int_0^\pi \sin t \sin (x \sin t)\, dt,$$

$$J_0''(x) = -\frac{1}{\pi} \int_0^\pi (\sin t)^2 \cos (x \sin t)\, dt,$$

etc.   The integrands of all integrals which we obtain by differentiation are bounded in absolute value by 1.   Thus

$$|J_0^{(n)}(x)| \leq \frac{1}{\pi} \int_0^\pi 1\, dt = 1, \qquad n = 0, 1, 2, \ldots.$$

Problems

5. A table of a function of one variable is well suited for linear interpolation if the error due to interpolation does not exceed the rounding error of the entries. What is the greatest permissible step of such a "well interpolable" table of cos x as a function of the number of decimal places, (a) if x is given in radians; (b) if x is given in degrees? Make a survey of some tables accessible to you and decide whether they are well suited for linear interpolation.

6. What is the maximum value of the combined error due to linear interpolation and rounding of the formula

$$f(x) \sim \frac{x - x_0}{x_1 - x_0} f_1 + \frac{x_1 - x}{x_1 - x_0} f_0,$$

if f_1 and f_0 are known to N places, and if the products are rounded to N places? (It may be assumed that the fractions $(x - x_0)/(x_1 - x_0)$ and $(x_1 - x)/(x_1 - x_0)$ are exact decimal fractions.)

7. The function $\log_{10} (\sin x)$, where x is given in degrees, is tabulated to five decimal places with a step of $1/60$ of one degree. From what value of x on is this table well suited for linear interpolation?

8. The function $f(x) = \sqrt{x}$ is tabulated at the integers, $x = 1, 2, 3, \ldots$, giving four decimals. From what x on is this table well suited for linear interpolation?

9. The Bessel function of order n can be defined by

$$J_n(x) = \frac{1}{\pi} \int_0^{\pi} \cos (x \sin t - nt) \, dt.$$

How do we have to choose the step h of a table of J_n so that the error is less than 10^{-6}
(a) if linear interpolation is to be used?
(b) if cubic interpolation (as described in example 4) is to be used?

10. *Interpolation near the end of a table.* The fourth derivative of a function f is known to be bounded by M_4. Let P be the polynomial interpolating f at the points $x_k = kh$, $k = 0, 1, 2, 3$. Give the best possible bound for

$$\max_{x_0 \leq x \leq x_1} |f(x) - P(x)|.$$

11. Theorem 9.2 implies that if $|f^{(n+1)}(x)| \leq M_{n+1}$, $x \in I$, then, in the notation of theorem 9.2

$$|f(x) - P(x)| \leq \frac{M_{n+1}}{(n + 1)!} |L(x)|, \qquad x \in I.$$

Are there any functions f (and corresponding points in I) for which this inequality becomes an equality?

12. Let n be a positive integer. Somebody proposes to calculate the value of

e^{n+1} by constructing the polynomial P interpolating the function $f(x) = e^x$ at the points $x = 0, 1, \ldots, n$ and evaluating P for $x = n + 1$.

(a) Indicate a *lower* bound for the error $e^{n+1} - P(n + 1)$.

(b) Determine the number ξ_x of theorem 9.2, and thus obtain an exact expression for the error.

13. The function f is defined on $[0, 1]$ and is known to have a bounded second derivative. Its values are to be computed from a fixed interpolating polynomial using two interpolating points x_0 and x_1. How should one place the points x_0 and x_1 in the interval $[0, 1]$ in order to minimize the error due to interpolation?

9.3 Convergence of Sequences of Interpolating Polynomials

Let the function f be defined for $-\infty < x < \infty$, and let it and all its derivatives be bounded by one and the same constant,

$$|f^{(n)}(x)| \leqq M, \qquad n = 0, 1, \ldots; \quad -\infty < x < \infty.$$

Assume one wishes to calculate $f(x)$ in the interval $[0, h]$ by means of the interpolating polynomial P of degree $2n - 1$ using the interpolating points

$$x_0 = 0, \qquad\qquad x_1 = h,$$
$$x_2 = -h, \qquad\qquad x_3 = 2h,$$
$$\ldots$$
$$x_{2n-2} = -(n-1)h, \qquad x_{2n-1} = nh.$$

For what values of h does the error tend to zero as $n \to \infty$?

Obviously, the above procedure cannot be effective for unrestricted values of h, as the example $f(x) = \sin x$, $h = \pi$ shows. (All interpolating polynomials are zero in this case.) By theorem 9.2, the interpolation error of P is bounded by $M|L(x)|/(2n)!$, where

$$L(x) = [x + (n-1)h][x + (n-2)h]\ldots[x - nh].$$

The maximum of the function $|L(x)|$ on the interval $[0, h]$ occurs at the point $x = h/2$. At this point,

$$\frac{1}{(2n)!}\left|L\left(\frac{h}{2}\right)\right| = \frac{[(n-\frac12)h(n-\frac32)h\ldots\frac12 h]^2}{(2n)!}$$

$$= \frac{[(2n-1)(2n-3)\ldots3\cdot1]^2 h^{2n}}{2^{2n}\cdot(2n)!}$$

$$= \left[\frac{(2n)!}{2^n n!}\right]^2 \frac{h^{2n}}{2^{2n}(2n)!}$$

$$= \frac{(2n)!}{2^{4n}(n!)^2} h^{2n}.$$

Using Stirling's asymptotic formula for $n!$,

$$n! \sim \sqrt{2\pi n} \left(\frac{n}{e}\right)^n,$$

(see Buck [1956], p. 159) we find

$$\frac{1}{(2n)!} \left| L\left(\frac{h}{2}\right) \right| \sim \frac{\sqrt{2\pi \cdot 2n} \left(\frac{2n}{e}\right)^{2n}}{2^{4n} 2\pi n \left(\frac{n}{e}\right)^{2n}} \cdot h^{2n}$$

$$= \left(\frac{h}{2}\right)^{2n} \cdot \frac{1}{\sqrt{\pi n}}.$$

The last expression tends to zero if $n \to \infty$ if and only if $|h| \leq 2$. Thus the convergence of the interpolation process described above can be guaranteed only if $|h| \leq 2$.

The sequence of interpolating polynomials constructed above looks somewhat unnatural in view of the fact that we use interpolating points farther and farther removed from the interval where we wish to approximate the function f. The following question, however, is very natural: Let f be continuous on the interval $[0, 1]$ and denote by $P_n(x)$ the polynomial interpolating f at the points

$$x_k = \frac{k}{n}, \qquad k = 0, 1, \ldots, n.$$

Is it true that

(9-7) $$\lim_{n \to \infty} P_n(x) = f(x)$$

for all $x \in [0, 1]$? An important result due to Runge states that there are continuous functions for which (9-7) does not hold. (A simple example is $f(x) = |x - \tfrac{1}{2}|$.) Actually, the relation (9-7) even fails to hold for some functions which have derivatives of all orders.

It is important, however, to understand Runge's result correctly. The result does not mean that a continuous function cannot always be approximated by polynomials. In fact, a famous theorem due to Weierstrass (see Buck [1956], p. 39) states that every f continuous on a closed finite interval I can be approximated by polynomials to any desired accuracy. Runge's result merely states that these approximating polynomials can in general not be obtained by interpolation at uniformly spaced points.

Problems

14. *Missing entry in a table.* A function f is defined on the whole real line and satisfies

$$|f^{(m)}(x)| \leq M^m, \qquad -\infty < x < \infty; \qquad m = 0, 1, 2, \ldots$$

for some constant M. For $n = 1, 2, \ldots$ let P_{2n-1} denote the polynomial interpolating f at the points $-n, -n + 1, \ldots, -1, 1, \ldots, n - 1, n$. Prove that

$$\lim_{n \to \infty} P_{2n-1}(0) = f(0)$$

holds provided that $M < 2$.

15. A function f is defined for $x \geq 0$ and satisfies

$$|f^{(m)}(x)| \leq 1, \qquad x \geq 0, \qquad m = 0, 1, 2, \ldots.$$

For a fixed value of h, let P_n denote the polynomial interpolating f at the points $0, h, 2h, \ldots, nh$. For what values of h can you guarantee that

$$\lim_{n \to \infty} P_n(x) = f(x)$$

for every fixed value of $x > 0$?

16*. Let the function f be continuous on the interval $[0, 1]$. From the fact that such a function is *uniformly continuous* (see Buck [1956], p. 34) one can easily prove that f can be approximated to arbitrary accuracy by the piecewise linear function coinciding with f at suitable points $x_0, x_1, \ldots, x_n$. Thus, in a sense, f can be approximated arbitrarily well by linear interpolation. Why does this not contradict Runge's theorem?

17. A function f is defined on $[0, 1]$, and its derivatives satisfy

$$|f^{(m)}(x)| \leq m!, \qquad m = 0, 1, 2, \ldots, 0 \leq x \leq 1.$$

(Example: $f(x) = (1 + x)^{-1}$.) Let P_n denote the polynomial interpolating f at the points $1, q, q^2, \ldots, q^n$, where q is some number such that $0 < q < 1$. Show that

$$\lim_{n \to \infty} P_n(0) = f(0).$$

9.4   How to Approximate a Polynomial of degree n by One of Degree $n - 1$

Let Q be a polynomial of degree n with leading coefficient 1,

$$Q(x) = x^n + a_{n-1}x^{n-1} + \cdots + a_0.$$

We wish to interpolate Q in the interval $[-1, 1]$ by a polynomial P of degree $n - 1$ such that the maximum of the error $|Q(x) - P(x)|$ is minimized. How do we have to choose the interpolating points $x_0, x_1, x_2, \ldots, x_{n-1}$, and how large is the smallest possible maximum error? From the general error formula (9-5) we find, since $Q^{(n)}(x) = n!$,

(9-8) $$Q(x) - P(x) = L(x)$$

where

$$L(x) = (x - x_0)(x - x_1)\ldots(x - x_{n-1}) = x^n + \cdots.$$

Our problem is thus equivalent to the problem of selecting the points $x_0, x_1, \ldots, x_{n-1}$ in such a manner that the quantity

$$\max_{-1 \leq x \leq 1} |(x - x_0)(x - x_1)\ldots(x - x_{n-1})|$$

is minimized. Although seemingly difficult this problem can be solved explicitly.

Theorem 9.4   The best choice of the interpolating points $x_0, x_1, \ldots,$ x_{n-1} for the approximation of the polynomial $Q(x) = x^n + \cdots$ in the interval $-1 \leq x \leq 1$ by a polynomial of degree $n - 1$ is the choice for which

$$(9\text{-}9) \qquad\qquad L(x) = \frac{1}{2^{n-1}} T_n(x),$$

where T_n denotes the nth Chebyshev polynomial,†

$$T_n(x) = \cos{(n \text{ arc cos } x)}.$$

Proof.   Let us first convince ourselves that the function L defined by (9-9) really is a polynomial of degree n with leading coefficient 1.   From the difference equation satisfied by the Chebyshev polynomials,

$$T_n(x) = 2xT_{n-1}(x) - T_{n-2}(x)$$

and from the fact that $T_0(x) = 1$, $T_1(x) = x$ it readily follows that T_n is a polynomial in x with leading coefficient 2^{n-1}.   Hence our assertion on L follows immediately.

Since we are interested in minimizing the maximum of $|L(x)|$, let us calculate the extrema of $L(x)$.   The extrema of $\cos x$ occur for $x = k\pi$, where k is an integer, hence the extrema of $L(x)$ in the interval $[-1, 1]$ occur at the points where $n \text{ arc cos } x = k\pi$, i.e., for $x = t_k$, where

$$t_k = \cos\frac{k\pi}{n}, \qquad k = 0, 1, \ldots, n.$$

The values of L at these points are

$$L(t_k) = (-1)^k 2^{-n+1},$$

i.e., the extrema all have the same absolute value 2^{-n+1}, but oscillate in sign (see Fig. 9.4).

---

† See example 3, chapter 6.

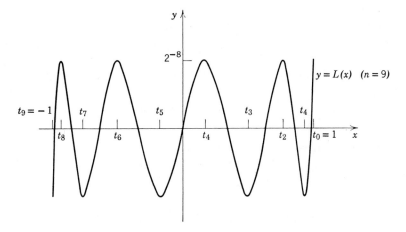

Figure 9.4 The function $L(x)$ $(n = 9)$.

Now suppose there exists another polynomial $M(x) = x^n + \cdots$ for which $|M(x)|$ has a smaller maximum $m < 2^{-n+1}$ in $[-1, 1]$. Then the difference polynomial

$$D(x) = L(x) - M(x)$$

would at the $n + 1$ points t_k have the same sign as L, i.e.,

$$D(t_k) \begin{cases} > 0, & k \text{ even}, \\ < 0, & k \text{ odd}. \end{cases}$$

Since $t_0 > t_1 > t_2 > \cdots > t_n$, it would follow that D has at least n distinct zeros, namely one in each interval $[t_{k+1}, t_k]$. However, since both L and M have leading coefficient 1, it follows that D is a polynomial of degree $\leq n - 1$, hence D cannot have n distinct zeros without vanishing identically. The assumption of the existence of a polynomial M with a maximum deviation from 0 smaller than 2^{-n+1} thus has led to a contradiction.

The interpolating points x_k for the best approximation of Q by a polynomial of lower degree are the zeros of L, that is the points $x = x_k$ satisfying

$$n \arccos x = (k + \tfrac{1}{2})\pi, \qquad k = 0, 1, \ldots, n - 1.$$

It follows that the interpolating points are given by

$$x_k = \cos\left(\frac{2k + 1}{2n}\pi\right), \qquad k = 0, 1, \ldots, n - 1.$$

EXAMPLE

6. How can we best approximate the function

$$f(x) = x^2 + ax + b$$

in $-1 \leq x \leq 1$ by a straight line? This is the case $n = 2$ of theorem 9.4. We have

$$x_0 = \cos \frac{\pi}{4} = \frac{\sqrt{2}}{2}, \qquad x_1 = \cos \frac{3\pi}{4} = -\frac{\sqrt{2}}{2}.$$

The interpolating polynomial is given by

$$P(x) = \frac{f(x_0)(x_1 - x) + f(x_1)(x_0 - x)}{x_1 - x_0}$$

$$= \frac{(x_0^2 + ax + b)(x - x_1) + (x_1^2 + ax + b)(x_0 - x)}{x_0 - x_1}$$

$$= ax + b + \tfrac{1}{2}.$$

The maximum deviation from $x^2 + ax + b$ in $[-1, 1]$ is $\tfrac{1}{2}$, as predicted by the theory.

It is not necessary to use the interpolation points x_k in order to construct the polynomial P of best approximation. From the error formula (9-8) we find, if L is given by (9-9),

(9-10) $$P(x) = Q(x) - \frac{1}{2^{n-1}} T_n(x).$$

EXAMPLE

7. We consider once more the problem of example 6. From the recurrence relation we easily find $T_2(x) = 2x^2 - 1$, hence

$$P(x) = x^2 + ax + b - \tfrac{1}{2}(2x^2 - 1)$$
$$= ax + b + \tfrac{1}{2},$$

in accordance with the earlier result.

Relation (9-8) shows that the error curve for the best approximation of a polynomial $Q(x) = x^n + \cdots$ by a polynomial of lower degree is given by $2^{-n+1}T_n(x)$. The discussion in the proof of theorem 9.4 revealed that this curve has $n + 1$ extrema in $[-1, 1]$ with alternating signs, but all of the same absolute value. This property is shared by the polynomial P minimizing

$$\max_{-1 \leq x \leq 1} |f(x) - P(x)|$$

where f is any continuous function. The theory of such minimizing polynomials was initiated by Chebyshev (1821–1894); it plays an outstanding role in modern numerical computation.

Problems

18. Determine the polynomial of degree $\leq n - 1$ that best approximates

$$Q(x) = a_0x^n + a_1x^{n-1} + \cdots + a_n$$

on an arbitrary interval $a \leq x \leq b$, and show that the least value of the maximum deviation is given by

$$\frac{1}{2^{n-1}}\left(\frac{b-a}{2}\right)^n a_0.$$

[Hint: Reduce the problem to the special case considered above by introducing a new variable x^* by setting

$$x = \frac{b+a}{2} + \frac{b-a}{2}x^*.]$$

19. Determine a polynomial of degree ≤ 4 that provides the best approximation to the function $f(x) = x^5$ on the interval $[0, 4]$.
20. Approximate $f(x) = x^3$ on the interval $[0, 1]$ by a polynomial of degree 2.
21. Approximate $f(x) = x^3$ on the interval $[0, 1]$ by a polynomial of degree 1 by approximating the approximating polynomial of problem 20 by a linear polynomial.
22. Determine directly (by calculus) a linear polynomial $P(x) = ax + b$ such that the quantity

$$\max_{0 \leq x \leq 1} |x^3 - ax - b|$$

is minimized.
23. Prove the uniqueness of the solution (9-10) of the approximation problem considered at the beginning of §9.4.

Recommended Reading

A more general treatment of the error of Lagrangian interpolation is given in Ostrowski [1960], chapter 1. For a discussion of the convergence of sequences of interpolating polynomials see Hildebrand [1956], pp. 114–118. A first introduction to the theorem of Weierstrass and to the approximation of continuous functions by polynomials in general is given in Todd [1963].

Research Problems

1. Assuming that f is sufficiently differentiable, how well does the derivative of an interpolating polynomial approximate the derivative of f? (For a partial answer, see §12.1.)
2. By extending the procedure outlined in problem 21, how well can you *at least* approximate a polynomial of degree n by one of arbitrary degree $m < n$?

chapter 10 construction of the interpolating polynomial: methods using ordinates

After considering the more theoretical aspects of the interpolating polynomial in chapter 9, we shall now discuss some algorithms for actually constructing the polynomial. Many such algorithms have been devised, frequently with some special purpose in mind. There are two main categories of such algorithms: In the first category, the function f enters through its values (or "ordinates") at all interpolation points. In the second category, f enters through its value at one point, and through differences of the function values. Here we are concerned with algorithms of the first category.

10.1 Muller's Method†

For some purposes the interpolating polynomial is calculated most conveniently from the Lagrangian formula (9-4). The Lagrangian formula is especially convenient if the polynomial is to be subjected to algebraic manipulations. As an example of an application of the Lagrangian representation, we shall discuss in more detail Muller's method for solving the equation $f(x) = 0$ mentioned in §4.11.

The reader will recall that the essence of Muller's method is as follows. Assuming that three distinct approximations x_{n-2}, x_{n-1}, x_n to the desired solution s are available, we gain a new approximation by interpolating the function f at the points x_{n-2}, x_{n-1}, x_n by a (normally) quadratic polynomial P. Of the (normally) two zeros of P, one closest to x_n is selected as the new approximation x_{n+1}. The process then is continued with (x_{n-1}, x_n, x_{n+1}) in place of (x_{n-2}, x_{n-1}, x_n) and terminated as soon as $|x_{n+1} - x_n|/|x_{n+1}|$ becomes less than some preassigned number.

† This section may be omitted at first reading.

The Lagrangian representation of P is

$$P(x) = \frac{(x - x_{n-1})(x - x_{n-2})}{(x_n - x_{n-1})(x_n - x_{n-2})}f_n + \frac{(x - x_n)(x - x_{n-2})}{(x_{n-1} - x_n)(x_{n-1} - x_{n-2})}f_{n-1}$$

$$+ \frac{(x - x_n)(x - x_{n-1})}{(x_{n-2} - x_n)(x_{n-2} - x_{n-1})}f_{n-2}.$$

In order to write this in a more compact manner, we introduce the quantities

(10-1) $$h_n = x_n - x_{n-1}, \qquad h = x - x_n$$

and obtain

$$P(x) = P(x_n + h)$$

$$= \frac{(h + h_n)(h + h_n + h_{n-1})}{h_n(h_n + h_{n-1})}f_n - \frac{h(h + h_n + h_{n-1})}{h_n h_{n-1}}f_{n-1}$$

$$+ \frac{h(h + h_n)}{(h_n + h_{n-1})h_{n-1}}f_{n-2}.$$

Collecting terms involving like powers of h and writing

(10-2) $$q_n = \frac{h_n}{h_{n-1}}, \qquad q = \frac{h}{h_n},$$

we find

$$P(x) = P(x_n + qh_n)$$
$$= (1 + q_n)^{-1}(A_n q^2 + B_n q + C_n)$$

where

(10-3) $$\begin{cases} A_n = q_n f_n - q_n(1 + q_n)f_{n-1} + q_n^2 f_{n-2}, \\ B_n = (2q_n + 1)f_n - (1 + q_n)^2 f_{n-1} + q_n^2 f_{n-2}, \\ C_n = (1 + q_n)f_n. \end{cases}$$

Solving the quadratic equation $P(x_n + qh_n) = 0$, we find

$$x_{n+1} = x_n + h_n q_{n+1},$$

where

$$q_{n+1} = \frac{-B_n \pm \sqrt{B_n^2 - 4A_n C_n}}{2A_n}.$$

In order to avoid loss of accuracy due to forming differences, this formula is better written in the form

(10-4) $$q_{n+1} = -\frac{2C_n}{B_n \pm \sqrt{B_n^2 - 4A_n C_n}}.$$

Here the sign yielding the *smaller* value of q_{n+1}, i.e., the larger absolute value of the denominator, should be chosen.

It may happen, of course, that the square root in (10-4) becomes imaginary. If f is defined for real values only, the algorithm then breaks down, and a new start must be made. If f is a polynomial, the possibility of imaginary square roots is considered an advantage, since this will automatically lead to approximations to complex zeros.

Three starting values x_0, x_1, x_2, for the algorithm have to be provided from some other source. Muller recommends to start the algorithm by taking for P the Taylor polynomial of degree 2 of f at $x = 0$. If f is a polynomial,

$$f(x) = a_0 x^N + a_1 x^{N-1} + \cdots + a_N,$$

this can be achieved artificially by putting $x_0 = -1$, $x_1 = 1$, $x_2 = 0$, thus

$$(10\text{-}5) \qquad h_1 = 2, \qquad h_2 = -1, \qquad q_2 = -\tfrac{1}{2}$$

and setting

$$(10\text{-}6) \qquad \begin{cases} f_0 = a_N - a_{N-1} + a_{N-2}, \\ f_1 = a_N + a_{N-1} + a_{N-2}, \\ f_2 = a_N. \end{cases}$$

As soon as a zero of f has been determined, it is to be divided out by algorithm 3.4 in connection with theorem 4.10.

It follows from the work of Ostrowski ([1960], p. 86, although without reference to Muller's work) that Muller's method converges whenever the three initial approximations are sufficiently close to a simple zero of f. The degree of convergence lies somewhere between that of the regula falsi and of Newton's method. No convergence theorems in the large similar to those for the QD algorithm appear to be known. Nevertheless, the method is (in the United States) among the most popular for finding zeros of polynomials.

Problems

1. Use Muller's method to find all zeros of the polynomial

$$p(x) = 128x^4 - 256x^3 + 160x^2 - 32x + 1.$$

(Real arithmetic may be used here.)

2. Use complex arithmetic to determine all zeros of the polynomial

$$p(x) = x^4 - 8x^3 + 39x^2 - 62x + 51$$

by Muller's method.

10.2 The Lagrangian Representation for Equidistant Abscissas†

In the present section we assume that the points x_k, where the values of the function f are given, are equally spaced.   This is the case, for instance, for most mathematical tables.   If h denotes the distance between two consecutive interpolating points, we then have

(10-7) $$x_k = x_0 + kh.$$

where $k = 0, \pm 1, \pm 1, \pm 2, \ldots$.   We now introduce a new variable s by means of the relation

(10-8) $$x = x_0 + sh.$$

At $x = x_k$, s obviously has the value k.   The variable s thus measures x in units of h, starting at x_0.

We now consider the polynomial P of degree $n - m$ which interpolates the function f at the points $x_m, x_{m+1}, \ldots, x_n$.   Here m and n may be any two integers such that $n \geqq m$.   (Ordinarily, we have $m \leqq 0, n \geqq 0$.)   By (9-4), this polynomial is given by

$$P(x) = \sum_{k=m}^{n} L_k(x) f_k,$$

where

$$L_k(x) = \prod_{\substack{q=m \\ q \neq k}}^{n} \frac{x - x_q}{x_k - x_q}.$$

We now express P in terms of the variable s defined by (10-8).   Evidently

$$x - x_q = (s - q)h,$$

and in particular,

$$x_k - x_q = (k - q)h.$$

If $P(x) = P(x_0 + sh) = p(s)$, we thus have

(10-9) $$p(s) = \sum_{k=m}^{n} l_k(s) f_k,$$

where

$$l_k(s) = \prod_{\substack{q=m \\ q \neq k}}^{n} \frac{s - q}{k - q}.$$

The remarkable fact about this representation of the Lagrangian polynomial is the independence of the functions $l_k(s)$ from h.   These functions, which may be called the normalized Lagrangian interpolation coefficients, depend only on s (the relative location of x with respect to x_0

---

† This section may be omitted at first reading.

and x_1), and, of course, on the integers m and n, which define the set of interpolating points.

EXAMPLES

1. $m = 0, n = 1$.   We have

$$l_0(s) = \prod_{\substack{q=0 \\ q \neq 0}}^{1} \frac{s - q}{0 - q} = 1 - s,$$

$$l_1(s) = \prod_{\substack{q=0 \\ q \neq 1}}^{1} \frac{s - q}{1 - q} = \frac{s}{1 - 0} = s.$$

We get, of course, the formula for linear interpolation, expressed as

$$p(s) = (1 - s)f_0 + sf_1.$$

2. A case which is frequently used in practice is given by $m = -1$, $n = 2$.   Here we find

$$l_{-1}(s) = \prod_{\substack{q=-1 \\ q \neq -1}}^{2} \frac{s - q}{-1 - q} = \frac{s(s - 1)(s - 2)}{(-1)(-2)(-3)} = -\frac{s(s - 1)(s - 2)}{6},$$

$$l_0(s) = \frac{(s + 1)(s - 1)(s - 2)}{1(-1)(-2)} = \frac{(s + 1)(s - 1)(s - 2)}{2},$$

$$l_1(s) = -\frac{(s + 1)s(s - 2)}{2} = l_0(1 - s),$$

$$l_2(s) = \frac{(s + 1)s(s - 1)}{6} = l_{-1}(1 - s).$$

In view of the fact that the normalized Lagrangian interpolation coefficients depend on one continuous variable only, extensive tables for them have been prepared (National Bureau of Standards [1948]).   Such tables take into account symmetry properties such as the relation $l_0(s) = l_1(1 - s)$ noted above.

We note some interesting algebraic relations between the normalized interpolation coefficients.   Let us consider the general case, where the interpolating points are $x_m, x_{m+1}, \ldots, x_n$.   The polynomial

$$P(x) = \sum_{k=m}^{n} L_k(x)f_k,$$

using $n - m + 1$ points, will furnish an exact representation of the function f if f is a polynomial of degree $n - m$ or less.   Thus it will be exact, in particular, for the functions

$$f(x) = \left(\frac{x - x_0}{h}\right)^q, \qquad q = 0, 1, \ldots, n - m.$$

Since $f(x) = s^q$, we have $f_k = k^q$. Hence (10-9) yields the identities in s,

$$(10\text{-}10) \qquad \sum_{k=m}^{n} l_k(s)k^q = s^q, \qquad q = 0, 1, \ldots, n - m.$$

These identities can be regarded as a system of $n - m + 1$ equations for $n - m + 1$ unknowns $l_k(s)$. They thus could be used to calculate the l_k numerically.

EXAMPLE

3. For $m = 0$, $n = 2$ the relations (10-10) take the form

$$\begin{aligned} l_0(s) + l_1(s) \ \ + l_2(s) &= 1 \\ l_1(s)1 + l_2(s)2 &= s \\ l_1(s)1^2 + l_2(s)2^2 &= s^2. \end{aligned}$$

If f is not a polynomial of degree $\leq n - m$, the error formula (9-5) still stands. In the present situation,

$$L(x) = \prod_{k=m}^{n} (x - x_k)$$

and thus

$$l(s) = L(x_0 + sh) = h^{n-m+1} \prod_{k=m}^{n} (s - k).$$

Thus the error formula now appears in the form

$$(10\text{-}11) \quad f(x_0 + sh) - p(s) = h^{n-m+1} \frac{f^{(n-m+1)}(\xi_s)}{(n-m+1)!} \prod_{k=m}^{n} (s - k),$$

where ξ_s is a point between the largest and the smallest of the numbers x_m, x_n, x.

EXAMPLE

4. If linear interpolation is used as in example 1, we have for $0 \leq s \leq 1$

$$p(s) - f(x_0 + sh) = h^2 \frac{f''(\xi_s)}{2} s(1 - s),$$

where $x_0 \leq \xi_s \leq x_1$.

Problems

3. Use normalized Lagrangian interpolation coefficients to determine $J_0(2.4068)$ by interpolation from the following values:

x	$J_0(x)$
2.1	0.16661
2.3	0.05554
2.5	−0.04838
2.7	−0.14245

4. If the interpolating points are $x_m, x_{m+1}, \ldots, x_n$, prove that

$$l_k(s) = l_{n+m-k}(n + m - s), \qquad k = m, \quad m + 1, \ldots, n.$$

5. Make a general statement about the signs of the $l_k(s)$ as a function of m, n, k, and s.

6. If the interpolating points are $x_0, x_1, \ldots, x_n$, show that

$$l_k(s) = (-1)^{n-k} \binom{s}{n+1} \binom{n}{k} \frac{n+1}{s-k}, \qquad k = 0, 1, \ldots, n; \quad s \neq k.$$

7. Assuming that the interpolating points are $x_m, x_{m+1}, \ldots, x_n$, find a closed expression for the sum

$$\sum_{k=m}^{n} k^{n-m+1} l_k(s).$$

[Hint: Apply the error formula (10-11) to the function

$$f(x) = (x - x_0)^{n-m+1}.]$$

10.3 Aitken's Lemma

We now shall discuss certain algorithms that permit us to construct the interpolating polynomial recursively, without reference to the Lagrangian formula (9-4).   The basic tool is a lemma which enables us to represent an interpolating polynomial of degree $d + 1$ in terms of two such polynomials of degree d.

Some special notation will be required.   We again denote the points at which the function f is to be interpolated by $x_0, x_1, x_2, \ldots, x_n$, and by f_k the value of f at x_k.   We shall have to consider polynomials that interpolate f at some, but not all of the points $x_0, x_1, \ldots, x_n$.   If S is any nonempty subset of $\{x_0, x_1, \ldots, x_n\}$, we denote by P_S the polynomial interpolating f at those x which are in S.   Thus, if S contains $k + 1$ points, P_S is the unique polynomial of degree $\leq k$ such that

$$P_S(x_i) = f_i, \qquad x_i \in S.$$

EXAMPLES

5. If S contains just one point x_i, then $P_S = f_i$.

6. $P_{\{x_1, x_2, x_5\}}$ denotes the polynomial interpolating at the points x_1, x_2, x_5.

Denoting by W the set of all interpolating points, we can state the following lemma:

Lemma 10.3   Let S and T be two proper subsets of W having all but the two points $x_i \in S$ and $x_j \in T$ in common.   Then

(10-12) $\qquad P_{S \cup T}(x) = \dfrac{(x_i - x)P_T(x) - (x_j - x)P_S(x)}{x_i - x_j}$

identically in x.

Here, as usual, $S \cup T$ denotes the union of the sets S and T.

Proof. Let the sets S and T contain $m + 1$ points each. Both polynomials P_S and P_T interpolate at $m + 1$ points, hence are of degree $\leq m$. Denoting the expression on the right of (10-12) by P, we see that P has a degree $\leq m + 1$. Hence if we can show that P interpolates at all points of $S \cup T$, then theorem 9.1 implies that $P = P_{S \cup T}$.

Let x_k be a point of the intersection $S \cap T$ of S and T. By virtue of

$$P_S(x_k) = P_T(x_k) = f_k,$$

(10-12) yields

$$P(x_k) = \frac{(x_i - x_k)f_k - (x_j - x_k)f_k}{x_i - x_j}$$

$$= f_k,$$

as desired. For $x = x_i$ we have

$$P(x_i) = \frac{-(x_j - x_i)P_S(x_i)}{x_i - x_j} = f_i,$$

and similarly for $x = x_j$

$$P(x_j) = \frac{(x_i - x_j)P_T(x_j)}{x_i - x_j} = f_j.$$

Thus P has been shown to interpolate at all points in $S \cup T$, completing the proof.

EXAMPLE

7.   If $S = \{x_0, x_2\}$, $T = \{x_2, x_5\}$, we obtain

$$P_{\{x_0, x_2, x_5\}}(x) = \frac{(x_0 - x)P_{\{x_2, x_5\}}(x) - (x_5 - x)P_{\{x_0, x_2\}}(x)}{x_0 - x_5}.$$

8.   Lemma 10.3 is already familiar if the intersection $S \cap T$ is empty. We then have, using example 5,

$$P_{\{x_i, x_j\}}(x) = \frac{(x_i - x)P_{\{x_j\}}(x) - (x_j - x)P_{\{x_i\}}(x)}{x_i - x_j}$$

$$= \frac{(x_i - x)f_j - (x_j - x)f_i}{x_i - x_j}.$$

This is the familiar formula for *linear* interpolation.

Lemma 10.3 can be used in two ways. We may use it to get a formal *representation* for the interpolating polynomial, or we may use it to calculate the *value* of the polynomial for a given value of x. In the latter case formula (10-12) requires dividing a sum of products by a single number, an operation that can be performed on a desk computer without writing down intermediate results.

Problems

8. Obtain the Lagrangian formula for quadratic interpolation on the set $\{x_0, x_1, x_2\}$ from the formulas for linear interpolation on the sets $\{x_0, x_1\}$ and $\{x_0, x_2\}$.

9. Prove the following generalization of lemma 10.3: Let S be an arbitrary subset of $\{x_{m+1}, x_{m+2}, \ldots, x_n\}$ and let $S_k = \{x_k, S\}$, $k = 1, 2, \ldots, m$. If $L_k(x)$ $(k = 1, 2, \ldots, m)$ are the Lagrangian interpolating coefficients for interpolation on the set $\{x_1, x_2, \ldots, x_m\}$, then

$$P_{\cup S_k}(x) = \sum_{k=1}^{m} L_k(x) P_{S_k}(x).$$

10.4   Aitken's Algorithm

Lemma 10.3 enables us to generate the interpolating polynomials of higher degrees successively from polynomials of lower degrees. It still leaves us considerable freedom in the choice of the sets S and T used to finally obtain the polynomial P_W. Two standardized choices have become widely used, one named after Aitken, the other named after Neville. In both choices a triangular array of polynomials $P_{k,d}$ is generated. Here $P_{k,d}$ is a certain polynomial of degree d that interpolates on a set of $d + 1$ points depending on k. Aitken's scheme is as follows:

Algorithm 10.4   For $d = 0, 1, \ldots, n$, generate the polynomials $P_{k,d}$ as follows:

$$(10\text{-}13) \qquad P_{k,0}(x) = f_k, \qquad k = 0, 1, \ldots, n;$$

$$(10\text{-}14) \qquad P_{k,d+1}(x) = \frac{(x_k - x)P_{d,d}(x) - (x_d - x)P_{k,d}(x)}{x_k - x_d},$$

$$k = d + 1, d + 2, \ldots, n.$$

The arrangement of the polynomials $P_{k,d}$ is shown in scheme 10.4.

d	0	1	2	$\cdots$	n	
x_0	$P_{0,0}$					$x_0 - x$
x_1	$P_{1,0}$	$\underline{P_{1,1}}$				$x_1 - x$
x_2	$P_{2,0}$	$P_{2,1}$	$P_{2,2}$			$x_2 - x$
$\vdots$						
x_k	$P_{k,0}$	$\underline{P_{k,1}}$	$\underline{\underline{P_{k,2}}}$			$\underline{x_k - x}$
$\vdots$						
x_n	$P_{n,0}$	$P_{n,1}$	$P_{n,2}$	$\cdots$	$P_{n,n}$	$x_n - x$

Scheme 10.4

The doubly underlined entry in scheme 10.4 is obtained by crosswise multiplication of the simply underlined entries.

Theorem 10.4 In the notation of lemma 10.3,

$$P_{k,d} = P_{\{x_0, x_1, \ldots, x_{d-1}, x_k\}},$$
$$d = 0, 1, \ldots, n; \quad k = d, d+1, \ldots, n.$$

Proof. We use induction with respect to d. By (10-13), the assertion is true for $d = 0$. Assuming it to be true for some $d \geq 0$, lemma 10.3 shows that the polynomial defined by (10-14) interpolates on the union of the sets $\{x_0, x_1, \ldots, x_{d-1}, x_d\}$ and $\{x_0, x_1, \ldots, x_{d-1}, x_k\}$ that is, on the set $\{x_0, x_1, \ldots, x_d, x_k\}$ proving our assertion for d increased by one. For $d = k = n$ we obtain

Corollary 10.4 $P_{n,n} = P_W$.

The rightmost entry in scheme 10.4 is the polynomial that interpolates on the set of *all* points $x_0, x_1, \ldots, x_n$.

EXAMPLE

9. Let $f(x) = x^4$. We wish to calculate $f(3)$ by interpolation at the points $-4, -2, 0, 2, 4$. Scheme 10.4 looks as follows:

x_k	$P_{k,0}$	$P_{k,1}$	$P_{k,2}$	$P_{k,3}$	$P_{k,4}$	$x_k - x$
-4	256					-7
-2	16	-584				-5
0	0	-192	396			-3
2	16	-24	116	-24		-1
4	256	256	116	186	81	1

Problems

10. Use Aitken's algorithm to obtain a value of $\sin \pi/4$ from the following values of the function $f(x) = \sin x\pi/2$:

x	-2	-1	0	1	2	3
$f(x)$	0	-1	0	1	0	-1

11. Use algorithm 10.4 to determine $J_0(2.4068)$ by interpolation from the values given in problem 3.

10.5 Neville's Algorithm

In Neville's use of lemma 10.3, the polynomials $P_{k,d}$ are built up in such a manner that each polynomial interpolates on a set of points with $d + 1$ *consecutive* indices. The algorithm is as follows:

Algorithm 10.5 For $d = 0, 1, \ldots, n$, construct the polynomials $P_{k,d}$ as follows:

$$(10\text{-}15) \qquad P_{k,0}(x) = f_k, \qquad k = 0, 1, \ldots, n;$$

$$(10\text{-}16) \qquad P_{k,d+1}(x) = \frac{(x_k - x)P_{k-1,d}(x) - (x_{k-d-1} - x)P_{k,d}(x)}{x_k - x_{k-d-1}},$$

$$k = d + 1, d + 2, \ldots, n.$$

The arrangement of the polynomials $P_{k,d}$ is the same as in scheme 10.4, but the doubly underlined entry is now computed from asymmetrically located entries as shown in scheme 10.5:

d	0	1	2	$\cdots$	n	
x_0	$P_{0,0}$					$x_0 - x$
x_1	$P_{1,0}$	$P_{1,1}$				$x_1 - x$
x_2	$P_{2,0}$	$P_{2,1}$	$P_{2,2}$			$x_2 - x$
$\vdots$						
x_{k-2}	$P_{k-2,0}$	$P_{k-2,1}$	$P_{k-2,2}$			$x_{k-2} - x$
x_{k-1}	$P_{k-1,0}$	$\underline{P_{k-1,1}}$	$P_{k-1,2}$			$x_{k-1} - x$
x_k	$P_{k,0}$	$\underline{P_{k,1}}$	$\underline{\underline{P_{k,2}}}$			$x_k - x$
$\vdots$						
x_n	$P_{n,0}$	$P_{n,1}$	$P_{n,2}$	$\cdots$	$P_{n,n}$	$x_n - x$

Scheme 10.5

Theorem 10.5 In the notation of lemma 10.3, if the polynomials $P_{k,d}$ are generated by algorithm 10.5,

$$P_{k,d} = P_{\{x_{k-d}, x_{k-d+1}, \ldots, x_k\}},$$
$$d = 0, 1, \ldots, n; \quad k = d, d + 1, \ldots, n.$$

Proof. By (10-15), the assertion is true for $d = 0$. If true for some $d \geq 0$, then it follows from (10-16) by virtue of lemma 10.3 that $P_{k,d+1}$ interpolates on the union of the sets $\{x_{k-d-1}, x_{k-d}, \ldots, x_{k-1}\}$ and $\{x_{k-d}, x_{k-d+1}, \ldots, x_k\}$, i.e., on the set $\{x_{k-d-1}, x_{k-d}, \ldots, x_k\}$, proving the assertion with d increased by one.

For $d = k = n$ we have, in particular

Corollary 10.5 $P_{n,n} = P_W,$

thus again, the rightmost polynomial in Neville's scheme is the desired polynomial interpolating at *all* points $x_0, x_1, \ldots, x_n$.

EXAMPLE

10. We again consider $f(x) = x^4$ and calculate $f(3)$ from the values at $x = -4, -2, 0, 2, 4$, using Neville's algorithm. The following scheme results:

x_k	$P_{k,0}$	$P_{k,1}$	$P_{k,2}$	$P_{k,3}$	$P_{k,4}$	$x_k - x$
-4	256					-7
-2	16	-584				-5
0	0	-24	396			-3
2	16	24	36	-24		-1
4	256	136	108	96	81	1

Problems

12. Which entries in the schemes of the polynomials $P_{k,d}$ generated by the algorithms 10.4 and 10.5 are necessarily identical?

13. Find an approximate value of $\sqrt{2}$ by interpolation, using Neville's algorithm, from the values of the function $f(x) = 2^x$ at the points $x = -2, -1, 0, 1, 2, 3$.

14. Calculate an approximate value of the infinite series

$$1 + \frac{1}{2^2} + \frac{1}{3^2} + \frac{1}{4^2} + \cdots$$

in the following manner: Let

$$f\left(\frac{1}{n}\right) = 1 + \frac{1}{2^2} + \cdots + \frac{1}{n^2} \qquad (n = 1, 2, 3, \ldots)$$

and calculate $f(0)$ by extrapolation from $f(1), f(\frac{1}{2}), f(\frac{1}{3}), \ldots$, using Neville's algorithm.

10.6 Inverse Interpolation

Interpolation (approximately) solves the problem of finding the value of $y = f(x)$ when x is given. It does *not* solve the problem of finding x when $y = f(x)$ is given. (We could, of course, replace f by the interpolating polynomial P and solve the equation $y = P(x)$ for x, but in doing so we would merely replace one problem by another problem of comparable difficulty.) The problem can be easily solved, however, by interchanging the roles of x and y. Speaking abstractly, this amounts to interpolating the *inverse function* $f^{[-1]}$ instead of f itself. Speaking concretely, it means interchanging the roles of the x_k and the f_k. Since, even for equidistant x_k, the corresponding values f_k are not equidistant, it is essential that we are able to calculate the interpolating polynomial for nonequidistant interpolating points. As an example, we consider the problem of solving $f(x) = 0$ when the function f is known at $n + 1$ distinct points x_k. If the polynomial $P(y)$ interpolating the inverse function at the points $f(x_m) = f_m$ is constructed by Aitken's algorithm and evaluated at $y = 0$, we obtain the following algorithm:

Algorithm 10.6   Let

$$X_{m,0} = x_m \qquad (m = 0, 1, \ldots, n)$$

and form for $n = 0, 1, \ldots, m - 1$ the numbers

$$X_{m,n+1} = \frac{f_m X_{n,n} - f_n X_{m,n}}{f_m - f_n}.$$

The approximate solution is given by $X_{n,n}$. The arrangement of the triangular array of the values $X_{m,n}$ is as follows:

$$
\begin{array}{c|cccc}
f_0 & X_{0,0} \\
f_1 & X_{1,0} & X_{1,1} \\
f_2 & X_{2,0} & X_{2,1} & X_{2,2} \\
\vdots & \cdot & \cdot & \cdot & \cdot & \cdot \\
f_n & X_{n,0} & X_{n,1} & X_{n,2} & \cdots & X_{n,n}.
\end{array}
$$

Inverse interpolation is possible only if, in the range where interpolation is used, x is a single-valued function of y. In the example depicted in figure 10.6, where this condition is not satisfied, the interpolating polynomial bears no relationship to the inverse function.

The *error* of inverse interpolation obeys the same laws as the error of ordinary interpolation. It depends on the derivatives of the inverse function $f^{[-1]}(y)$. These derivatives can be calculated, in principle at least, from the derivatives of the function f. Differentiating the identity

$$f^{[-1]}(f(x)) = x$$

we obtain

(10-17) $$f^{[-1]\prime}(f(x))f'(x) = 1;$$

hence

$$f^{[-1]\prime}(f(x)) = \frac{1}{f'(x)}.$$

Higher derivatives can be obtained by repeatingly differentiating (10-17). For instance,

$$f^{[-1]\prime\prime}(f(x))[f'(x)]^2 + f^{[-1]\prime}(f(x))f''(x) = 0$$

shows that

$$f^{[-1]\prime\prime}(f(x)) = -\frac{f''(x)}{f'(x)^3}.$$

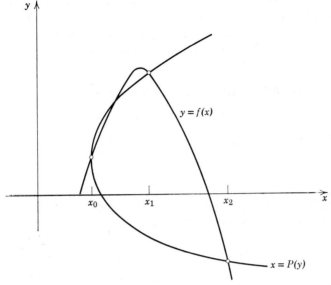

Figure 10.6

This process can be continued, but the results become more and more complicated, even if the derivatives of f are simple.

Problem

15. Using inverse interpolation, find an approximate value of the second zero of the Bessel function $J_0(x)$ if the following values are given:

x	$J_0(x)$
5.2	−0.1102904
5.4	−0.0412101
5.6	0.0269709
5.8	0.0917026

10.7   Iterated Inverse Interpolation

Solving an equation $f(x) = 0$ by simple inverse interpolation as described above is appropriate if the function f is known only at a set of discrete points x (e.g., if f is a tabulated function). If $f(x)$ can be calculated for arbitrary x, it is possible to test the result of the inverse interpolation procedure by evaluating f at the interpolated value of x. In general, $f(x)$ will not be exactly equal to 0. In this case, $f(x)$ and x are introduced as new entries in the interpolation table, and a new row of values of X is calculated by inverse interpolation. The Neville form of the interpolation table is especially appropriate here, because in it already the first entries in a new horizontal row are good approximations to the desired value of x. Beginning with two values of f, the Neville scheme is continued systematically row by row as follows:

Algorithm 10.7   Choose x_0 and x_1, and let

$$f_0 = f(x_0), \qquad X_{0,0} = x_0,$$

(10-18) $$\qquad f_1 = f(x_1), \qquad X_{1,0} = x_1,$$

$$X_{1,1} = \frac{f_1 X_{0,0} - f_0 X_{1,0}}{f_1 - f_0}.$$

Then form the triangular array of numbers $X_{m,n}$ ($m = 2, 3, \ldots$; $n = 0, 1, \ldots, m$) by means of the relations

$$X_{m,0} = X_{m-1,m-1}, \qquad f_m = f(X_{m,0}),$$

(10-19) $$\qquad X_{m,n+1} = \frac{f_m X_{m-1,n} - f_{m-n-1} X_{m,n}}{f_m - f_{m-n-1}}$$

$$n = 0, 1, \ldots, m - 1.$$

If f is sufficiently differentiable in a neighborhood of a solution s of $f(x) = 0$, if $f'(x) \neq 0$, and if x_0 and x_1 are sufficiently close to s, it is intuitively clear that the numbers $X_{n,n}$ converge to s for $n \to \infty$.

As described above, the table of the values $X_{m,n}$ extends farther and farther to the right with every new row. Ultimately, little extra accuracy will be gained from the entries far to the right, because they depend on values $X_{n,0}$ with small n which presumably are poor approximations to the desired solution. It is therefore advisable not to increase the degrees of the inverse interpolating polynomial beyond a certain degree d (say $d = 2$ or $d = 3$), that is, to truncate the table of values $X_{m,n}$ after the dth column. This means that the formulas (10-19) are only used for $m \leq d$, for $m > d$ they are to be replaced by the following:

(10-20) $$\begin{cases} X_{m,0} = X_{m-1,d}, \qquad f_m = f(X_{m,0}), \\ X_{m,n+1} = \dfrac{f_m X_{m-1,n} - f_{m-n-1} X_{m,n}}{f_m - f_{m-n-1}}, \qquad n = 0, 1, \ldots, d - 1. \end{cases}$$

The convergence of this modified version of algorithm 10.7 is, under suitable conditions, proved by Ostrowski ([1960], chapter 13).

The case $d = 2$ of the modified algorithm 10.7 is very similar to Muller's method discussed in §10.1, except that now the inverse function rather than the function itself is interpolated by a quadratic polynomial.   From the computational point of view it would even seem to be superior to Muller's method since it does not require the evaluation of square roots. However, just for this reason it lacks the advantage of automatically branching off into the complex domain if no real zeros are found.

Problems

16. Using repeated quadratic inverse interpolation, find the root of the polynomial

$$P(x) = 70x^4 - 140x^3 + 90x^2 - 20x + 1$$

located between 0.6 and 0.7.   (Use Horner's scheme to evaluate the polynomial.)

17. Show that iterated *linear* inverse interpolation is identical with the regula falsi (see §4.11).

Recommended Reading

The practical aspects of interpolation are dealt with in a volume issued by the Nautical Almanac Service [1956].   The theory of a number of processes for solving $f(x) = 0$ by methods based on interpolation is dealt with very thoroughly by Ostrowski [1960].

Research Problems

1.   How can the regula falsi be extended to the solution of systems of more than one equation?   (For some pertinent remarks, see Ostrowski [1960], p. 146.)

2.   Develop a theory for interpolating functions of two variables by bilinear polynomials of the form $a + bx + cy + dxy$.

chapter 11 construction of the interpolating polynomial: methods using differences

The representations of the interpolating polynomial discussed in chapter 10 were based directly on the values of the interpolated function. They do not convey any information, explicit or implicit, concerning the *error* of the interpolating polynomial. In this respect, the methods to be discussed in the present chapter do somewhat better. These representations are based on *differences* of the sequence of function values, and on certain properties of binomial coefficients.

11.1   Differences and Binomial Coefficients

Differences of a sequence of numbers were already defined in §4.4 and §6.9. We now introduce differences of a function f defined on a suitable interval. Let $h > 0$ be a constant. The function Δf whose value at x is given by

$$\Delta f(x) = f(x + h) - f(x)$$

is called the *first* (forward) *difference* of the function f. It obviously depends on the step h, although this fact is usually not made evident in the notation. Higher differences are defined inductively by the relation

$$\Delta^k f = \Delta(\Delta^{k-1}f), \qquad k = 2, 3, \ldots.$$

For instance,

$$\Delta^2 f(x) = f(x + 2h) - 2f(x + h) + f(x).$$

For symmetry we put

$$\Delta^0 f = f.$$

214

An induction argument entirely analogous to that employed in the proof of (6-27) shows that

$$\Delta^k f(x) = f(x + kh) - \binom{k}{1} f(x + (k - 1)h)$$

$$+ \binom{k}{2} f(x + (k - 2)h) + \cdots + (-1)^k f(x).$$

If, for integral k, $x_k = x_0 + kh$, and if we write

$$f(x_k) = f_k,$$

the differences thus introduced produce the same result as the difference operator Δ introduced in §4.4, if the latter is applied to the sequence of values $\{f_k\}$ of f.

It is to be expected that the differences of a function share many properties, and have many connections, with the derivatives of the function. For instance, the mean value theorem of differential calculus states that, for some ξ between x and $x + h$,

$$\Delta f(x) = h f'(\xi).$$

We shall soon become acquainted with a generalization of this relation to differences and derivatives of arbitrary order.

In differential calculus, a set of functions enjoying particularly simple properties with respect to differentiation is the set of monomials $x^n/n!$, $n = 0, 1, \ldots$. In fact,

(11-1)
$$\left(\frac{x^n}{n!}\right)' = \frac{x^{n-1}}{(n-1)!}.$$

In difference calculus, an analogous role is played by the binomial coefficients

(11-2)
$$\binom{s}{n} = \frac{s(s - 1)\ldots(s - n + 1)}{n!}.$$

Here s is any real (or even complex) number, and n is a positive integer. For $n = 0$, the symbol (11-2) is defined to be 1, for negative integers n, zero. It is always understood in the following that the operator Δ acts on the variable s, and that the step $h = 1$ is implied. With this understanding we have

(11-3)
$$\Delta \binom{s}{n} = \binom{s}{n - 1}.$$

This is trivially true for $n \leq 0$; for $n > 0$ we have

$$\Delta\binom{s}{n} = \binom{s+1}{n} - \binom{s}{n},$$

$$= \frac{(s+1)s(s-1)\ldots(s-n+2) - s(s-1)\ldots(s-n+1)}{n!}$$

$$= \frac{[s+1-(s-n+1)]s(s-1)\ldots(s-n+2)}{n(n-1)!}$$

$$= \binom{s}{n-1},$$

as desired.   By induction it follows immediately from (11-3) that

$$(11\text{-}4) \qquad \Delta^k\binom{s}{n} = \binom{s}{n-k}, \qquad k = 0, 1, 2, \ldots.$$

Another property of the monomials $x^n/n!$ also carries over to the binomial coefficients.   It is trivial that every polynomial of degree n can be expressed as a linear combination of the monomials

$$1, \frac{x}{1!}, \ldots, \frac{x^n}{n!}.$$

Similarly, such a polynomial can also be expressed as a linear combination of

$$1, \binom{s}{1}, \binom{s}{2}, \ldots, \binom{s}{n}.$$

In fact, a somewhat more general statement is true.   If $\{a_1, a_2, \ldots, a_n\}$ is an arbitrary set of real numbers, then a polynomial of degree n is expressible in terms of the generalized monomials

$$1, \frac{x + a_1}{1!}, \frac{(x + a_2)^2}{2!}, \ldots, \frac{(x + a_n)^n}{n!}.$$

This fact has the following analog:

Theorem 11.1   Let $a_1, a_2, \ldots, a_n$ be n arbitrary real numbers, and let p be any polynomial of degree n.   Then there exist constants A_0, $A_1, \ldots, A_n$ such that

$$(11\text{-}5) \quad p(s) = A_0 + A_1\binom{s+a_1}{1} + A_2\binom{s+a_2}{2} + \cdots + A_n\binom{s+a_n}{n}$$

identically in s.

Proof.   Evidently the statement of the theorem is true for $n = 0$; we proceed by induction with respect to n and assume that the theorem is true for some nonnegative integer $n - 1$. If

$$p(s) = b_0 s^n + b_1 s^{n-1} + \cdots + b_n,$$

then the polynomial

(11-6) $$q(s) = p(s) - b_0 n! \binom{s + a_n}{n}$$

is of degree $n - 1$, since the leading coefficient of

$$\binom{s + a_n}{n} \quad \text{is} \quad \frac{s^n}{n!}.$$

By the induction hypothesis, q can be represented in the form

$$q(s) = A_0 + A_1 \binom{s + a_1}{1} + \cdots + A_{n-1} \binom{s + a_{n-1}}{n-1}.$$

Solving (11-6) for $p(s)$ we obtain a representation of the desired form (11-5), where $A_n = b_0 n!$.

EXAMPLES

1. A special case of theorem 11.1 was used in example 13 of chapter 6, where we obtained the formula

$$s^3 = 6\binom{s}{3} + 6\binom{s}{2} + \binom{s}{1}.$$

2. The truth of corollary 6.8 (which was not proved in §6.8) follows from the special case $a_1 = a_2 = \cdots = a_n = 0$ of the above theorem 11.1.

Problems

1. Determine all functions f that are defined on the whole real line and satisfy

$$\Delta f(x) = hcf(x)$$

identically in x, where c is a constant.
2. If f and g are two differentiable functions, find a formula for $\Delta(fg)$ and derive the product rule for differentiation.
3. Formulate an algorithm for obtaining the representation (11-5) for a given polynomial $p(s) = b_0 s^n + \cdots + b_n$ and for given constants $a_1, a_2, \ldots, a_n$. (Determine A_n first.)
4. Represent the polynomials $p(s) = s^n$ ($n = 1, 2, \ldots$) in the form (11-5), where $a_n = 1, 2, \ldots$.
5. Find a closed expression for the differences

$$\Delta^k f_0, \quad k = 0, 1, 2, \ldots,$$

if $f(x) = e^x$ and $x_0 = 0$. Show that in this case

$$\lim_{h \to 0} h^{-k} \Delta^k f_0 = f'(0).$$

11.2   Finalized Representations of Sequences of Interpolating Polynomials

We now return to the problem of constructing the polynomials interpolating a function f on a given set of points. The interpolating points are a set of equidistant points

$$x_k = x_0 + kh,$$

where the integer k may be positive or negative. As usual we write

$$f_k = f(x_k).$$

It will be convenient to express the interpolating polynomial in terms of the variable

(11-7) $$s = \frac{x - x_0}{h}.$$

Thus, if S is a set of interpolating points x_k and P_S denotes the polynomial interpolating f at the points of S, we shall define p by

$$p(s) = P_S(x) = P_S(x_0 + sh).$$

The polynomial p is characterized by the property that

$$p(k) = f_k \quad \text{whenever} \quad x_k \in S;$$

for nonintegral values of s, $p(s)$ is to be regarded as an approximation to $f(x_0 + sh)$.

Actually, we are now not merely interested in constructing a single polynomial, interpolating f on a single set S. Instead, we shall try to determine sequences of interpolating polynomials that interpolate f on a sequence of sets $S_0, S_1, S_2, \ldots$ of interpolating points. These sets are defined as follows:

Let $m_0, m_1, m_2, \ldots$ be a nonincreasing sequence of integers such that

(11-8) $$m_k - 1 \leqq m_{k+1} \leqq m_k$$

for all $k = 0, 1, 2, \ldots$, and let

$$S_k = \{x_{m_k}, x_{m_k + 1}, \ldots, x_{m_k + k}\},$$

$k = 0, 1, 2, \ldots$.   The set S_k thus contains precisely $k + 1$ consecutive interpolating points, beginning with the point x_{m_k}. By virtue of (11-8), each set S_k contains the preceding set S_{k-1} and thus *all* preceding sets $S_{k-2}, \ldots, S_0$.

EXAMPLES

3.   Let $m_k = 0$, $k = 0, 1, 2, \ldots$.   Then

$$S_k = \{x_0, x_1, \ldots, x_k\}.$$

4. If $m_k = -k$, $k = 0, 1, 2, \ldots$, we have

$$S_k = \{x_{-k}, x_{-k+1}, \ldots, x_0\}.$$

5. Setting $m_0 = m_1 = 0$, $m_2 = m_3 = -1, \ldots$, we obtain the sets

$$S_{2k} = \{x_{-k}, x_{-k+1}, \ldots, x_k\},$$
$$S_{2k+1} = \{x_{-k}, x_{-k+1}, \ldots, x_{k+1}\}, \qquad k = 0, 1, 2, \ldots.$$

6. Setting $m_0 = 0$, $m_1 = m_2 = -1$, $m_3 = m_4 = -2, \ldots$, the sets S_k are given by

$$S_{2k} = \{x_{-k}, x_{-k+1}, \ldots, x_k\},$$
$$S_{2k+1} = \{x_{-k-1}, x_{-k}, \ldots, x_k\}.$$

By the fundamental theorem 9.1, there exists, for every $k = 0, 1, 2, \ldots$, a unique polynomial P_{S_k} of degree $\leq k$ such that

$$P_{S_k}(x_m) = f_m, \qquad x_m \in S_k.$$

If x and s are connected by (11-7), we shall write

$$p_k(s) = P_{S_k}(x).$$

We now wish to consider the following *problem*: Given a sequence of integers $\{m_k\}$ satisfying (11-8) (and hence a sequence of sets $S_0, S_1, \ldots$), determine two sequences of real numbers $\{a_k\}$ and $\{A_k\}$ such that, for *every* $n = 0, 1, 2, \ldots$,

$$(11\text{-}9) \qquad p_n(s) = A_0 + A_1\binom{s + a_1}{1} + \cdots + A_n\binom{s + a_n}{n}.$$

identically in s.

It is not clear at all that this problem has a solution. Theorem 11.1 merely tells us that, for an arbitrary sequence $\{a_k\}$ and for every fixed integer n, constants $A_0, A_1, \ldots, A_n$ can be found such that (11-9) holds. It is to be expected, however, that if n is replaced by $n + 1$, the constants $A_0, A_1, \ldots, A_n$ already found will have to be replaced by other constants. If we wish to obtain a "finalized" representation of the sequence $\{p_n\}$ with the property that the constants A_k, once determined, remain unchanged, we can hope to do so only by a judicious choice of the sequence $\{a_n\}$.

EXAMPLE

7. Let $a_n = n$, $n = 1, 2, \ldots$. The function $f(s) = s$ is interpolated at $s = 0$ by $p_0(s) = 0$. This is a representation of the form (11-9) with $A_0 = 0$. In order to interpolate f at $s = 0$ and $s = 1$, we must take

$$p_1(s) = s = -1 + \binom{s + 1}{1}.$$

The last expression is again of the form (11-9), but we now have $A_0 = -1$. The coefficient $A_0 = 0$ is preserved if and only if we choose $a_1 = -0$.

Let us now investigate the properties which the sequence $\{a_k\}$ must have if the above problem is to have a solution. For an arbitrary integer $n \geq 0$, consider the two polynomials

$$p_n(s) = A_0 + A_1 \binom{s + a_1}{1} + \cdots + A_n \binom{s + a_n}{n}$$

and

$$p_{n+1}(s) = A_0 + A_1 \binom{s + a_1}{1} + \cdots + A_n \binom{s + a_n}{n} + A_{n+1} \binom{s + a_{n+1}}{n + 1}.$$

The polynomial p_n interpolates on the set S_n, p_{n+1} on the set S_{n+1}. Since S_n is contained in S_{n+1}, both polynomials interpolate on the set S_n. Both thus have identical values for x_s in the set S_n. This means that the last term

$$A_{n+1} \binom{s + a_{n+1}}{n + 1}$$

must vanish whenever s is equal to one of the integers

(11-10) $$m_n, m_n + 1, \ldots, m_n + n.$$

If f is such that p_{n+1} has degree $n + 1$, then $A_{n+1} \neq 0$, and the required condition is

(11-11) $$\binom{s + a_{n+1}}{n + 1} = 0$$

for all said integers. The binomial coefficient in (11-11) is zero if and only if s is one of the numbers

$$-a_{n+1}, \; -a_{n+1} + 1, \ldots, \; -a_{n+1} + n.$$

Evidently the set of these numbers coincides with the set (11-10) if and only if

(11-12) $$a_{n+1} = -m_n, \qquad n = 0, 1, 2, \ldots.$$

This condition fully determines the sequence $\{a_1, a_2, \ldots\}$ as a function of the sequence $\{m_n\}$.

There remains the problem of determining the constants A_k. For a fixed value of n, there certainly exist, by theorem 11.1, constants A_0, $A_1, \ldots, A_n$ such that

$$p_n(s) = A_0 + A_1 \binom{s - m_0}{1} + \cdots + A_n \binom{s - m_{n-1}}{n}.$$

The question is whether these A_k are independent of n. In order to determine A_k for $0 \leq k \leq n$, we form the kth difference of $p_n(s)$. By (11-4), we get

$$(11\text{-}13) \quad \Delta^k p_n(s) = A_k + A_{k+1}\binom{s - m_k}{1} + \cdots + A_n\binom{s - m_{n-1}}{n - k}.$$

In this identity we set $s = m_k$. The values of p involved in forming the difference $\Delta^k p_n(s)$ then are the values $p(s)$ for $s = m_k, m_k + 1, \ldots, m_k + k$, i.e., those values of s for which $x_s \in S_k$. Since $S_k \subset S_n$, we have

$$p_n(s) = f_s$$

for these values, and hence

$$\Delta^k p_n(m_k) = \Delta^k f_{m_k}.$$

In the expression on the right of (11-13), all binomial coefficients are zero for $s = m_k$, as it follows from (11-8) that

$$0 \leqq m_k - m_{k+l} \leqq l, \qquad l = 0, 1, 2, \ldots.$$

We thus obtain

$$A_k = \Delta^k f_{m_k},$$

and it turns out that A_k is independent of n, the degree of the interpolating polynomial, as we had hoped. The polynomials p_n solving the problem posed initially thus are given by

$$(11\text{-}14) \qquad p_n(s) = \sum_{k=0}^{n} \Delta^k f_{m_k}\binom{s - m_{k-1}}{k}.$$

They can be generated recursively by the following simple algorithm:

Algorithm 11.2 If $\{m_k\}$ is a sequence of integers satisfying (11-8), let $p_0(s) = f_{m_0}$, and for $n = 0, 1, 2, \ldots$,

$$(11\text{-}15) \qquad p_{n+1}(s) = p_n(s) + \Delta^{n+1} f_{m_{n+1}}\binom{s - m_n}{n + 1}.$$

By construction, we have

$$p_n(k) = f(x_k),$$

$k = m_n, m_n + 1, \ldots, m_n + n$. We wish to find an expression for the error of $p_n(s)$ if s is not equal to one of the above values of k. This is easily possible in our new notation. According to theorem 9.2 the difference

$$f(x) - P_{S_n}(x) = f(x) - p_n(s)$$

can, if f has a continuous derivative of order $n + 1$, be written in the form

$$\frac{(x - x_{m_n})(x - x_{m_n+1})\ldots(x - x_{m_n+n})}{(n + 1)!} f^{(n+1)}(\xi_x),$$

where ξ_x is some point in the smallest interval containing x, x_{m_n}, and x_{m_n+n}. We have

$$x - x_{m_n+k} = h[s - (m_n + k)], \qquad k = 0, 1, \ldots, n.$$

The product of the factors $x - x_{m_n+k}$ appearing above thus can be written

$$h^{n+1}(n + 1)!\binom{s - m_n}{n + 1},$$

and we obtain

Theorem 11.2 Let the function f have a continuous $(n + 1)$st derivative on an interval containing the points $x = x_0 + sh$, x_{m_n}, and x_{m_n+n}. If $p_n(s)$ is defined by algorithm 11.2, then for a suitable point ξ_x of that interval

$$(11\text{-}16) \qquad f(x) = p_n(s) + h^{n+1}f^{(n+1)}(\xi_x)\binom{s - m_n}{n + 1}.$$

A comparison of the equations (11-15) and (11-16) shows that the correction term which has to be added to $p_n(s)$ in order to obtain the exact value of f is of the same form as the term that has to be added in order to pass from p_n to p_{n+1}, with the exception that

$$\Delta^{n+1}f_{m_n+1} \text{ is to be replaced by } h^{n+1}f^{(n+1)}(\xi_x).$$

If the function $f^{(n+1)}$ does not change very rapidly, these two terms are of the same order of magnitude (see the problems 11 and 12). One can thus say with some justification that the error of p_n is of the order of the first omitted term in the sum (11-14). Thus, if the degree of the interpolating polynomial is not fixed beforehand, one may hope to obtain an accurate representation of f (to within rounding errors) by extending the sum (11-14) through such a value of n that the omitted terms are insignificant.

EXAMPLES

We shall construct the sequences of interpolating polynomials corresponding to the sequences $\{m_k\}$ considered in the examples 3, 4, 5, and 6.

8. For $m_n = 0$, $n = 0, 1, 2, \ldots$ we obtain the polynomials

$$p_n(s) = f_0 + \binom{s}{1}\Delta f_0 + \binom{s}{2}\Delta^2 f_0 + \cdots + \binom{s}{n}\Delta^n f_0,$$

interpolating on the sets $\{x_0, x_1, \ldots, x_n\}$.

9. For $m_n = -n$, $n = 0, 1, 2, \ldots$ we obtain

$$p_n(s) = f_0 + \binom{s}{1}\Delta f_{-1} + \binom{s+1}{2}\Delta^2 f_{-2} + \cdots + \binom{s+n-1}{n}\Delta^n f_{-n}.$$

These polynomials p_n interpolate on the set $\{x_0, x_{-1}, x_{-2}, \ldots, x_{-n}\}$.

The formulas obtained in examples 8 and 9 are known, respectively, as the *Newton forward* and the *Newton backward formula*.

10. Letting

$$\{m_0, m_1, m_2, \ldots\} = \{0, 0, -1, -1, \ldots\},$$

we get the polynomials

$$p_n(s) = f_0 + \binom{s}{1}\Delta f_0 + \binom{s}{2}\Delta^2 f_{-1}$$

$$+ \binom{s+1}{3}\Delta^3 f_{-1} + \binom{s+1}{4}\Delta^4 f_{-2} + \cdots$$

$$(n+1 \text{ terms})$$

interpolating on the sets

$$\{x_{-k}, x_{-k+1}, \ldots, x_k\} \quad \text{for } n = 2k$$

and

$$\{x_{-k}, x_{-k+1}, \ldots, x_{k+1}\} \quad \text{for } n = 2k+1.$$

11. Taking $\{m_0, m_1, m_2, \ldots\} = \{0, -1, -1, -2, -2, \ldots\}$, we obtain the polynomials

$$p_n(s) = f_0 + \binom{s}{1}\Delta f_{-1} + \binom{s+1}{2}\Delta^2 f_{-1}$$

$$+ \binom{s+1}{3}\Delta^3 f_{-2} + \binom{s+2}{4}\Delta^4 f_{-2} + \cdots$$

$$(n+1 \text{ terms}).$$

They interpolate on the sets $\{x_{-k}, x_{-k+1}, \ldots, x_k\}$ for $n = 2k$ and $\{x_{-k-1}, x_{-k}, \ldots, x_k\}$ for $n = 2k+1$.

The formulas obtained in the examples 10 and 11 are known, respectively, as the *Gauss forward* and the *Gauss backward formula*.

Problems

6. Forming differences of the values of the function J_0 given in problem 15, chapter 10, find $J_0(5.5)$ by the four interpolation formulas given above.

7. Using the fact that

$$\binom{s}{m} = (-1)^m \binom{m-s-1}{m}$$

and expressing forward differences by backward differences (see §6.9), show that the Newton backward formula can be written in the form

$$p_n(s) = f_0 - \binom{-s}{1} \nabla f_0 + \binom{-s}{2} \nabla^2 f_0 - \cdots + (-1)^n \binom{-s}{n} \nabla^n f_0.$$

8. *Expressing ordinates in terms of differences.* Establish the formula

$$f_k = \sum_{m=0}^{k} \binom{k}{m} \Delta^m f_0, \qquad k = 0, 1, 2, \ldots$$

(a) by induction; (b) by considering it as a special case of the Newton formula.

9. Using the fact that the identity of problem 8 must hold for arbitrary values of $f_0, f_1, \ldots$, show that for arbitrary integers n and k such that $0 \leq n \leq k$

$$\sum_{m=n}^{k} (-1)^{m-n} \binom{k}{m} \binom{m}{n} = \begin{cases} 1, & n = k \\ 0, & n < k. \end{cases}$$

10. Study the convergence of the infinite series

$$\sum_{k=0}^{\infty} \Delta^k f_0 \binom{s}{k}$$

(Newton's formula extended to infinitely many terms), where $f(x) = e^x$, as it depends on x and h. (Use problem 5 and apply the ratio test.)

11. Let f be n times continuously differentiable on a suitable interval. Show that for some $\xi \in (x_0, x_n)$

$$\Delta^n f_0 = h^n f^{(n)}(\xi).$$

[Differentiate the function

$$f(x) - \sum_{k=0}^{n} \binom{s}{k} \Delta^k f_0, \qquad s = \frac{x - x_0}{h}$$

n times with respect to x and apply Rolle's theorem.]

12. As an application of the preceding problem, show that

$$\lim_{h \to 0} h^{-n} \Delta^n f_0 = f^{(n)}(x_0)$$

for any sufficiently differentiable function f.

13. Assume that the values of f_n are known only up to rounding errors ε_n, where $|\varepsilon_n| \leq \varepsilon$. Show that the maximum error in $\Delta^k f_n$ can be as large as $2^k \varepsilon$.

11.3   Some Special Interpolation Formulas

In spite of their basic simplicity the interpolation formulas of Newton and Gauss given in §11.2 are not frequently used in practice, mainly because of their lack of formal symmetry.   More frequently used in practical interpolation are certain formulas named after Stirling, Bessel, and Everett.

The elegant formulation of these formulas requires the introduction of two new operators μ and δ in addition to the forward difference operator Δ. The operator μ is defined by

(11-17)
$$\mu f(x) = \frac{1}{2}\left[f\left(x + \frac{h}{2}\right) + f\left(x - \frac{h}{2}\right)\right]$$

and consequently is called the *averaging operator*. The operator δ is defined by

(11-18)
$$\delta f(x) = f\left(x + \frac{h}{2}\right) - f\left(x - \frac{h}{2}\right)$$

and is called the *central difference operator*. Note that δ can always be expressed in terms of Δ, and vice versa. For instance,

$$\Delta^{2k}f_{-k} = \delta^{2k}f_0, \qquad k = 0, 1, 2, \ldots.$$

A further identity to be noted is

$$\mu\delta f_0 = \tfrac{1}{2}(\Delta f_{-1} + \Delta f_0) = \tfrac{1}{2}(f_1 - f_{-1}).$$

Stirling's formula. For a given even integer $n = 2k$ both the Gauss forward and the Gauss backward formula yield the interpolating polynomial corresponding to the set $\{x_{-k}, x_{-k+1}, \ldots, x_k\}$. Thus also their arithmetic mean must yield the same polynomial. The resulting polynomial

$$p_{2k}(s) = f_0 + \binom{s}{1}\frac{1}{2}[\Delta f_{-1} + \Delta f_0]$$

$$+ \frac{1}{2}\left[\binom{s}{2} + \binom{s+1}{2}\right]\Delta^2 f_{-1} + \binom{s+1}{3}\frac{1}{2}[\Delta^3 f_{-1} + \Delta^3 f_{-2}]$$

$$+ \cdots + \frac{1}{2}\left[\binom{s+k-1}{2k} + \binom{s+k}{2k}\right]\Delta^{2k}f_{-k}$$

can by virtue of the identity

$$\binom{s+k-1}{2k} + \binom{s+k}{2k} = \frac{s}{k}\binom{s+k-1}{2k-1} \qquad (k = 1, 2, \ldots)$$

be written in the form

(11-19)
$$p_{2k}(s) = f_0 + \binom{s}{1}\left[\mu\delta f_0 + \frac{s}{2}\delta^2 f_0\right] + \cdots$$

$$+ \binom{s+k-1}{2k-1}\left[\mu\delta^{2k-1}f_0 + \frac{s}{2k}\delta^{2k}f_0\right].$$

This formula, called Stirling's formula, expresses the polynomial interpolating at an odd number of equidistant points in terms of central differences at the center point. It is preferably used for interpolation near that center point.

Bessel's formula. If n is odd, $n = 2k + 1$, the Gaussian forward formula at x_0 and the Gaussian backward formula at x_1 interpolate on the same set of points $S_{2k+1} = \{x_{-k}, x_{-k+1}, \ldots, x_{k+1}\}$. The Gaussian backward formula centered at x_1 is given by

$$p_{2k+1}(s) = f_1 + \binom{s'}{1} \Delta f_0 + \binom{s'+1}{2} \Delta^2 f_0 + \cdots + \binom{s'+k}{2k+1} \Delta^{2k+1} f_{-k},$$

where

$$s' = \frac{x - x_1}{h} = \frac{x - x_0 - h}{h} = s - 1.$$

Averaging this expression with the Gaussian forward formula at x_0 we get, after some simplification,

(11-20) $p_{2k+1}(s) = \mu f_{1/2} + (s - \tfrac{1}{2}) \delta f_{1/2}$

$$+ \binom{s}{2} [\mu \delta^2 f_{1/2} + \tfrac{1}{3}(s - \tfrac{1}{2}) \delta^3 f_{1/2}] + \cdots$$

$$+ \binom{s+k-1}{2k} \left[\mu \delta^{2k} f_{1/2} + \frac{1}{2k+1} (s - \tfrac{1}{2}) \delta^{2k+1} f_{1/2} \right].$$

This formula is known as *Bessel's formula.* It expresses the interpolating polynomial for an even number of consecutive points in terms of central differences. It is preferably used for interpolation halfway between the two center points.

Everett's formula. We start from Gauss' forward formula, where n is odd, $n = 2k + 1$. Eliminating differences of odd order by use of the formula

$$\Delta^{2k+1} f_{-k} = \Delta^{2k} f_{-k+1} - \Delta^{2k} f_{-k},$$

we obtain

$$p_{2k+1}(s) = f_0 + \binom{s}{1}(f_1 - f_0) + \binom{s}{2} \Delta^2 f_{-1}$$

$$+ \binom{s+1}{3}[\Delta^2 f_0 - \Delta^2 f_{-1}] + \binom{s+1}{4} \Delta^4 f_{-2} + \cdots$$

$$+ \binom{s+k-1}{2k} \Delta^{2k} f_{-k} + \binom{s+k}{2k+1}[\Delta^{2k} f_{-k+1} - \Delta^{2k} f_{-k}].$$

Collecting equal differences, expressing them in terms of the central difference operator and using the identity

$$\binom{s + k - 1}{2k} - \binom{s + k}{2k + 1} = \binom{s + k - 1}{2k}\left(1 - \frac{s + k}{2k + 1}\right)$$

$$= \binom{t + k}{2k + 1},$$

where

$$t = 1 - s,$$

we obtain the formula

$$(11\text{-}21) \quad p_{2k+1}(s) = \binom{t}{1} f_0 + \binom{t + 1}{3} \delta^2 f_0 + \cdots + \binom{t + k}{2k + 1} \delta^{2k} f_0$$

$$+ \binom{s}{1} f_1 + \binom{s + 1}{3} \delta^2 f_1 + \cdots + \binom{s + k}{2k + 1} \delta^{2k} f_1$$

due to the British astronomer Everett.

Of the three formulas given above, the highly symmetrical and elegant formula due to Everett has found great favor in practice. It has the advantage of using only differences of even order, and of furnishing an interpolation polynomial whose degree (and consequently, accuracy) is higher than the order of the highest difference employed. For instance, with a column of second differences alone we can calculate the cubic interpolating polynomial, whereas the application of all other formulas requires three difference columns. For these reasons many tables of higher transcendental functions, if they give any differences at all, give second differences only. No special tables of the Everett interpolation coefficients

$$\binom{s + k}{2k + 1}, \qquad k = 1, 2, \ldots,$$

are required, as these coefficients are identical with the extreme (first and last) Lagrangian interpolation coefficients interpolating on the set $x_{-k}, x_{-k+1}, \ldots, x_{k+1}$ (see §10.2).

11.4 Throwback†

Throwback of higher differences into lower differences is an extremely simple but ingenious device, due to Comrie, which enhances the accuracy of interpolation formulas without increasing the required numerical work. The idea is quite general; we explain it in the simplest possible situation.

---

† This section may be omitted at first reading.

Everett's formula for the interpolating polynomial of degree 5 may be written

$$p_5(s) = tf_0 + \binom{t+1}{3}\left\{\delta^2 f_0 + \frac{(t+2)(t-2)}{4 \cdot 5}\,\delta^4 f_0\right\}$$

plus a similar term with t replaced by s and the subscript 0 replaced by 1. In the interval $0 \leq t \leq 1$ the factor

$$\frac{(t+2)(t-2)}{4 \cdot 5} = \frac{t^2 - 4}{20}$$

varies between the narrow limits $-(4/20)$ and $-(3/20)$. Thus if we define modified second differences $\delta^2 f_k^*$ by the formula

$$\delta^2 f_k^* = \delta^2 f_k - \tfrac{7}{40}\,\delta^4 f_k$$

then Everett's formula with fourth differences can be approximated by the formula

$$p_5^*(s) = tf_0 + \binom{t+1}{3}\delta^2 f_0^* + sf_1 + \binom{s+1}{3}\delta^2 f_1^*$$

where $t = 1 - s$. This formula can be used in the same way as the formula involving second differences only. The error committed in replacing p_5 by p_5^* is equal to

$$\binom{t+1}{3}\left[-\frac{7}{40} - \frac{t^2-4}{20}\right]\delta^4 f_0 + \binom{s+1}{3}\left[-\frac{7}{40} - \frac{s^2-4}{20}\right]\delta^4 f_1$$

$$= \frac{1}{40}\left\{(1 - 2t^2)\binom{t+1}{3}\delta^4 f_0 + (1 - 2s^2)\binom{s+1}{3}\delta^4 f_1\right\}.$$

For $0 \leq s \leq 1$ (and consequently $0 \leq t \leq 1$) this turns out to be less than

$$0.00122 \max\{|\delta^4 f_0|, |\delta^4 f_1|\}.$$

Thus if one unit in the least significant digit carried in the computation is denoted by u, we have

$$|p_5^*(s) - p_5(s)| \leq \frac{u}{2}$$

already if the fourth differences are less than $400u$.

Many mathematical tables giving second differences print modified instead of ordinary second differences. In forming the modified differences, the factor $-(7/40)$ is frequently replaced by -0.184, a value suggested by Comrie from a consideration of Bessel's formula. Tables with modified second differences make it possible to calculate (with an error of less than one round-off error) the fifth degree interpolating polynomial from only one difference column, with an amount of work comparable to that for the third degree polynomial.

EXAMPLE

12. According to the tables of Jahnke and Emde [1945], $x = 11.620$ is a zero of the Bessel function $J_2(x)$. We check this statement by evaluating $J_2(11.620)$, using the following values given in the British Association tables:

x	$J_2(x)$	δ^2*
11.4	0.05118808	
11.5	0.02793593	
11.6	0.00461559	15622
11.7	−0.01854910	37729
11.8	−0.04133747	
11.9	−0.06353402	

With $x_0 = 11.6$, we have $s = 0.2$, $t = 0.8$

$$\binom{s+1}{3} = -0.032, \qquad \binom{t+1}{3} = -0.048,$$

yielding $J_2(11.620) = 0.00003692$. The derivative of $J_2(x)$ satisfies

$$J_2'(x) = \tfrac{1}{2}[J_1(x) - J_3(x)]$$

and thus, again from tables, has near the suspected zero the approximate value -0.23. From Newton's formula, we thus expect

$$11.62 - \frac{-0.00003692}{-0.23} = 11.61984$$

to be a more accurate value of the desired zero and indeed, Watson [1944] lists the desired zero as 11.6198412.

Problems

14. For the interpolation problem discussed in example 12, estimate
   (a) the error $J_2(x) - p_5(s)$ due to pure interpolation;
   (b) the error $p_5^*(s) - p_5(s)$ due to throwback.
   (For (a), use the integral representation of problem 9, chapter 9, to estimate the high derivative of J_2 required.)

15. Proceeding as in the derivation of Everett's formula, obtain an interpolation formula that uses only zeroth, third, sixth, ... differences.

16. Devise a method for "throwing back the second differences into the zeroth" in Everett's formula, and estimate the error involved. Why is the method less efficient than the one discussed in the text? When would it be useful?

17. The zeros of the Bessel function $J_0(x)$ are known to approach the points

$$x = (n - \tfrac{1}{4})\pi \qquad n = 1, 2, 3, \ldots.$$

Verify this statement by evaluating $J_0((n - \tfrac{1}{4})\pi)$, $n = 1, 2, 3, \ldots$, using a table of Bessel functions.

18. Check the values of the modified differences given in example 12 by forming ordinary second and fourth differences.

Recommended Reading

Finite difference techniques have been pushed to an especially high level in Great Britain. Books such as Fox [1957] as well as the volumes issued by the National Physical Laboratory [1961] and by the Nautical Almanac Service [1956] contain much excellent advice on interpolation by differences. On the whole, however, the subject of interpolation of tables is somewhat in the eclipse due to the fact that most tables have been replaced by prestored programs in digital computers.

chapter 12 numerical differentiation

Above we have used the interpolating polynomial to approximate values of a function f at points where f is not known. Another use of the interpolating polynomial, of equal or even higher importance in practice, is the imitation of the fundamental operations of calculus. In all these applications the basic idea is extremely simple: Instead of performing the operation on the function f, which may be difficult or—in cases where f is known at discrete points only—impossible, the operation is performed on a suitable interpolating polynomial. In the present chapter this program is carried out for the operation of differentiation.

12.1 The Error of Numerical Differentiation

Let f be a function defined on an interval I containing the set of points $S = \{x_0, x_1, \ldots, x_n\}$ (not necessarily equidistant) and let P_S be the polynomial interpolating f at the points of the set S. We seek to approximate

$$f'(x) \quad \text{by} \quad P_S'(x), \quad x \in I,$$

and wish to derive a formula for the error that must be expected in this approximation.

It seems natural to obtain an expression for the error $f' - P_S'$ by differentiating the error formula (9-5). If f has a continuous derivative of order $n + 1$ in I and if $x \in I$, it was shown in §9.2 that

(12-1) $$f(x) - P_S(x) = L(x)g(x),$$

where

(12-2) $$L(x) = (x - x_0)(x - x_1)\ldots(x - x_n),$$

(12-3) $$g(x) = \frac{1}{(n + 1)!}f^{(n+1)}(\xi_x),$$

ξ_x being some unspecified point in the smallest interval containing x and all the points x_i, $i = 0, 1, \ldots, n$. Corollary 9.2 shows that, although no assertion can be made about the continuity or differentiability of ξ_x as a function of x, the function g can be extended to a function that is continuous on I. A similar consideration shows that the extended function is even n times continuously differentiable on I. Hence we obtain by differentiating (12-1)

$$(12\text{-}4) \qquad f'(x) - P_S'(x) = L'(x)g(x) + L(x)g'(x).$$

If x is arbitrary, this expression is not of much use for the purpose of estimating $f' - P_S'$, since we lack a convenient explicit representation for g' such as (12-3). However, if $x = x_j$ ($j = 0, 1, \ldots, n$) in (12-4), we obtain by virtue of $L(x_j) = 0$

$$f'(x_j) - P_S'(x_j) = L'(x_j)g(x_j).$$

Recalling that

$$L'(x_j) = \lim_{x \to x_j} \frac{\prod\limits_{i=0}^{n} (x - x_i)}{x - x_j} = \lim_{x \to x_j} \prod\limits_{\substack{i=0 \\ i \neq j}}^{n} (x - x_i) = \prod\limits_{\substack{i=0 \\ i \neq j}}^{n} (x_j - x_i)$$

we can state the above result as follows:

> **Theorem 12.1** Let the function f be continuous and $n + 1$ times continuously differentiable on an interval I containing the $n + 1$ distinct points $x_0, x_1, \ldots, x_n$, and let P_S denote the polynomial of degree $\leq n$ interpolating f at these points. Then for each j, $j = 0, 1, \ldots, n$ the interval spanned by the largest and the smallest of the points x_i contains a point ξ_j such that
>
> $$(12\text{-}5) \quad f'(x_j) - P_S'(x_j) = \frac{1}{(n + 1)!} f^{(n+1)}(\xi_j) \prod\limits_{\substack{i=0 \\ i \neq j}}^{n} (x_j - x_i).$$

Problems

1. For a sufficiently differentiable function f, $f'(0)$ is approximated by differentiating the polynomial P interpolating f at the points $0, h, 2h, \ldots, nh$. Give a formula for the error $f'(0) - P'(0)$.

2. Indicate a *lower* bound for the error if the derivative $f'(0)$ of $f(x) = e^x$ is replaced by the derivative of the polynomial interpolating f at the points $0, 1, \ldots, n$.

3. Give a formula for the error of numerically differentiating f at $x = 0$ by differentiating the interpolating polynomial using the points $-h, 0, h$. Obtain the same error formula by subtracting the Taylor expansion for $f(-h)$ from that for $f(h)$, both terminated with the term involving h^3.

12.2   Numerical Differentiation Formulas for Equidistant Abscissas

We now shall assume that the points x_k are equidistant, $x_k = x_0 + kh$, $k = 0, \pm 1, \pm 2, \ldots$. For given m and $n > m$,

$$\left| \prod_{\substack{i=m \\ i \neq j}}^{n} (x_i - x_j) \right|$$

is smallest when x_j lies halfway between the extreme points x_m and x_n. This suggests the use of Stirling's formula (11-19)

$$p_{2k}(s) = f_0 + \binom{s}{1}\left[\mu\delta f_0 + \frac{s}{2}\delta^2 f_0\right] + \cdots$$
$$+ \binom{s + k - 1}{2k - 1}\left[\mu\delta^{2k-1}f_0 + \frac{s}{2k}\delta^{2k}f_0\right],$$

where $s = (x - x_0)/h$, for numerical differentiation. The derivative with respect to x at x_0 equals h^{-1} times the derivative with respect to s at $s = 0$. The derivatives of the coefficients

$$\frac{s}{2k}\binom{s + k - 1}{2k - 1}, \qquad k = 1, 2, \ldots$$

are zero, since they contain the factor s^2. For the remaining coefficients we find

$$\frac{d}{ds}\binom{s + k - 1}{2k - 1}\bigg|_{s=0}$$
$$= \lim_{s \to 0} \frac{1}{s}\binom{s + k - 1}{2k - 1}$$
$$= \lim_{s \to 0} \frac{(s + k - 1)(s + k - 2)\ldots(s + 1)(s - 1)\ldots(s - k + 1)}{(2k - 1)!}$$
$$= (-1)^{k-1}\frac{[(k - 1)!]^2}{(2k - 1)!}.$$

The derivative of the polynomial $P_{2k}(x) = p_{2k}((x - x_0)h^{-1})$ at the central point x_0 can thus be written

(12-6) $P'_{2k}(x_0) = \frac{1}{h}\left\{\mu\delta f_0 - \frac{(1!)^2}{3!}\mu\delta^3 f_0 + \cdots \right.$
$$\left. + (-1)^{k-1}\frac{[(k - 1)!]^2}{(2k - 1)!}\mu\delta^{2k-1}f_0\right\}$$

or, evaluating the numerical coefficients

$$P'_{2k}(x_0) = \frac{1}{h}\{\mu\delta f_0 - \tfrac{1}{6}\mu\delta^3 f_0 + \tfrac{1}{30}\mu\delta^5 f_0 - \tfrac{1}{140}\mu\delta^7 f_0 + \cdots\}.$$

Formulas in terms of ordinates can be found by expressing the central differences $\delta^{2k-1}f_{-1/2}$ and $\delta^{2k-1}f_{1/2}$ in the form $\sum c_n f_n$ and arranging the coefficients c_n in a table, as follows:

Table 12.2a

n	-1	0	1
$\delta f_{-1/2}$	-1	1	0
$\delta f_{1/2}$	0	-1	1
$\mu \delta f_0$	$-\frac{1}{2}$	0	$\frac{1}{2}$

Table 12.2b

n	-2	-1	0	1	2
$\delta^3 f_{-1/2}$	-1	3	-3	1	0
$\delta^3 f_{1/2}$	0	-1	3	-3	1
$\mu \delta^3 f_0$	$-\frac{1}{2}$	1	0	-1	$\frac{1}{2}$

The average of the coefficients in each column gives us the corresponding coefficient for $\mu \delta^{2k-1}f_0$. For the second and fourth degree polynomials we find in this manner

(12-7)
$$P_2'(x_0) = \frac{1}{2h}(f_1 - f_{-1}),$$

(12-8) $P_4'(x_0) = \frac{1}{h}\{\frac{1}{2}(f_1 - f_{-1}) - \frac{1}{12}(f_2 - 2f_1 + 2f_{-1} - f_{-2})\}$

$$= \frac{1}{12h}(-f_2 + 8f_1 - 8f_{-1} + f_{-2}).$$

An expression for the error $f'(x_0) - P_{2k}'(x_0)$ is easily determined from (12-5). With $n = 2k$, and the interpolating points arranged equidistantly and symmetrically about x_0, we find

$$\frac{1}{(2k+1)!}L'(x_0) = (-1)^k \frac{(k!)^2}{(2k+1)!}h^{2k}.$$

It thus follows that

(12-9) $f'(x_0) - P_{2k}'(x_0) = \frac{(-1)^k}{h}\frac{(k!)^2}{(2k+1)!}h^{2k+1}f^{(2k+1)}(\xi),$

where $x_{-k} < \xi < x_k$. As in the calculation of approximate function values by algorithm 11.2, it is thus true that the error committed in differentiating Stirling's polynomial in place of the function f equals the first omitted term in the difference formula, provided that $\mu\delta^{2k+1}f_0$ is replaced by $h^{2k+1}f^{(2k+1)}(\xi)$.

Problems

4. By using the Newton forward formula, devise a formula for numerical differentiation at the beginning of a table and obtain an error formula similar to (12-9).

5. Making use of Bessel's formula, obtain a formula for numerical differentiation halfway between two entries in a table, and write the formulas resulting from the polynomials of degree one and three in terms of ordinates.

6. By differentiating Stirling's form of the interpolating polynomial twice, show that

$$f''(x_0) \approx \frac{1}{h^2}\{\delta^2 f_0 - \tfrac{1}{12}\delta^4 f_0 + \cdots\}.$$

   Obtain a general formula for the coefficients on the right.

7. Suppose that, due to rounding, the values f_n are known only up to errors ε_n, where $|\varepsilon_n| \leq \varepsilon$. What is the maximum error resulting therefrom in the formulas (12-7) and (12-8), if all arithmetic operations are performed without rounding error?

8. Suppose we calculate $J_0'(x)$ by means of (12-7) from a table of $J_0(x)$ giving six decimals. What is the smallest value of h for which we can guarantee that the maximum possible discretization error (due to replacing J_0 by P) is not exceeded by the maximum possible error due to rounding?

12.3 Extrapolation to the Limit

The formulas derived in §12.2 can also be obtained by an entirely different method which recalls the fundamental principle of numerical analysis applied already in the derivation of Aitken's Δ^2-method: Improve the accuracy of an approximation using any (possibly incomplete) knowledge of the asymptotic behavior of the error.

The simplest numerical differentiation procedure consists in

$$\text{replacing } f'(x_0) \quad \text{by} \quad \frac{1}{2h}(f_1 - f_{-1})$$

(see formula (12-7) above). We introduce an abbreviation for the expression on the right that emphasizes its dependence on the step h by defining the *basic differentiation operator* D_h as follows:

(12-10) $$D_h f(x_0) = \frac{1}{2h}[f(x + h) - f(x - h)].$$

The error formula (12-9) then states in the special case $k = 1$ that

(12-11) $D_h f(x_0) - f'(x_0) = \tfrac{1}{6}h^2 f^{(3)}(\xi),$

where $x_0 - h < \xi < x_0 + h$.

In order to obtain a more accurate statement about the error of $D_h f$, we use Taylor's expansion with remainder term. If the function f is sufficiently differentiable, we have for $k = 0, 1, \ldots$

$$f(x + h) = f(x) + \frac{h}{1!}f'(x) + \cdots + \frac{h^k}{k!}f^{(k)}(x) + \frac{h^{k+1}}{(k+1)!}f^{(k+1)}(\xi),$$

where ξ is a point between x and $x + h$. Writing down this expression for $k = 2m$ and $x = x_0$ and then subtracting from it the same expression with h replaced by $-h$, we obtain after dividing by $2h$

(12-12) $D_h f(x_0) = f'(x_0) + a_1 h^2 + a_2 h^4 + \cdots + a_{m-1} h^{2m-2} + R_m(h),$

where

$$a_k = \frac{1}{(2k+1)!} f^{(2k+1)}(x_0), \qquad k = 1, 2, \ldots, m - 1,$$

$$R_m(h) = \frac{h^{2m}}{(2m+1)!} \frac{f^{(2m+1)}(\xi_1) + f^{(2m+1)}(\xi_2)}{2}.$$

Here ξ_1 is a point between x and $x + h$ and ξ_2 is between x and $x - h$. If $f^{(2m+1)}$ is continuous, it assumes every value between

$$f^{(2m+1)}(\xi_1) \quad \text{and} \quad f^{(2m+1)}(\xi_2).$$

Thus in particular for some ξ between ξ_1 and ξ_2 (and thus between $x - h$ and $x + h$)

$$f^{(2m+1)}(\xi) = \tfrac{1}{2}[f^{(2m+1)}(\xi_1) + f^{(2m+1)}(\xi_2)].$$

It follows that the remainder R_m in (12-12) can be expressed in the simpler form

(12-13) $R_m(h) = \dfrac{h^{2m}}{(2m+1)!} f^{(2m+1)}(\xi).$

With this form of the remainder, (12-12) for $m = 1$ reduces to (12-11).

In numerical applications the values of $f'(x_0)$ and of the constants $a_1, a_2, \ldots$ are of course unknown. But the mere fact that a formula of type (12-12) holds can be used to improve the accuracy of a numerical differentiation. Suppose the operator D_h is applied with two different values of h, say h and qh, where $q \neq 0, 1$. (If f is tabulated at equally spaced intervals, $q = \tfrac{1}{2}$ is a natural choice.) We then have from (12-12)

$$D_h f(x_0) = f'(x_0) + a_1 h^2 + 0(h^4),$$
$$D_{qh} f(x_0) = f'(x_0) + a_1 (qh)^2 + 0(h^4),$$

since $0(qh) = 0(h)$. Eliminating from this pair of equations the constant a_1 and solving for $f'(x_0)$, we find

(12-14) $$f'(x_0) = \frac{D_{qh}f(x_0) - q^2 D_h f(x_0)}{1 - q^2} + 0(h^4).$$

Using two values of the basic differentiation operator D_h (which ordinarily has an error $0(h^2)$) we thus have succeeded in deriving a differentiation formula with an error of only $0(h^4)$.

Without a more careful study of the error term we cannot assert that (12-14) is more accurate than (12-10) for any given h. However, if $f^{(5)}$ is continuous, then (12-14) is always more accurate for *sufficiently small values* of h.

In the special case $q = 2$ (12-14) takes the form

$$f'(x_0) = \frac{1}{3} \left\{ \frac{4}{2h} (f_1 - f_{-1}) - \frac{1}{2 \cdot 2h} (f_2 - f_{-2}) \right\} + 0(h^4).$$

We thus obtain the approximate differentiation formula

$$f'(x_0) = \frac{-f_2 + 8f_1 - 8f_{-1} + f_{-2}}{12h} + 0(h^4),$$

which turns out to be identical with (12-8).

Another way to look at formula (12-14), more in the spirit of Aitken's Δ^2-process, is to consider it as a device to speed up the convergence of the basic differentiation operator D_h as $h \to 0$. The speeded up operator may be written in the form

(12-15) $$D_{qh}^* = \frac{D_{qh} - q^2 D_h}{1 - q^2}.$$

EXAMPLE

1. To find the first derivative of $J_0(x)$ at $x = 2$.

Table 12.3

h	$D_h J_0(2)$	$D_h^* J_0(2)$	$D_h^{**} J_0(2)$
0.40	−0.56611 8105		
0.20	−0.57406 0360	−0.57670 7779	
0.10	−0.57605 7896	−0.57672 3741	−0.57672 4805
0.05	−0.57655 8030	−0.57672 4742	−0.57672 4808

The exact value, to nine places, is $J_0'(2) = -0.57672\,4808$.

Nothing prevents us from trying to further speed up the sequence of the values $D_h^* f(x_0)$. The appropriate formula can be obtained by considering the structure of the error of $D_h^* f(x_0)$. From (12-12) we find

$$D_{qh} f(x_0) = f'(x_0) + a_1 q^2 h^2 + a_2 q^4 h^4 + \cdots + a_{m-1} q^{2m-2} h^{2m-2} + R_m(qh).$$

It follows that

$$D_{qh}^* f(x_0) = f'(x_0) + a_2^*(qh)^4 + \cdots + a_{m-1}^*(qh)^{2m-2} + R_m^*(qh),$$

where

$$a_k^* = \frac{1 - q^{2-2k}}{1 - q^2} a_k,$$

$$R_m^*(qh) = \frac{R_m(qh) - q^2 R_m(h)}{1 - q^2}.$$

Thus, in particular,

$$D_h^* f(x_0) = f'(x_0) + a_2^* h^4 + 0(h^6),$$
$$D_{qh}^* f(x_0) = f'(x_0) + a_2^* q^4 h^4 + 0(h^6).$$

Eliminating a_2^* yields

(12-16) $f'(x_0) = D_{qh}^{**} f(x_0) + 0(h^6),$

where

(12-17) $D_{qh}^{**} = \dfrac{D_{qh}^* - q^4 D_h^*}{1 - q^4}.$

Still using the same basic differentiation operator D_h, we thus have obtained a differentiation formula with an error $0(h^6)$. Obviously the procedure can be continued, the only limits being those set by the accuracy of the values $f(x_0 + h)$ themselves. The effectiveness of the procedure is illustrated by the last column of table 12.3, where the derivative has been obtained to nine significant digits from only six values of the function.

Problems

9. Using Taylor's expansion, show that

$$\frac{1}{h^2} \delta^2 f_0 = f''(x_0) + \frac{h^2}{12} f^{(4)}(\xi),$$

where $x - h < \xi < x + h$.

10. By eliminating a_1 and a_2 in the formulas

$$D_{kh} f(x_0) = f'(x_0) + a_1(kh)^2 + a_2(kh)^4 + 0(h^6), \qquad k = 1, 2, 3,$$

obtain a differentiation formula that uses the abscissas $f_{-3}, f_{-2}, \ldots, f_3$ and has an error $0(h^6)$. Compare the result with the formula (12-6) for $k = 3$.

11. Suppose that, due to rounding, the values $D_h f(x_0)$ are known only up to rounding errors not exceeding ε/h. If the values $D_h^* f(x_0)$ and $D_h^{**} f(x_0)$ are formed with $q = \frac{1}{2}$, what are their errors due to the inaccuracies in $D_h f(x_0)$?

12.4 Extrapolation to the Limit: The General Case

Extrapolation to the limit as applied to numerical differentiation in §12.3 is a mere special case of the following general situation: An unknown quantity a_0 is approximated by a calculable quantity $A(y)$ $(y > 0)$ such that

$$(12\text{-}18) \qquad\qquad \lim_{y \to 0} A(y) = a_0,$$

and it is known that there exist constants $a_1, a_2, \ldots$ and $C_1, C_2, \ldots$ such†
that for $k = 1, 2, 3, \ldots$

$$(12\text{-}19) \qquad A(y) = a_0 + a_1 y + a_2 y^2 + \cdots + a_{k-1} y^{k-1} + R_k(y),$$

where

$$|R_k(y)| < C_k y^k, \qquad y > 0.$$

A triangular array of numbers $A_{m,n}$ $(m = 0, 1, 2, \ldots; n \leq m)$ is now formed in the following manner.

Algorithm 12.4 For two fixed constants r and y_0 $(0 < r < 1, y_0 > 0)$, let for $m = 0, 1, 2, \ldots$

$$A_{m,0} = A(r^m y_0),$$

$$A_{m,n+1} = \frac{A_{m,n} - r^{n+1} A_{m-1,n}}{1 - r^{n+1}}, \qquad n = 0, 1, \ldots, m-1.$$

The manner in which the numbers $A_{m,n}$ depend on each other is indicated in scheme 12.4.

$$
\begin{array}{lllll}
A_{0,0} \\
A_{1,0} & A_{1,1} \\
A_{2,0} & A_{2,1} & A_{2,2} \\
A_{3,0} & A_{3,1} & A_{3,2} & A_{3,3} \\
\vdots & \vdots & \vdots & \vdots & \ddots
\end{array}
$$

Scheme 12.4

This scheme has the following interesting property.

---

† In technical language the hypothesis (12-19) means that $A(y)$ admits an asymptotic expansion for $y \to 0+$. We do *not* assume that the infinite series $a_0 + a_1 y + a_2 y^2 + \cdots$ converges.

Theorem 12.4   Let $A = A(y)$ satisfy the relations (12-19).   Then for each n such that $a_{n+1} \neq 0$, the $(n + 1)$st column in the scheme 12.4 converges faster to a_0 than the nth column, in the sense that

$$(12\text{-}20) \qquad \lim_{m \to \infty} \frac{A_{m, n+1} - a_0}{A_{m, n} - a_0} = 0.$$

More generally, for each fixed value of $n \geq 0$,

$$(12\text{-}21) \quad A_{m,n} = a_0 + (-1)^n a_{n+1} r^{-n(n+1)/2}(r^m y_0)^{n+1} + 0((r^m y_0)^{n+2})$$

as $m \to \infty$.

Here $0(z)$ denotes a quantity that remains bounded for $z \to 0$ when divided by z.

Proof.   We shall establish the following proposition, which is stronger than (12-21): For each fixed n, and for each $p > n$,

$$(12\text{-}22) \qquad A_{m,n} = a_0 + a_{n+1, n}(r^m y_0)^{n+1} + a_{n+2, n}(r^m y_0)^{n+2}$$
$$+ \cdots + a_{p, n}(r^m y_0)^p + 0((r^m y_0)^{p+1})$$

as $m \to \infty$, where for $q = n + 1, n + 2, \ldots$

$$(12\text{-}23) \qquad a_{q, n} = \frac{(1 - r^{1-q})(1 - r^{2-q})\ldots(1 - r^{n-q})}{(1 - r)(1 - r^2)\ldots(1 - r^n)} a_q.$$

(Empty products are to be interpreted as 1.)

The proof is by induction with respect to n.   For $n = 0$, the formulas (12-22) and (12-23) are a mere restatement of our hypothesis (12-19). Assuming that they are valid for some $n \geq 0$, it follows from algorithm 12.4 that $A_{m, n+1}$ has a representation of the form (12-22) where the coefficient of $(r^m y_0)^q$ is

$$\frac{1 - r^{n+1-q}}{1 - r^{n+1}} a_{q, n}.$$

This is zero for $q = n + 1$ and equals $a_{q, n+1}$ for $q > n + 1$, verifying (12-22) with n increased by one.

Setting $q = n + 1$ in (12-23), we obtain

$$a_{n+1, n} = \frac{(1 - r^{-n})(1 - r^{-n+1})\ldots(1 - r^{-1})}{(1 - r)(1 - r^2)\ldots(1 - r^n)} a_{n+1}$$

$$= r^{-(1+2+\cdots+n)} \frac{(r^n - 1)(r^{n-1} - 1)\ldots(r - 1)}{(1 - r)(1 - r^2)\ldots(1 - r^n)} a_{n+1}$$

$$= (-1)^n r^{-n(n+1)/2} a_{n+1}$$

in view of the well-known formula

$$1 + 2 + \cdots + n = \frac{n(n + 1)}{2}.$$

Letting $p = n + 1$ in (12-22) and using the above value of $a_{n+1,n}$, we thus obtain (12-21).

If $a_{n+1} \neq 0$, relation (12-21) may be written

$$A_{m,n} - a_0 = (-1)^n a_{n+1} r^{-n(n+1)/2}(r^m y_0)^{n+1}[1 + 0(r^m y_0)].$$

Thus,

$$\frac{A_{m,n+1} - a_0}{A_{m,n} - a_0} = -\frac{a_{n+2}}{a_{n+1}} r^{-n-1}(r^m y_0)[1 + 0(r^m y_0)],$$

which implies (12-20).

The technique described in §12.3, and in particular in example 1, corresponds to the special case $A(h^2) = D_h$, $r = q^2$ of algorithm 12.4.

Problems

12. A numerical computation furnished the following approximations $A(2^{-n})$ to a quantity a_0:

n	$A(2^{-n})$
0	0.000000
1	0.250000
2	0.316406
3	0.343609
4	0.356074
5	0.362055
6	0.364987
7	0.366438

Determine a_0 as accurately as you can. (Exact value:

$$a_0 = e^{-1} = 0.367879.)$$

13. Show that the hypotheses of theorem 12.4 are satisfied for the function

$$A(y) = (1 + y)^{1/y} \qquad (y > 0).$$

14. Show that the scheme generated by algorithm 12.4 is identical with the scheme that would be obtained by calculating $A(0)$ by Neville interpolation (algorithm 10.5) using the interpolating points $y_0, ry_0, r^2 y_0, \ldots$

15. Suppose the values of the function $A(y)$ used in algorithm 12.4 are known only up to errors $\leq \varepsilon$. How large is the resulting uncertainty in the columns $A_{m,n}$? Give numerical values for the "noise amplification factor" for $n = 1, 2, 3$ and $r = 0.5$, $r = 0.25$.

16. The following relations hold for the symbol $0(z)$ introduced after theorem 12.4:

$$0(z) + 0(z) = 0(z);$$
$$0(cz) = 0(z) \quad \text{for any constant } c.$$

From what theorems about limits do they follow?

12.5   Calculating Logarithms by Differentiation†

In this section we shall discuss a further application of algorithm 12.4 to a problem of numerical differentiation.   Let $a > 1$ be a given number. We consider the function

$$f(x) = a^x$$

and wish to evaluate $f'(0)$.   By the rules of calculus, this of course equals

$$f'(0) = \log a.$$

By carrying out the differentiation numerically, we may thus hope to obtain a method for calculating the natural logarithm of a given number.

We use the following one-sided approximation to the first derivative, viz.:

$$S(h) = \frac{f(h) - f(0)}{h}$$

$$= \frac{1}{h}(a^h - 1).$$

By the definition of the derivative,

$$\log a = f'(0) = \lim_{h \to 0} S(h).$$

This limit relation also holds if we restrict h to the discrete set of values $h = 2^{-n}$, $n = 0, 1, 2, \ldots$.   Setting

$$(12\text{-}24) \qquad s_n = S(2^{-n}) = 2^n(a^{2^{-n}} - 1)$$

we thus have

$$\log a = \lim_{n \to \infty} s_n.$$

The values $a^{2^{-n}}$ required to form the numbers s_n can easily be generated recursively by successively evaluating square roots, as follows:

$$a^{2^{-0}} = a, \qquad a^{2^{-n-1}} = \sqrt{a^{2^{-n}}}.$$

We thus have, in principle, solved our problem.

Readers who have studied §8.4 will immediately observe, however, that the algorithm thus defined is numerically unstable.   The quantity $a^{2^{-n}} - 1$ approaches zero as $n \to \infty$.   If the numbers $a^{2^{-n}}$ are computed with a fixed number of decimals, the *relative* error of $a^{2^{-n}} - 1$ will ultimately

---

† This section may be omitted at first reading.

become quite large.   By rule (ii) of §8.4, the relative error of s_n will likewise be large, and, since the s_n approach a nonzero limit, the absolute
error, too, will grow rapidly.

A procedure that is more stable numerically can be defined if we
generate the s_n in a different way.   Solving (12-24) for $a^{2^{-n}}$ we get

(12-25) $$a^{2^{-n}} = 1 + 2^{-n}s_n.$$

On the other hand,

$$s_{n-1} = 2^{n-1}(a^{2^{-n+1}} - 1)$$
$$= 2^{n-1}(a^{2^{-n}} + 1)(a^{2^{-n}} - 1).$$

Substituting for $a^{2^{-n}}$ the value (12-25), we have

$$s_{n-1} = 2^{n-1}(2 + 2^{-n}s_n)2^{-n}s_n$$
$$= s_n + 2^{-n-1}s_n^2.$$

Solving the quadratic for s_n and observing that $s_n > 0$, we find

$$s_n = 2^n(\sqrt{1 + 2^{-n+1}s_{n-1}} - 1).$$

To avoid loss of accuracy, we write this in the form

(12-26) $$s_n = \frac{2s_{n-1}}{1 + \sqrt{1 + 2^{-n+1}s_{n-1}}}.$$

Together with the initial condition $s_0 = a - 1$ this relation may be used
to generate the sequence $\{s_n\}$ in a stable manner.†

For numerical purposes the convergence of the algorithm thus obtained,
even though stable, is intolerably slow (see problem 16).   If it is to be put
to any practical use, we must be able to speed it up.   Let us check whether
the function $S(h)$ satisfies the hypotheses of theorem 12.4.   Since $a^h = e^{h \log a}$ we have, using the exponential series,

$$S(h) = \log a + \frac{1}{2!}(\log a)^2 h + \frac{1}{3!}(\log a)^3 h^2 + \cdots.$$

Since this series converges for all values of h, condition (12-19) certainly
holds for $A(h) = S(h)$, and theorem 12.4 is applicable.   We thus obtain

---

† Relation (12-26) also provides us with an example of those relatively rare non-linear
difference equations that have an explicit solution.

Algorithm 12.5   For $a > 1$, let $A_{0,0} = a - 1$, and calculate for $m = 1, 2, \ldots$

$$A_{m,0} = \frac{2A_{m-1,0}}{1 + \sqrt{1 + 2^{-m+1}A_{m-1,0}}},$$

$$A_{m,n+1} = \frac{2^{n+1}A_{m,n} - A_{m-1,n}}{2^{n+1} - 1},$$

where

(12-27) $n = 0, 1, 2, \ldots, m - 1.$

Theorem 12.4 yields in the present special case

Theorem 12.5   All columns of the scheme generated in algorithm 12.5 converge to $\log a$, and each column converges faster than the preceding one.

One can also show that the sequence $\{A_{n,n}\}$ of diagonal elements converges to $\log a$, and converges faster than any column.

EXAMPLE

2.   For $a = 6$ algorithm 12.5 yields the following triangular array:

Table 12.5

5.000000						
2.898980	0.797959					
2.260338	1.621697	1.896277				
2.008267	1.756196	1.801029	1.787422			
1.895937	1.783606	1.792743	1.791559	1.791835		
1.842872	1.789806	1.791873	1.791748	1.791761	1.791759	
1.817076	1.791281	1.791773	1.791759	1.791759	1.791759	1.791759

The table shows that $\log 6 = 1.791759$ can be obtained to seven significant digits by the evaluation of six square roots.

If higher accuracy is desired, it may be necessary to limit the number of columns of the scheme in order to avoid build-up of rounding errors. Condition (12-27) should then be replaced by

$$n = 0, 1, \ldots, \min(m, N) - 1$$

where N is the number of columns desired.

Problems

17.   Using Taylor's formula, find a *lower* bound for the error $|S(h) - \log a|$. Thus find a lower bound for the number n necessary that

$$|s_n - \log a| < 0.5 \times 10^{-6},$$

if $a = 6$.

18. Use algorithm 12.5 to calculate log e to six decimals.

19. Let $1 \leqq a \leqq e$. By obtaining explicit values for the constants C_k in (12-19) from Taylor's formula, find bounds for the errors of the quantities $A_{m,1}$ and $A_{m,2}$, considered as approximations to log a.

Recommended Reading

The principle of extrapolation to the limit was first clearly stated by Richardson [1927]. Repeated extrapolation to the limit is discussed by Bauer *et al.* [1963]. Concerning its application to numerical differentiation, see also Rutishauser [1963].

Research Problem

Since every positive number can be represented in the form $e^n z$, where $e^{-1/2} \leqq z < e^{1/2} = 1.648\ldots$, it is only necessary to know log x in the interval $(1, e^{1/2})$. Make a comparative study of the computation of log x in that interval (a) by algorithm 12.5, (b) by Taylor's series.

chapter 13 numerical integration

We now turn to the problem of numerical evaluation of definite integrals. The method is the same as in chapter 12. Instead of performing the integration on the function f, which may be difficult, we perform the integration on a polynomial interpolating f at suitable points. We begin by giving a theoretical appraisal of the error committed in this approximation.

13.1 The Error in Numerical Integration

Our starting point is once again the general error formula proved in §9.2. There it was shown that if P_S interpolates f on the set of points $S = \{x_0, x_1, \ldots, x_n\}$, then

$$(13\text{-}1) \qquad f(x) - P_S(x) = L(x)g(x),$$

where

$$L(x) = (x - x_0)(x - x_1)\ldots(x - x_n),$$

and where g is a continuous function that can be expressed in the form

$$(13\text{-}2) \qquad g(x) = \frac{1}{(n + 1)!} f^{(n+1)}(\xi_x),$$

ξ_x being a suitable number contained between the largest and the smallest of the numbers $x_0, \ldots, x_n$ and x. If we integrate (13-2) between two arbitrary limits a and b, we evidently have

$$\int_a^b f(x)\, dx = \int_a^b P_S(x)\, dx + R_S^{(a,b)},$$

where the integration error $R_S^{(a,b)}$ can be expressed in the form

$$(13\text{-}3) \qquad R_S^{(a,b)} = \int_a^b L(x)g(x)\, dx.$$

246

Let us now assume that the polynomial L does not change its sign in the interval (a, b). This will evidently be the case if and only if no interpolating point x_k is contained in the interior of (a, b). We then can apply to the integral representation (13-3) the second mean value theorem of the integral calculus (see Buck [1956], p. 58), with the result that

$$R_S^{(a, b)} = g(t) \int_a^b L(x)\, dx,$$

where t is a suitable point in (a, b). In view of the definition of g, we thus have the following result:

Theorem 13.1 Let the function f and the polynomial P satisfy the conditions of theorem 9.2. If the interval (a, b) contains no interpolating point in its interior, then

$$(13\text{-}4) \qquad \int_a^b f(x)\, dx - \int_a^b P(x)\, dx = \frac{f^{(n+1)}(\xi)}{(n+1)!} \int_a^b L(x)\, dx,$$

where ξ is a number contained between the largest and the smallest of the numbers $x_0, \ldots, x_n, a, b$.

It is clear the theorem 13.1 cannot be true if (a, b) contains an interpolating point, since it then may happen that the integral on the right of (13-4) vanishes.

EXAMPLE

1. $x_0 = -1,\ x_1 = 0,\ x_2 = 1,\ L(x) = x(x^2 - 1),\ a = -b,$

$$\int_a^b L(x)\, dx = 0.$$

Problems

1. The integral $\int_0^\pi \sin x\, dx$ is evaluated by interpolating the function $f(x) = \sin x$ at the points $x = 0$ and $x = \pi$. Calculate the bound for the error given by theorem 13.1 (using the fact that all derivatives are bounded by 1), and compare it with the actual error.

2. Dropping the assumption that the interval (a, b) contains no interpolating point, show that

$$\left| \int_a^b f(x)\, dx - \int_a^b P(x)\, dx \right| \leq \frac{M_{n+1}}{(n+1)!} \int_a^b |L(x)|\, dx$$

where $M_{n+1} = \max_{x \in I} |f^{(n+1)}(x)|$.

13.2   Numerical Integration using Backward Differences

Once again we specialize to the situation where the interpolating points x_k are equidistant, $x_k = x_0 + kh$, $k = \pm 1, \pm 2, \ldots$. We begin by deriving an integration formula which appears to be lacking in symmetry, but which for this very reason will be of use in the numerical integration of differential equations.

Let us assume that the function f is interpolated at the points x_0, $x_{-1}, \ldots, x_{-k}$, and that the integral of f is desired between the limits x_0 and x_1. Setting as before

$$s = \frac{x - x_0}{h},$$

we use the Newton backward polynomial in the form given in problem 7, chapter 11,

$$P_k(x) = \sum_{m=0}^{k} (-1)^m \binom{-s}{m} \nabla^m f_0.$$

Introducing s as variable of integration and observing that the differences are independent of s, we find

$$\int_{x_0}^{x_1} P_k(x)\, dx = h \int_0^1 \left\{ \sum_{m=0}^{k} (-1)^m \binom{-s}{m} \nabla^m f_0 \right\} ds$$

$$= h \sum_{m=0}^{k} c_m \nabla^m f_0,$$

where

(13-5) $$c_m = (-1)^m \int_0^1 \binom{-s}{m} ds, \qquad m = 0, 1, 2, \ldots.$$

We also have, in the present case

$$\frac{1}{(k+1)!} L(x) = \frac{(x - x_0)(x - x_{-1})\ldots(x - x_{-k})}{(k+1)!}$$

$$= (-1)^{k+1} h^{k+1} \binom{-s}{k+1}.$$

Since L does not vanish in the interval (x_0, x_1), and since

$$\frac{1}{(k+1)!} \int_{x_0}^{x_1} L(x)\, dx = h^{k+2} c_{k+1},$$

theorem 13.1 thus yields the formula

(13-6) $$\int_{x_0}^{x_1} f(x)\, dx = h\{c_0 f_0 + c_1 \nabla f_0 + c_2 \nabla^2 f_0 + \cdots$$

$$+ c_k \nabla^k f_0 + c_{k+1} h^{k+1} f^{(k+1)}(\xi)\},$$

where $x_{-k} < \xi < x_1$, and where the constants c_m are given by (13-5). As on several previous occasions it is seen that the error due to interpolation equals the first omitted term in the difference formula, provided that the difference is replaced by the corresponding derivative, suitably normalized (compare similar statements in §11.2 and §12.2).

From the definition (13-5) we find

$$c_0 = \int_0^1 1\, ds = 1, \qquad c_1 = -\int_0^1 \binom{-s}{1} ds = \tfrac{1}{2}.$$

For larger values of m, however, the c_m are more easily calculated by a method to be explained in §13.4.

In an entirely similar manner we can derive the formula

$$(13\text{-}7) \quad \int_{x_{-1}}^{x_0} f(x)\, dx = h\{c_0^* f_0 + c_1^* \nabla f_0 + c_2^* \nabla^2 f_0 + \cdots$$
$$+ c_k^* \nabla^k f_0 + c_{k+1}^* h^{k+1} f^{(k+1)}(\xi)\},$$

where $x_{-k} < \xi < x_0$, and

$$(13\text{-}8) \qquad c_m^* = (-1)^m \int_{-1}^0 \binom{-s}{m} ds, \qquad m = 0, 1, 2, \ldots.$$

In particular we find

$$c_0^* = 1, \qquad c_1^* = -\tfrac{1}{2}.$$

Again a similar statement about the error applies.

Problem

3. Find a formula that integrates between $x_{-1/2}$ and $x_{1/2}$, using backward differences at x_0. What can you say about the error?

13.3 Numerical Integration using Central Differences

If values of f are available on both sides of the interval of integration, it seems preferable to perform the integration on a polynomial that takes into account all these values. Bessel's formula (11-20) recommends itself for the purpose of integrating between the limits x_0 and x_1. We again introduce s as a variable of integration. If s is replaced by $1 - s$, the binomial coefficients

$$\binom{s + m - 1}{2m}$$

remain unchanged, but the factors $s - \tfrac{1}{2}$ change sign. It follows that all integrals of coefficients of the form

$$(s - \tfrac{1}{2})\binom{s + m - 1}{2m}$$

vanish. If P_{2k+1} denotes the polynomial interpolating f on the set $x_{-k}, x_{-k+1}, \ldots, x_{k+1}$, we thus find

$$\int_{x_0}^{x_1} P_{2k+1}(x)\, dx = \int_{x_0}^{x_1} \left\{ \sum_{m=0}^{k} \binom{s+m-1}{2m} \mu \delta^{2m} f_{1/2} \right\} dx$$

$$= h \sum_{m=0}^{k} b_m \mu \delta^{2m} f_{1/2},$$

where

(13-9) $$b_m = \int_0^1 \binom{s+m-1}{2m}\, ds, \qquad m = 0, 1, 2, \ldots.$$

Since, in the present case,

$$\frac{1}{(2k+2)!} L(x) = \frac{(x-x_{-k})(x-x_{-k+1})\ldots(x-x_{k+1})}{(2k+2)!}$$

$$= h^{2k+2} \binom{s+k}{2k+2}$$

and consequently

$$\frac{1}{(2k+2)!} \int_{x_0}^{x_1} L(x)\, dx = h^{2k+3} b_{k+1},$$

theorem 13.1 yields the formula

(13-10) $$\int_{x_0}^{x_1} f(x)\, dx = h\{b_0 \mu f_{1/2} + b_1 \mu \delta^2 f_{1/2} + \cdots$$

$$+ b_k \mu \delta^{2k} f_{1/2} + b_{k+1} h^{2k+2} f^{(2k+2)}(\xi)\},$$

where $x_{-k} < \xi < x_{k+1}$.

We easily find $b_0 = 1$, $b_1 = -\frac{1}{12}$. Thus the case $k = 0$ of (13-10) yields

$$\int_{x_0}^{x_1} f(x) = \frac{h}{2} [f_0 + f_1] - \frac{h^3}{12} f''(\xi),$$

where $x_0 < \xi < x_1$. This is the familiar *trapezoidal rule* of numerical integration. More about it in §13.6!

Problems

4. By integrating Stirling's formula (11-19) between the limits x_{-1} and x_1, obtain an approximate integration formula of the form

$$\int_{x_{-1}}^{x_1} f(x)\, dx = h\{s_0 f_0 + s_1 \delta^2 f_0 + \cdots + s_k \delta^{2k} f_0 + R_{k+1}\},$$

and show that the remainder R_{k+1} is of the order of h^{2k+2}.   (Theorem 13.1 is not applicable in this case, why?)   In particular, show that $s_0 = 2$, $s_1 = \frac{1}{3}$ and hence obtain the approximate integration formulas

$$\int_{x_{-1}}^{x_1} f(x)\,dx = 2hf_0 + 0(h^3) \qquad \text{(Midpoint rule)};$$

$$\int_{x_{-1}}^{x_1} f(x)\,dx = \frac{h}{3}(f_{-1} + 4f_0 + f_1) + 0(h^5) \qquad \text{(Simpson's rule)}.$$

5. Using Taylor's expansion, show that

$$\int_{x_{-1}}^{x_1} f(x)\,dx = 2hf_0 + \tfrac{1}{3}h^3 f''(\xi), \qquad x_{-1} < \xi < x_1.$$

13.4   Generating Functions for Integration Coefficients†

We left unresolved the problem of finding numerical values of the coefficients c_m, c_m^*, and b_m introduced in the two preceding sections. Such values are conveniently found by the method of *generating functions*. The generating function of a sequence of coefficients c_m, for instance, is the function C defined by the power series

$$C(t) = \sum_{m=0}^{\infty} c_m t^m.$$

If we succeed in determining a closed formula for C, we may hope to find numerical values of the c_m in a very simple manner.

We exemplify the method with the coefficients c_m^* defined by (13-8). Their generating function is

$$C^*(t) = \sum_{m=0}^{\infty} c_m^* t^m = \sum_{m=0}^{\infty} (-1)^m t^m \int_{-1}^{0} \binom{-s}{m}\,ds.$$

It is easily seen that $|c_m^*| \leq 1$, $m = 0, 1, 2, \ldots$.   Hence the power series converges uniformly for $|t| \leq \frac{1}{2}$, say.   Interchanging summation and integration we find

$$C^*(t) = \int_{-1}^{0} \sum_{m=0}^{\infty} (-1)^m \binom{-s}{m} t^m\,ds.$$

By the general form of the binomial theorem (see Taylor [1959], p. 479),

$$\sum_{m=0}^{\infty} (-1)^m \binom{-s}{m} t^m = (1 - t)^{-s}.$$

† This section may be omitted at first reading.

We thus find

$$C^*(t) = \int_{-1}^{0} (1 - t)^{-s}\, ds.$$

The integration can be carried out by observing that

$$(1 - t)^{-s} = e^{-s \log (1 - t)}.$$

We thus find

(13-11)
$$C^*(t) = \frac{1}{-\log (1 - t)} [e^{-s \log (1 - t)}]_{-1}^{0}$$

$$= \frac{t}{-\log (1 - t)}.$$

Observing

$$-\frac{\log (1 - t)}{t} = 1 + \tfrac{1}{2}t + \tfrac{1}{3}t^2 + \cdots,$$

we thus have the identity

$$1 = C^*(t) \frac{-\log (1 - t)}{t}$$

$$= (c_0^* + c_1^* t + c_2^* t^2 + \cdots)(1 + \tfrac{1}{2}t + \tfrac{1}{3}t^2 + \cdots).$$

Comparing coefficients of like powers of t on both sides of this identity, we find

$$c_0^* = 1,$$

$$c_m^* + \tfrac{1}{2}c_{m-1}^* + \cdots + \frac{1}{m + 1} c_0^* = 0, \qquad m = 1, 2, \ldots.$$

These relations can readily be used to calculate the values of the coefficients c_m^* recursively.

Table 13.4a

m	0	1	2	3	4	5
c_m^*	1	$-\dfrac{1}{2}$	$-\dfrac{1}{12}$	$-\dfrac{1}{24}$	$-\dfrac{19}{720}$	$-\dfrac{3}{160}$

In a like manner, we obtain for the generating function

(13-12)
$$C(t) = \sum_{m=0}^{\infty} c_m t^m$$

of the coefficients c_m defined by (13-5) the closed expression

(13-13) $$C(t) = -\frac{t}{(1-t) \log (1-t)},$$

the recurrence relation $c_0 = 1$,

$$c_m + \tfrac{1}{2}c_{m-1} + \tfrac{1}{3}c_{m+2} + \cdots + \frac{1}{m+1}\,c_0 = 1, \qquad m = 1, 2, \ldots$$

and from it the numerical values

Table 13.4b

m	0	1	2	3	4	5
c_m	1	$\dfrac{1}{2}$	$\dfrac{5}{12}$	$\dfrac{3}{8}$	$\dfrac{251}{720}$	$\dfrac{95}{288}$

In order to obtain a recurrence relation for the coefficients b_m defined by (13-9), we write their generating function in the form

$$B(t) = \sum_{m=0}^{\infty} b_m (2t)^{2m} = \int_0^1 \sum_{m=0}^{\infty} \binom{s + m - 1}{2m} (2t)^{2m}\, ds.$$

To evaluate it, we require the formula

(13-14) $$\sum_{m=0}^{\infty} \binom{s + m - 1}{2m} (2t)^{2m} = \frac{(T + t)^{1 - 2s} + (T - t)^{1 - 2s}}{2T},$$

where

$$T = \sqrt{1 + t^2}.$$

This is a result from the theory of Legendre functions (see Erdelyi [1953], equation 3.2 (14) in connection with 3.5 (12)) whose derivation cannot be given here. Some manipulation yields

$$B(t) = \sum_{m=0}^{\infty} b_m (2t)^{2m} = \frac{t}{T \log (T + t)}.$$

Since

$$\frac{d}{dt} \log (T + t) = \frac{1}{T} = 1 + \binom{-\frac{1}{2}}{1} t^2 + \binom{-\frac{1}{2}}{2} t^4 + \cdots$$

we easily find

$$t^{-1} \log (T + t) = 1 + \frac{1}{3} \binom{-\frac{1}{2}}{1} t^2 + \frac{1}{5} \binom{-\frac{1}{2}}{2} t^4 + \cdots.$$

Hence we have

$$\frac{\log{(T + t)}}{t} B(t) = \frac{1}{T}$$

or

$$\left[1 + \frac{1}{3}\binom{-\frac{1}{2}}{1}t^2 + \frac{1}{5}\binom{-\frac{1}{2}}{2}t^4 + \cdots\right][b_0 + 4b_1 t^2 + 16b_2 t^4 + \cdots]$$

$$= 1 + \binom{-\frac{1}{2}}{1}t^2 + \binom{-\frac{1}{2}}{2}t^4 + \cdots.$$

Comparing coefficients, we get

$$(13\text{-}15) \quad 4^m b_m + \frac{4^{m-1}}{3}\binom{-\frac{1}{2}}{1}b_{m-1} + \cdots + \frac{1}{2m+1}\binom{-\frac{1}{2}}{m}b_0 = \binom{-\frac{1}{2}}{m},$$

$$m = 0, 1, 2, \ldots.$$

The following values are obtained:

Table 13.4c

m	0	1	2	3
b_m	1	$-\dfrac{1}{12}$	$\dfrac{11}{720}$	$-\dfrac{191}{60480}$

Problems

6. Find the generating function for the coefficients in the formula

$$f'(x_0) = \frac{1}{h}[a_1 \Delta f_0 + a_2 \Delta^2 f_0 + \cdots]$$

obtained by differentiating the Newton forward interpolating polynomial at $x = x_0$.

7*. Obtain the generating function for the coefficients introduced in problem 4.

8. Using generating functions, show that

$$c_m^* = c_m - c_{m-1}, \quad m = 1, 2, \ldots.$$

13.5   Numerical Integration over Extended Intervals

In §13.2 and §13.3 we have considered the problem of evaluating the integral of a function f over an interval of length h in terms of differences calculated with the step h.   Here we shall study the equally important problem of evaluating

$$I = \int_a^b f(x)\, dx,$$

where $[a, b]$ is a fixed, not necessarily short, interval as accurately as possible. It is assumed that f is continuous on $[a, b]$ and can be evaluated at arbitrary points of that interval. Several procedures offer themselves; we shall be able to dismiss two of them very briefly.

(*i*) *Newton-Cotes formulas.* The most natural idea that offers itself seems to select a certain number of interpolating points within $[a, b]$, to interpolate f at these points, and to approximate the integral of f by the integral of the interpolating polynomial. If the interpolating points divide $[a, b]$ into N equal parts, we arrive in this.manner at certain integration formulas which are called the *Newton-Cotes* formulas. Unfortunately these formulas have, for large values of N, some very undesirable properties. In particular, it turns out that there exist functions, even analytic ones, for which the sequence of the integrals of the interpolating polynomials does not converge towards the integral of the function f. Also, the coefficients in these formulas are large and alternate in sign, which is undesirable for the propagation of rounding error. For these reasons, the Newton-Cotes formulas are rarely used for high values of N. For $N = 2, 3, 4$ the formulas are identical with certain well-known integration formulas which we shall discuss below from a different point of view.

(*ii*) *Gaussian quadrature.* One may try to avoid some of the shortcomings of the Newton-Cotes formulas by relinquishing the equal spacing of the interpolating points. Gauss discovered that by a proper choice of the interpolating points one can construct integration formulas which, using $N + 1$ interpolating points, give the accurate value of the integral if f is a polynomial of degree $2N + 1$ or less. These formulas turn out to be numerically stable, and they are in successful use at a number of computation laboratories. The formulas suffer from the disadvantage, however, that the interpolating points as well as the corresponding weights are irregular numbers that have to be stored. This practically (although not theoretically) limits the applicability of these highly interesting formulas.

In the following two sections, we shall discuss two integration schemes that are easy to use in practice. They are (*iii*) *the Trapezoidal rule with end correction*, and (*iv*) *Romberg integration*.

Both can be regarded as more sophisticated forms of the trapezoidal rule discussed in §13.3.

13.6   Trapezoidal Rule with End Correction

As in the discussion of the Newton-Cotes formula, let us divide the interval $[a, b]$ into N equal parts of length

$$h = \frac{b - a}{N}.$$

We write

$$x_n = a + nh, \qquad n = 0, 1, \ldots, N \quad (x_N = b)$$

and evaluate the integral of f over each of the subintervals $[x_{n-1}, x_n]$ separately, using a different interpolating polynomial each time. Using the symmetrical formula (13-10), we find for the integral over the nth subinterval, if f has a continuous derivative of order $2k + 2$,

$$(13\text{-}16) \quad \int_{x_{n-1}}^{x_n} f(x)\, dx = h\{b_0 \mu f_{n-1/2} + b_1 \mu \delta^2 f_{n-1/2} + \cdots$$
$$+ b_k \mu \delta^{2k} f_{n-1/2} + b_{k+1} h^{2k+2} f^{(2k+2)}(\xi_n)\},$$

where $x_{n-1-k} < \xi_n < x_{n+k}$. Adding up the integrals (13-16) for $n = 1, 2, \ldots, N$, we clearly obtain, in view of $x_0 = a$, $x_N = b$

$$(13\text{-}17) \quad I = \int_a^b f(x)\, dx$$

$$= h\Bigg\{ b_0 \sum_{n=1}^N \mu f_{n-1/2} + b_1 \sum_{n=1}^N \mu \delta^2 f_{n-1/2} + \cdots$$

$$+ b_k \sum_{n=1}^N \mu \delta^{2k} f_{n-1/2} + b_{k+1} h^{2k+2} \sum_{n=1}^N f^{(2k+2)}(\xi_n) \Bigg\}$$

where $b_0 = 1$, $b_1 = -\frac{1}{12}, \ldots$, are the constants defined by (13-9) and tabulated in §13.4. Let us consider the sums on the right separately. In view of

$$\mu f_{n-1/2} = \tfrac{1}{2}(f_{n-1} + f_n)$$

the first term

$$h \sum_{n=1}^N \mu f_{n-1/2} = h[\tfrac{1}{2}(f_0 + f_1) + \tfrac{1}{2}(f_1 + f_2) + \cdots + \tfrac{1}{2}(f_{N-1} + f_N)]$$

clearly reduces to

$$(13\text{-}18) \quad T_N = h[\tfrac{1}{2} f_0 + f_1 + f_2 + \cdots + f_{N-2} + f_{N-1} + \tfrac{1}{2} f_N].$$

We shall call this term the *N-point trapezoidal value* of the desired integral. It is the result of evaluating the integral by the trapezoidal rule familiar from elementary calculus (see Taylor [1959], p. 515).

Even more drastic simplifications occur in the other sums on the right of (13-17). Recalling that

$$\delta^{2m} f_{n-1/2} = \delta^{2m-1} f_n - \delta^{2m-1} f_{n-1}$$

and hence

$$\mu \delta^{2m} f_{n-1/2} = \mu \delta^{2m-1} f_n - \mu \delta^{2m-1} f_{n-1}$$

we find that the remaining sums in (13-17) "telescope" as follows:

$$\sum_{n=1}^{N} \mu\delta^{2m}f_{n-1/2} = \sum_{n=0}^{N} (\mu\delta^{2m-1}f_n - \mu\delta^{2m-1}f_{n-1})$$

$$= \mu\delta^{2m-1}f_N - \mu\delta^{2m-1}f_0$$

$(m = 1, 2, \ldots, k)$. Thus, each of the sums multiplying $b_1, b_2, \ldots, b_k$ reduces to a difference of two central differences at the endpoints of the interval of integration.

In order to simplify the last term, let us denote by M the maximum and by m the minimum of the function $f^{(2k+2)}(x)$ for $x_{-k-1} \leqq x \leqq x_{N+k+2}$. The sum

$$\sum_{n=1}^{N} f^{(2k+2)}(\xi_n),$$

having $N = (b - a)/h$ terms, is then contained between the limits

$$\frac{b-a}{h} m \quad \text{and} \quad \frac{b-a}{h} M.$$

Since the continuous function $f^{(2k+2)}$ assumes all values between its extreme values, there must be a ξ in the above-mentioned interval such that

$$\frac{b-a}{h} f^{(2k+2)}(\xi) = \sum_{n=1}^{N} f^{(2k+2)}(\xi_n).$$

For that value of ξ, we thus have

$$h^{2k+2}b_{k+1} \sum_{n=1}^{N} f^{(2k+2)}(\xi_n) = (b-a)h^{2k+1}b_{k+1}f^{(2k+2)}(\xi).$$

Gathering together the above results, we have

Theorem 13.6   Let N and k be any two positive integers, and let f have a continuous derivative of order $2k+2$ on the interval $J = [a - kh, b + kh]$. Then

(13-19) $\int_a^b f(x)\,dx = T_N + C_N^{(k)} + R_N^{(k)}$,

where T_N denotes the trapezoidal value of the integral defined by (13-18), $C_N^{(k)}$ denotes the "end correction"

(13-20) $C_N^{(k)} = h\{b_1(\mu\delta f_N - \mu\delta f_0) + b_2(\mu\delta^3 f_N - \mu\delta^3 f_0) + \cdots$
$$+ b_k(\mu\delta^{2k-1}f_N - \mu\delta^{2k-1}f_0)\},$$

and where, for some suitable $\xi \in J$,

(13-21) $R_N^{(k)} = h^{2k+2}b_{k+1}(b-a)f^{(2k+2)}(\xi)$.

In order to formulate the integration procedure implied in theorem 13.6 in algorithmic terms, one would have to devise a systematic method for forming the sequence of end corrections $C_N^{(k)}$ ($k = 1, 2, \ldots$). This would best be done by first forming, in the neighborhood of the points x_0 and x_N, the two-step differences

$$\mu \delta f_n = \tfrac{1}{2}(f_{n+1} - f_{n-1})$$

and then using the identities

$$\mu \delta^{2m-1} f_n = \delta^{2m-2}(\mu \delta f_n)$$

$$= \delta^{2m-2} \left(\frac{f_{n+1} - f_{n-1}}{2} \right) \qquad (n = 0, N).$$

The coefficients b_m can be generated recursively from (13-15).

EXAMPLE

2. To evaluate $\int_0^{1.6} J_0(x)\, dx$. We choose the step $h = 0.1$, leading to $N = 16$. From a table of Bessel functions we find

$$T_{16} = 1.28934\ 6003.$$

Since $J_0(-x) = J_0(x)$, the contributions to the end correction at $x_0 = 0$ is zero. At $x_{16} = 1.6$ we find

$$
\begin{array}{ll}
\mu \delta f_{16} = -0.05692\ 1406, & hb_1 \mu \delta f_{16} = 0.00047\ 4345, \\
\mu \delta^3 f_{16} = 0.00040\ 8458, & hb_2 \mu \delta^3 f_{16} = 0.00000\ 0624, \\
\mu \delta^5 f_{16} = -0.00000\ 3327, & hb_3 \mu \delta^5 f_{16} = 0.00000\ 0001,
\end{array}
$$

yielding the more accurate value

$$\int_0^{1.6} J_0(x)\, dx = 1.28982\ 0973,$$

correct to the number of places given.

Problems

9. Evaluate

$$\int_0^\pi \frac{\sin x}{x}\, dx$$

with an error less than 10^{-8}.

10. Devise a method, similar to the one discussed above, based on the formula obtained in problem 4. (Divide into an even number of subintervals.)

11. Using the integration formula (13-6) and a similar formula involving forward differences, obtain an end correction for the trapezoidal rule that involves values f_n satisfying $0 \leq n \leq N$ only.

12. Show that if the function f is periodic with period T, and if formula (13-19) is used to evaluate the integral of f over a full period, then

$$C_N^{(k)} = 0 \quad \text{for all } k \text{ and } N.$$

13. Problem 12 shows that for the integration of a sufficiently differentiable periodic function over a full period,

$$\int_a^b f(x)\,dx - T_N = 0(h^{2k})$$

for every k.   Does it follow that the trapezoidal value is exact?

14. Show that the trapezoidal rule T_N yields for $N > 1$ the exact values of the integrals

$$\int_0^{2\pi} \cos x\,dx, \qquad \int_0^{2\pi} \sin x\,dx.$$

(Use problem 19, chapter 6.)

13.7 Romberg Integration

Theorem 13.6 states, in effect, that the trapezoidal approximation to $\int_a^b f(x)\,dx$ calculated with the step h,

$$A(h) = T_{N(h)},$$

where $N(h) = (b - a)h^{-1}$, satisfies

$$A(h) = a_0 - C_{N(h)}^{(k)} + 0(h^{2k+2}),$$

where a_0 is the exact value of the integral, and where $C_{N(h)}^{(k)}$ is defined by (13-20).   If we could show that for every positive integer k

(13-22) $\qquad C_{N(h)}^{(k)} = a_1 h^2 + a_2 h^4 + \cdots + a_k h^{2k} + 0(h^{2k+2}),$

then $A(h)$ would satisfy the hypotheses for successful $k$fold extrapolation to the limit, that is, it would be of the form (12-19) where $y = h^2$.   We might expect then to speed up the convergence of the trapezoidal values by an application of algorithm 12.4.

By expanding the differences appearing in (13-20) in powers of h, it is easy to see that $C_{N(h)}^{(k)}$ indeed can be expanded in the form (13-22).   For instance,

$$h\mu\delta f_0 = \frac{h}{2}(f_1 - f_{-1})$$

$$= h^2 f'(x_0) + \frac{h^4}{6} f'''(x_0) + \cdots,$$

$$h\mu\delta^3 f_0 = \frac{h}{2}(f_2 - 2f_1 + 2f_{-1} - f_{-2})$$

$$= h^4 f'''(x_0) + \cdots.$$

We thus may apply algorithm 12.4, keeping in mind that $y = h^2$. The choices of the values y_0 and r are dictated by the obvious procedure to begin with a single subinterval ($N = 1$), and then to double the number of subintervals at each step. Since $y = h^2$ this means that the ratio between consecutive values of y is $r = \frac{1}{4}$. The algorithm thus results in the triangular array

$$
\begin{array}{llll}
A_{0,0} & & & \\
A_{1,0} & A_{1,1} & & \\
A_{2,0} & A_{2,1} & A_{2,2} & \\
A_{3,0} & A_{3,1} & A_{3,2} & A_{3,3} \\
\cdot\;\;\cdot & \cdot\;\;\cdot\;\;\cdot & \cdot\;\;\cdot\;\;\cdot & \cdot\;\;\cdot\;\;\cdot
\end{array}
$$

defined by the recurrence relations

(13-23) $A_{m,0} = T_{2^m}, \qquad m = 0, 1, 2, \ldots,$

(13-24) $A_{m,n+1} = \dfrac{4^{n+1} A_{m,n} - A_{m-1,n}}{4^{n+1} - 1}, \qquad n = 0, 1, \ldots, m - 1.$

In order to define an economical algorithm, we note that of the ordinates necessary to compute T_{2^m} only those whose index is odd have to be freshly calculated. The others are known from $T_{2^{m-1}}$. The computation is most easily arranged by introducing besides the trapezoidal values T_N the *midpoint values*

$$
M_N = h(f_{1/2} + f_{3/2} + \cdots + f_{N-1/2})
$$

$$
= h \sum_{n=1}^{N} f(a + (n - \tfrac{1}{2})h) \qquad \left(h = \frac{b - a}{N}\right).
$$

Like the trapezoidal value, the midpoint value is an approximation to the desired integral that is in error by $O(h^2)$. We clearly have

$$
T_{2N} = \frac{b - a}{2N} [\tfrac{1}{2}f_0 + f_{1/2} + f_1 + \cdots + f_{N-1} + f_{N-1/2} + \tfrac{1}{2}f_N]
$$

$$
= \tfrac{1}{2}(T_N + M_N).
$$

Thus in the present situation algorithm 12.4 can be formulated as follows:

Algorithm 13.7 If the function f is defined on the interval $[a, b]$, generate the triangular array of numbers $A_{m,n}$ by means of the recurrence relation (13-24) and

$$
A_{0,0} = T_1 = \tfrac{1}{2}(b - a)[f(a) + f(b)],
$$

$$
A'_{m,0} = M_{2^m} = h \sum_{n=1}^{2^m} f(a + (n - \tfrac{1}{2})h), \qquad h = 2^{-m}(b - a),
$$

$$
A_{m+1,0} = \tfrac{1}{2}(A_{m,0} + A'_{m,0}), \qquad m = 0, 1, 2, \ldots.
$$

The hypotheses of theorem 12.4 are satisfied when f satisfies the hypotheses of theorem 13.6 for every k when N is sufficiently large. This is the case if f has derivatives of all orders on an interval containing the interval $[a, b]$ in its interior. This condition, in turn, is satisfied if f is *analytic* on the closed interval $[a, b]$ (see Buck [1956], p. 78).

Theorem 13.7   Let the function f be analytic on the closed interval $[a, b]$. Then all columns of the array generated by algorithm 13.7 converge to $\int_a^b f(x)\, dx$. If none of the coefficients a_k in (13-22) vanishes, each column converges faster than the preceding one.

The scheme generated by algorithm 13.7 is commonly known as the *Romberg* scheme (see Bauer *et al.* [1963]).

EXAMPLE

3. For the problem considered in example 2 the Romberg scheme is as follows:

Table 13.7

1.16432 1734			
1.25919 0749	1.29081 3754		
1.28220 7763	1.28988 0101	1.28981 7857	
1.28792 0410	1.28982 4626	1.28982 0927	1.28982 0976
1.28934 6003	1.28982 1201	1.28982 0973	1.28982 0973

The accuracy of extrapolation to the limit with 16 steps is about the same as applying T_{16} with "end correction." The advantage of the Romberg procedure is that no decision has to be made in advance concerning the best stepsize.

Problem 13 shows that it may happen that some or even all coefficients in the expansion (13-22) are zero. In such cases extrapolation to the limit will not speed up the convergence of the trapezoidal rule.

Problems

15. Verify experimentally that extrapolation to the limit does not speed up the convergence of the trapezoidal rule for the integral

$$\int_0^1 \sqrt{x}\, dx.$$

Can you explain why?

16. Evaluate

$$\int_0^\pi \frac{\sin x}{x}\, dx$$

with an error of less than 10^{-8}, using repeated extrapolation to the limit.

17. Show that the values $A_{m,1}$ obtained by algorithm 13.7 are identical with the values obtained by applying Simpson's rule (see problem 4) to 2^{m-1} subintervals.

18. Express the values $A_{m,2}$ in terms of the ordinates f_n. Can you recognize a familiar integration rule?

Recommended Reading

Hildebrand [1956] gives an excellent account of Gaussian quadrature in chapter 8. Many examples of the use of generating functions to obtain integration coefficients are given in Henrici [1962], chapter 5. Bauer *et al.* [1963] contains a definite treatment of Romberg integration. Milne [1949] gives numerous integration formulas not considered here and also has a general treatment of the error in numerical integration.

Research Problems

1. Make a comparative study of the effectiveness of the end correction versus repeated extrapolation to the limit.

2. Study the connection of the end correction with the Euler-MacLaurin summation formula.

chapter 14 numerical solution of differential equations

The mathematical formulation of many problems in science and engineering leads to a relation between the values of an unknown function and the value of one or several of its derivatives at the same point.   Such relations are called *differential equations*.   The present chapter is devoted to the problem of finding numerical values of solutions of differential equations.

14.1   Theoretical Preliminaries

Let $f = f(x, y)$ be a real-valued function of two real variables defined for $a \leq x \leq b$, where a and b are finite, and for all real values of y.   The equation

$$(14\text{-}1) \qquad\qquad y' = f(x, y)$$

is called an *ordinary differential equation* of the first order; it symbolizes the following problem: To find a function $y = y(x)$, continuous and differentiable for $x \in [a, b]$, such that

$$y'(x) = f(x, y(x))$$

for all $x \in [a, b]$.   A function y with this property is called a *solution* of the differential equation (14-1).

EXAMPLE

1.   The differential equation $y' = -y$ has the solutions $y(x) = Ce^{-x}$, $C = $ const.

As the above example shows, a given differential equation may have many solutions.   In order to pin down a solution, one must specify its value at one point, say at $x = a$.   The problem of finding a solution y of

263

(14-1) such that $y(a) = s$, where s is a given number, is called an *initial value problem* and is schematically described by the equations

(14-2)
$$\begin{cases} y' = f(x, y), \\ y(a) = s. \end{cases}$$

EXAMPLE

2. The initial value problem $y' = -y$, $y(0) = 1$ has the solution $y(x) = e^{-x}$.

In courses on differential equations much stress is placed on differential equations whose solutions can be expressed in terms of elementary functions, or of indefinite integrals of such functions. It is shown, for instance, that any differential equation of the form $y' = g(x)y + p(x)$ can be solved in this manner. The emphasis on explicitly solvable differential equations tends to create the impression that almost any differential equation can be solved in explicit form, if only the proper trick is found. Nothing could be farther from the truth, however. Most differential equations that occur in actual practice are much too complicated to admit an explicit solution; if numerical values of the solution are desired, they can only be found by special numerical methods.

EXAMPLE

3. The differential equation $y' = 1-y$ can be solved explicitly; for the equation $y' = 1-y + \varepsilon y^3 \sin x$, which for small values of ε differs only little from it, no such solution can be found.

The fact that a differential equation does not possess an explicit solution does not mean that the solution does not exist, in the mathematical sense. The following existence theorem is proved in courses on differential equations (see Coddington [1961], p. 217).

Theorem 14.1 Let the function $f = f(x, y)$ be continuous for $a \leq x \leq b$, $-\infty < y < \infty$, and let there exist a constant L such that, for any two numbers y, z and all $x \in [a, b]$,

(14-3)
$$|f(x, y) - f(x, z)| \leq L|y - z|.$$

Then, whatever the initial value s, the initial value problem (14-2) possesses a unique solution $y = y(x)$ for $x \in [a, b]$.

This theorem can be proved, for instance, by the method of successive approximations. Although this is, in a sense, a constructive method for proving the existence of a solution, the method is not suitable for the numerical computation of the solution, because it requires the evaluation of infinitely many indefinite integrals. The methods to be given below are far more economical and are therefore preferred in practice.

Condition (14-3) resembles condition (4-4) imposed on functions suitable for iteration, with the difference that the function f now depends on the additional parameter x. The condition is again called a *Lipschitz condition*.

Problems

1. Which of the following functions f satisfy a Lipschitz condition?

(a) $\qquad\qquad f(x, y) = \dfrac{y^2}{1 + y^2};$

(b) $\qquad\qquad f(x, y) = \dfrac{x^2}{1 + x^2} y, \qquad 0 \le x \le 1;$

(c) $\qquad\qquad f(x, y) = y^2;$

(d) $\qquad\qquad f(x, y) = \sqrt{|y|}.$

2. Show that the following condition is sufficient for f to satisfy (14-3): The partial derivative f_y exists and is continuous for $a \le x \le b$, $-\infty < y < \infty$, and there exists a constant M such that

$$|f_y(x, y)| \le M$$

for all values of x and y in the indicated domain.

14.2   Numerical Integration by Taylor's Expansion

Throughout the balance of the present chapter we shall assume that the function f not only satisfies the conditions of the basic existence theorem 14.1, but also that it possesses continuous derivatives with respect to both x and y of all orders required to justify the analytical operations to be performed.   We shall refer to this property by the phrase "f is sufficiently differentiable."

It is a basic fact in the theory of ordinary differential equations that if f is sufficiently differentiable, all derivatives of a solution of the differential equation (14-1) are expressible in terms of the function f and its derivatives. In fact, this statement is obviously true for the first derivative.   Assuming its truth for the nth derivative, we write

(14-4) $\qquad\qquad y^{(n)} = f^{(n-1)}(x, y),$

where $f^{(n-1)}$, the $(n-1)st$ total derivative of f with respect to x, is a certain combination of derivatives of f.   Differentiating the identity

$$y^{(n)}(x) = f^{(n-1)}(x, y(x))$$

once more with respect to x, we have

$$y^{(n+1)}(x) = f_x^{(n-1)}(x, y(x)) + f_y^{(n-1)}(x, y(x))y'(x)$$

and thus, in view of $y'(x) = f(x, y(x))$,

$$y^{(n+1)}(x) = f^{(n)}(x, y(x)),$$

where

(14-5) $f^{(n)}(x, y) = f_x^{(n-1)}(x, y) + f_y^{(n-1)}(x, y)f(x, y).$

We thus have

> **Algorithm 14.2**   If f is sufficiently differentiable, the nth derivative of a solution y of (14-1) is expressible in the form
>
> $$y^{(n)}(x) = f^{(n-1)}(x, y(x)), \qquad n = 0, 1, 2, \ldots$$
>
> where the functions $f^{(m)}$ can be calculated by means of the recurrence relation
>
> $$f^{(0)} = f, \qquad f^{(m+1)} = f_x^{(m)} + f_y^{(m)}f, \qquad m = 0, 1, 2, \ldots.$$

In place of $f^{(1)}$ and $f^{(2)}$ we shall also write f' and f''.

EXAMPLES

4.
$$f' = f_x + f_y f,$$
$$f'' = f_{xx} + f_{xy}f + f_y f_x + (f_{xy} + f_{yy}f + f_y^2)f$$
$$= f_{xx} + 2f_{xy}f + f_{yy}f^2 + (f_x + f_y f)f_y.$$

5.   Let $f(x, y) = x^2 + y^2$.   We find

$$f'(x, y) = 2x + 2y(x^2 + y^2),$$
$$f''(x, y) = 2 + 4xy + (2x^2 + 6y^2)(x^2 + y^2).$$

In view of the fact that the derivatives of the solution of a differential equation can be determined whenever the derivatives of f can be calculated (which is, in principle, always the case if f is an elementary function of x and y), one might think of approximating the solution y of the initial value problem (14-2) by its Taylor series at $s = a$.   In view of the initial condition $y(a) = s$ this series takes the form

(14-6) $y(x) = s + \dfrac{x-a}{1!}f(a, s) + \dfrac{(x-a)^2}{2!}f'(a, s)$
$$+ \dfrac{(x-a)^3}{3!}f''(a, s) + \cdots.$$

However, the above examples suggest that unless the function f is very simple, the higher total derivatives of f rapidly increase in complexity. This circumstance makes it necessary to truncate the infinite series (14-6) already after very few terms.   This necessarily restricts the range of values of x over which the truncated series (14-6) can be expected to define a good approximation of the solution y. One therefore will use the truncated series with a small value of $x - a$, $x - a = h$ say, and re-evaluate the derivatives $f, f', f'', \ldots$, at the point $x = a + h$.

Problems

3. Calculate the nth total derivative of f if

(a) $f(x, y) = y + e^x$,   (b) $f(x, y) = e^x y$.

4. If $f(x, y) = (2y/x) - 1$, show that f'' is identically zero. What is the explanation of this fact?

14.3   The Taylor Algorithm

In order to formalize the procedure outlined in the preceding section we introduce the following notation. Let $h > 0$ be a constant, and set

$$x_n = a + nh, \qquad n = 0, 1, 2, \ldots.$$

We shall always denote by y_n a number intended to approximate $y(x_n)$, the value of the exact solution of the initial value problem (14-1) at $x = x_n$. If the method of Taylor expansion is used, these values are calculated according to the following scheme: Let, for some fixed integer $p \geq 1$,

(14-7) $T_p(x, y; h) = \dfrac{1}{1!} f(x, y) + \dfrac{h}{2!} f'(x, y) + \cdots + \dfrac{h^{p-1}}{p!} f^{(p-1)}(x, y).$

The numbers y_n are then calculated successively as follows:

> **Algorithm 14.3**   For a fixed value of h, generate the sequence of numbers $\{y_n\}$ by the recurrence relation
>
> (14-8a) $y_0 = s,$
>
> (14-8b) $y_{n+1} = y_n + hT_p(x_n, y_n; h), \qquad n = 0, 1, 2, \ldots.$

This scheme is known as the *Taylor algorithm* of order p.

EXAMPLE

6.   The Taylor algorithm of order one is particularly simple. In view of $T_1(x, y; h) = f(x, y)$, (14-8b) then becomes

(14-9) $y_{n+1} = y_n + hf(x_n, y_n).$

This is also known as the *Euler method*, or as the *Euler-Cauchy* method.

Naturally we expect the values defined by (14-8) to approximate the corresponding values $y(x_n)$ of the exact solution, particularly when h is small. To see whether this is the case, let us examine the purely academic example

$$(14\text{-}10) \qquad\qquad y(0) = 1, \qquad y' = y.$$

The exact solution, of course, is $y(x) = e^x$. In view of

$$f(x, y) = f'(x, y) = f''(x, y) = \cdots = y,$$

the function T_p takes the simple form

$$T_p(x, y; h) = \left(\frac{1}{1!} + \frac{h}{2!} + \cdots + \frac{h^{p-1}}{p!}\right) y.$$

The recurrence relation (14-8) thus becomes

$$y_{n+1} = \left(1 + \frac{h}{1!} + \frac{h^2}{2!} + \cdots + \frac{h^p}{p!}\right) y_n.$$

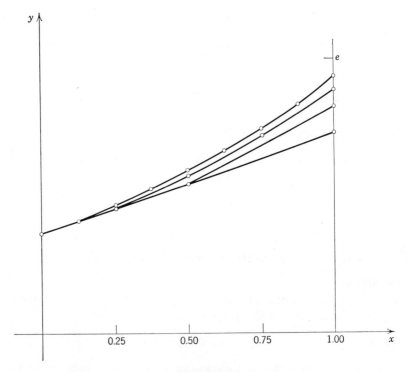

Figure 14.3

This is a difference equation of order 1 (see chapter 3); the solution satisfying the initial condition $y_0 = 1$ is given by

$$(14\text{-}11) \quad y_n = \left(1 + \frac{h}{1!} + \frac{h^2}{2!} + \cdots + \frac{h^p}{p!}\right)^n, \qquad n = 0, 1, 2, \ldots.$$

We wish to examine how closely y_n approximates $y(x_n) = e^{x_n}$ at a fixed point x of the interval of integration, if the approximation y_n is calculated by successively smaller integration steps h. We thus must let $h \to 0$ and $n \to \infty$ simultaneously in such a manner that $x = x_n = nh$ remains fixed. By Taylor's formula with remainder term (see Taylor [1959], p. 471)

$$e^h = 1 + \frac{h}{1!} + \frac{h^2}{2!} + \cdots + \frac{h^p}{p!} + \frac{h^{p+1}}{(p+1)!} e^{\theta h},$$

where θ is some unspecified number between 0 and 1. We now use the fact that for any two real numbers A and B such that $A \geq B \geq 0$ and for $n = 1, 2, \ldots$

$$A^n - B^n = (A - B)(A^{n-1} + A^{n-2}B + \cdots + AB^{n-2} + B^{n-1})$$
$$\leq (A - B)nA^{n-1}.$$

In particular, taking

$$A = e^h,$$
$$B = 1 + \frac{h}{1!} + \frac{h^2}{2!} + \cdots + \frac{h^p}{p!},$$

we obtain

$$y(x_n) - y_n = e^{nh} - \left(1 + \frac{h}{1!} + \cdots + \frac{h^p}{p!}\right)^n$$
$$\leq n \frac{h^{p+1}}{(p+1)!} e^{\theta h} e^{(n-1)h}$$

or, in view of $nh = x_n$, $0 < \theta < 1$,

$$(14\text{-}12) \qquad |y_n - y(x_n)| \leq \frac{h^p}{(p+1)!} x_n e^{x_n}.$$

This relation shows that, at a fixed point $x_n = x$, the error in the approximation of the exact solution y by the Taylor algorithm of order p tends to zero like h^p as the stepsize h tends to zero.

The convergence if the integration steps are successively halved is illustrated in figure 14.3.

We shall now derive a similar bound for the error $y_n - y(x_n)$ of the Taylor approximation y_n to the solution y of an arbitrary initial value problem (14-2). To this end we make the following assumptions:

(*i*) Not only the function *f*, but also the function T_p defined by (14-7) satisfies a Lipschitz condition with Lipschitz constant *L*:

(14-13) $$|T_p(x, y; h) - T_p(x, z; h)| \leq L|y - z|$$

for any *y*, *z* and all *x* and *h* such that $x \in [a, b]$, $x + h \in [a, b]$;

(*ii*) The $(p + 1)$st derivative of the exact solution *y* of (14-2) is continuous and hence bounded on the closed interval $[a, b]$, and

(14-14) $$|y^{(p+1)}(x)| \leq Y_{p+1}, \qquad x \in [a, b].$$

We then have:

Theorem 14.3   Under the above assumptions (*i*) and (*ii*), the error $z_n = y_n - y(x_n)$ of the Taylor algorithm of order *p* is bounded as follows:

(14-15) $$|z_n| \leq h^p \frac{Y_{p+1}}{(p + 1)!} \frac{e^{L(x_n - a)} - 1}{L}.$$

The result shows that z_n again tends to zero at least like h^p as $x_n = x$ is fixed and $h \to 0$.

Proof.   By Taylor's formula with remainder, the exact solution satisfies

$$y(x_{n+1}) = y(x_n + h)$$
$$= y(x_n) + hT_p(x_n, y(x_n); h) + \frac{h^{p+1}}{(p + 1)!} y^{(p+1)}(\xi_n),$$

where ξ_n is some point between x_n and x_{n+1}. Subtracting the last relation from (14-8b), we get

(14-16) $$z_{n+1} = z_n + h[T_p(x_n, y_n; h) - T_p(x_n, y(x_n); h)]$$
$$- \frac{h^{p+1}}{(p + 1)!} y^{(p+1)}(\xi_n)$$

and hence

$$|z_{n+1}| \leq |z_n| + h|T_p(x_n, y_n; h) - T_p(x_n, y(x_n); h)|$$
$$+ \frac{h^{p+1}}{(p + 1)!} |y^{(p+1)}(\xi_n)|.$$

Using (14-13) and (14-14), the expression on the right can be simplified, with the following result:

(14-17) $$|z_{n+1}| \leq |z_n| + hL|z_n| + h^{p+1} \frac{Y_{p+1}}{(p + 1)!}.$$

We now make use of the following auxiliary result:

Lemma 14.3   Let the elements of a sequence $\{w_n\}$ satisfy the inequalities

(14-18) $$w_{n+1} \leqq (1 + a)w_n + B, \qquad n = 0, 1, 2, \ldots$$

where a, and B are certain positive constants, and let $w_0 = 0$. Then

(14-19) $$w_n \leqq B \frac{e^{na} - 1}{a}, \qquad n = 0, 1, 2, \ldots.$$

Proof of lemma 14.3.   The relation (14-19) is evidently true for $n = 0$. Assuming its truth for some nonnegative integer n we have from (14-18), in view of $1 + a \leqq e^a$,

$$w_{n+1} \leqq (1 + a) \frac{e^{na} - 1}{a} B + B$$

$$= B \frac{(1 + a)e^{na} - 1}{a} \leqq B \frac{e^{(n+1)a} - 1}{a},$$

establishing (14-19) with n increased by one.   The truth of the lemma now follows by the principle of induction.

Returning to the proof of theorem 14.3, we apply the lemma to the relation (14-17), setting $w_n = |z_n|$, $a = hL$, $B = h^{p+1} Y_{p+1}/(p + 1)!$. By virtue of (14-8a) and (14-2), $|z_0| = 0$.   Thus the conclusion (14-19) applies, yielding (14-15).

Problems

5. Solve the initial value problem

$$y' = 1 - y, \qquad y(0) = 0,$$

by the Taylor algorithm of order $p = 2$, using the steps $h = 0.5, h = 0.25$, and $h = 0.125$, and compare the values of the numerical solution at $x_n = 1$ with the values of the exact solution.

6. Find an analytical expression for the values y_n defined by (14-9) for the initial value problem

$$y' = -\frac{2y}{1 + x}, \qquad y(0) = 1,$$

and verify the statement of theorem 14.3.   [Hint: Use theorem 3.3 to solve the difference equation involved.]

14.4   Extrapolation to the Limit

The estimate for the error $z_n = y_n - y(x_n)$ given in theorem 14.3 should not be regarded as a realistic indication of the actual size of the error.   It

merely serves to prove the convergence of the Taylor algorithm and to
indicate the order of magnitude of the error. The estimate states, in
effect, that there exists a constant K so that

(14-20)
$$|z_n| \leq Kh^p$$

for all $x_n \in [a, b]$. This relation shows that the error tends to zero as
$h \to 0$. Can we make a statement about the manner in which the error
tends to zero? As we have observed repeatedly, such knowledge could be
helpful for the purpose of speeding up the convergence of the method as
$h \to 0$.

We shall prove:

Theorem 14.4 If f is sufficiently differentiable, the errors $z_n = y_n$
$- y(x_n)$ of the values defined by (14-8) satisfy

(14-21)
$$z_n = h^p z(x_n) + 0(h^{p+1}),$$

where z denotes the solution of the initial value problem

(14-22)
$$\begin{cases} z(a) = 0, \\ z' = f_y(x, y(x))z - \dfrac{1}{(p+1)!} y^{(p+1)}(x). \end{cases}$$

Proof. We start from relation (14-16) above. By Taylor's theorem,

$$T_p(x_n, y_n; h) - T_p(x_n, y(x_n); h) = \frac{\partial T_p}{\partial y}(x_n, y(x_n); h)z_n + 0(z_n^2)$$

$$= \frac{\partial T_p}{\partial y}(x_n, y(x_n); 0)z_n + 0(z_n^2) + 0(hz_n)$$

$$= f_y(x_n, y(x_n))z_n + 0(h^{p+1}),$$

since $T_p(x, y; 0) = f(x, y)$. Also, $y^{(p+1)}(\xi_n) = y^{(p+1)}(x_n) + 0(h)$. We
thus have

$$z_{n+1} = z_n + hf_y(x_n, y(x_n))z_n - \frac{h^{p+1}}{(p+1)!} y^{(p+1)}(x_n) + 0(h^{p+2}).$$

From this we subtract h^p times the relation

$$z(x_{n+1}) = z(x_n) + hz'(x_n) + 0(h^2)$$

$$= z(x_n) + h\left[f_y(x_n, y(x_n))z(x_n) - \frac{1}{(p+1)!} y^{(p+1)}(x_n)\right] + 0(h^2)$$

which follows from the definition of z. Setting

$$w_n = z_n - h^p z(x_n) \quad (n = 0, 1, 2, \ldots)$$

we obtain

$$|w_{n+1}| \leq |w_n| + hL|w_n| + 0(h^{p+1}).$$

Since $w_0 = 0$, lemma 14.3 now yields the relation $w_n = O(h^{p+1})$ equivalent to (14-21).

In order to make explicit the dependence of the numerical values y_n on the step h with which they are calculated, we shall denote (in this section only) by $y(x; h)$ the approximation to $y(x)$ calculated with the step h. Relation (14-21) can then be written more explicitly as follows:

$$(14\text{-}23) \qquad y(x, h) = y(x) + h^p z(x) + O(h^{p+1}).$$

In general, both quantities $y(x_n)$ and $z(x_n)$ are unknown in this relation, the latter because it depends on the solution of a differential equation involving the unknown function y. However, the mere fact that a relation of the form (14-21) holds can be made the basis for an extrapolation to the limit procedure. Eliminating $z(x_n)$ between (14-23) and the similar relation

$$y(x, r^{-1}h) = y(x) + r^{-p}h^p z(x) + O(h^{p+1})$$

for the numerical approximation calculated with the step $r^{-1}h$, where $r \neq 1$, we find that the quantity

$$(14\text{-}24) \qquad y^*(x, h) = \frac{y(x, h) - r^p y(x, r^{-1}h)}{1 - r^p}$$

differs from $y(x)$ by only $O(h^{p+1})$. As $h \to 0$, the values $y^*(x, h)$ will thus converge to $y(x)$ faster than the values $y(x, h)$. Here r is, in principle, arbitrary; for practical reasons one usually chooses $r = \frac{1}{2}$.

Under the assumption that (14-23) holds in the generalized form

$$(14\text{-}25) \qquad \begin{aligned} y(x; h) = y(x) &+ a_p(x)h^p + a_{p+1}(x)h^{p+1} + \cdots \\ &+ a_k(x)h^k + O(h^{k+1}) \end{aligned}$$

for every $k > p$, we can speed up the values $y^*(x, h)$ once more, using (14-24) with p replaced by $p + 1$. Continuing the process systematically in the manner of algorithm 12.4, we obtain

Algorithm 14.4   For a fixed x in $[a, b]$, form the triangular array of numbers $A_{m, n}$ as follows:

$$A_{m, 0} = y\left(x; \frac{x - a}{2^m}\right), \qquad m = 0, 1, 2, \ldots,$$

$$A_{m, n+1} = \frac{2^{p+n} A_{m, n} - A_{m-1, n}}{2^{p+n} - 1}, \qquad n = 0, 1, \ldots, m - 1.$$

The fact that (14-25) holds for sufficiently differentiable functions f was established by Gragg [1963]. If this is taken for granted, we can deduce as in theorem 12.4 that each succeeding column of the array $A_{m, n}$ converges faster to $y(x)$.

EXAMPLE

7. We integrate the initial value problem

$$y' = x - y^2, \qquad y(0) = 0$$

by Euler's method, using the steps $h = 0.4, 0.2, 0.1, 0.05$. The table below shows the resulting values at $x = 0.8$, and the values obtained by repeated extrapolation to the limit.

Table 14.4

h	$A_{m,0}$	$A_{m,1}$	$A_{m,2}$	$A_{m,3}$	$A_{m,4}$
0.8	0.00000				
0.4	0.16000	0.32000			
0.2	0.23682	0.31363	0.31151		
0.1	0.27201	0.30720	0.30505	0.30413	
0.05	0.28871	0.30541	0.30481	0.30478	0.30482

Problems

7. Let the initial value problem

$$y' = -y, \qquad y(0) = 1$$

be solved by the Taylor algorithm of order 1. Where does the absolute value of the error (approximately) attain its maximum value? Verify your result numerically by performing a numerical integration using the step $h = 0.1$.

8. Improve the values obtained in problem 5 by extrapolation to the limit.

9. Devise a method for extrapolation to the limit if p, the order of the method, is not known.

14.5   Methods of Runge-Kutta Type

The practical application of the Taylor algorithm described in the preceding sections is frequently tedious because of the necessity of evaluating the functions $f', f'', \ldots$ at each integration step. It therefore is a remarkable fact that formulas exist that produce values y_n of the same accuracy as some of the Taylor algorithms of order $p \geq 2$, but without requiring the evaluation of any derivatives. We mention only a few of the many available formulas of this type.

(*i*) *The simplified Runge-Kutta method.* Here we replace the function T_2 in algorithm 14.3 by K_2, where

$$(14\text{-}26) \qquad K_2(x, y; h) = \tfrac{1}{2}[f(x, y) + f(x + h, y + hf(x, y))].$$

Clearly, no evaluation of f' is necessary. Instead, the function f is evaluated twice at each step. It can be shown that

(14-27) $$K_2(x, y; h) - T_2(x, y; h) = 0(h^2),$$

and that the accumulated error satisfies an inequality of the form (14-20), and, with a suitable definition of z, a relation of the form of (14-21), where $p = 2$.

(ii) *The classical Runge-Kutta method.* Here the function T_4 in algorithm 14.3 is replaced by K_4, a sum of four values of the function f, defined as follows:

(14-28) $$K_4(x, y; h) = \tfrac{1}{6}[k_1 + 2k_2 + 2k_3 + k_4],$$

where

$$k_1 = f(x, y),$$

$$k_2 = f\left(x + \frac{h}{2}, y + \frac{h}{2}k_1\right),$$

$$k_3 = f\left(x + \frac{h}{2}, y + \frac{h}{2}k_2\right),$$

$$k_4 = f(x + h, y + hk_3).$$

It can be shown that

$$K_4(x, y; h) = T_4(x, y; h) + 0(h^4)$$

and that, as a consequence, the relations (14-20) and (14-21) now hold with $p = 4$. (The proofs of these statements are extremely complicated). In view of this fact, the classical Runge-Kutta method is one of the most widely used methods for the numerical integration of differential equations.

(iii) *A mixed Runge-Kutta-Taylor method.* If the first total derivative of f is easily evaluated, we may use in place of T_3 in algorithm 14.3 the function $G_3(x, y; h)$ defined by

(14-29) $$G_3(x, y; h) = f(x, y) + \frac{h}{2}f'\left(x + \frac{h}{3}, y + \frac{h}{3}f(x, y)\right).$$

It can be shown that $G_3(x, y; h) - T_3(x, y; h) = 0(h^3)$ and that, as a consequence, (14-20) and (14-21) are true for $p = 3$.

Problems

10. Solve the initial value problem

$$y' = x - y^2, \qquad y(0) = 0$$

for $0 \leq x \leq 0.8$ by the classical Runge-Kutta method, using the step $h = 0.1$. Also, perform the same integration with the steps $h = 0.2$ and $h = 0.4$, and perform an extrapolation to the limit.

11. Establish the relation (14-27).

12. Show that for the differential equation $y' = Ay$ ($A = $ const.) the values y_n produced by the methods K_2 and K_4 are identical, respectively, with the values produced by T_2 and T_4.

13. Prove that G_3 differs from T_3 by $O(h^3)$.

14.6   Methods Based on Numerical Integration: The Adams-Bashforth Method

All the methods discussed so far are based directly or indirectly on the idea of expanding the exact solution in a Taylor series. A different approach to the problem can be based on the idea of *numerical integration*. A solution y of (14-1) by definition satisfies the identity

$$y'(x) = f(x, y(x)).$$

Integrating between the limits x_n and x_{n+1} we obtain

(14-30) $$y(x_{n+1}) = y(x_n) + \int_{x_n}^{x_{n+1}} f(x, y(x))\, dx.$$

Let us now suppose that we have somehow already obtained approximate values $y_0, y_1, \ldots, y_n$ of the solution y at the points $x_0, x_1, \ldots, x_n$, where we again assume the points x_m to be equally spaced, $x_m = a + mh$. The approximate values of $f(x, y(x))$ at these points then are

$$f_m = f(x_m, y_m), \qquad m = 0, 1, \ldots, n.$$

If $k \leq n$, we can use these values to approximate $f(x, y(x))$ by the Newton backward polynomial

$$P_k(x) = \sum_{m=0}^{k} (-1)^m \binom{-s}{m} \nabla^m f_n, \qquad s = \frac{x - x_n}{h}.$$

Performing the integration in (14-30) on P_k in place of f, we obtain the following algorithm for computing a new value y_{n+1}:

Algorithm 14.6 (The Adams-Bashforth method.) For a given function $f = f(x, y)$, a given constant $h > 0$, a given integer $k \geq 0$, and given values $y_0, y_1, \ldots, y_k$, let $f_m = f(x_m, y_m)$ ($m = 0, 1, 2, \ldots, k$), and generate the sequence $\{y_n\}$ recursively from the formulas

(14-31) $$\begin{cases} x_{n+1} = x_n + h, \\ y_{n+1} = y_n + h[c_0 f_n + c_1 \nabla f_n + c_2 \nabla^2 f_n + \cdots + c_k \nabla^k f_n], \end{cases}$$

and

$$\begin{aligned} f_{n+1} &= f(x_{n+1}, y_{n+1}) = \nabla^0 f_{n+1}, \\ \nabla^m f_{n+1} &= \nabla^{m-1} f_{n+1} - \nabla^{m-1} f_n, \qquad m = 1, 2, \ldots, k. \end{aligned}$$

where $n = k, k+1, \ldots$.

Here the coefficients c_m are defined by (13-5); for numerical values see table 13.4b. The working of algorithm 14.6 is illustrated by scheme 14.6 for $k = 3$.

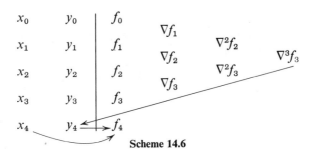

Scheme 14.6

As soon as y_{n+1} is known, f_{n+1} can be calculated and a new diagonal of differences may be formed. The process requires exactly one evaluation of f per step, no matter how many differences are carried. This compares favorably with methods such as the classical Runge-Kutta method, which requires four evaluations.

Algorithm 14.6 does not say how to obtain the starting values $y_0, y_1, \ldots,$ y_k. This can be done by any of the methods discussed in §14.3 or §14.5; other starting methods, more in the spirit of the algorithm itself, are also known (see Collatz [1960], p. 81).

Let us now study the error $z_n = y_n - y(x_n)$ of the values defined by (14-31) under the assumption that the starting values are in error by at most δ:

(14-32) $|z_m| = |y_m - y(x_m)| < \delta, \qquad m = 0, 1, \ldots, k.$

Expressing differences in terms of ordinates, (14-31) appears in the form

$$y_{n+1} = y_n + h\{b_{k0}f_n + b_{k1}f_{n-1} + \cdots + b_{kk}f_{n-k}\},$$

where the b_{km} are certain constants, involving the c_m, that need not be specified. The corresponding relation for the exact solution is, by (13-6),

$$y(x_{n+1}) = y(x_n) + h\{b_{k0}y'(x_n) + b_{k1}y'(x_{n-1}) + \cdots + b_{kk}y'(x_{n-k})\}$$
$$+ c_{k+1}h^{k+2}y^{(k+2)}(\xi_n),$$

where ξ_n is a point between x_{n-k} and x_{n+1}. We now subtract the last two relations from each other. Observing that

$$|f_m - y'(x_m)| = |f(x_m, y_m) - f(x_m, y(x_m))|$$
$$\leq L|y_m - y(x_m)| = L|z_m|,$$

where L denotes the Lipschitz constant of the function f, and assuming that

(14-33) $$|y^{(k+2)}(x)| \leq Y_{k+2}, \qquad a \leq x \leq b,$$

we obtain

$$|z_{n+1}| \leq |z_n| + hL\{|b_{k0}||z_n| + |b_{k1}||z_{n-1}| + \cdots + |b_{kk}||z_{n-k}|\} + |c_{k+1}|h^{k+2}Y_{k+2}.$$

Our aim is to find an explicit bound for the quantities $|z_n|$. An induction argument shows that

$$|z_n| \leq w_n, \qquad n = 0, 1, 2, \ldots$$

where $\{w_n\}$ is any solution of the difference equation

(14-34) $$w_{n+1} = w_n + hL\{|b_{k0}|w_n + |b_{k1}|w_{n-1} + \cdots + |b_{kk}|w_{n-k}\} + |c_{k+1}|h^{k+2}Y_{k+2}$$

satisfying

(14-35) $$|w_n| \geq \delta, \qquad n = 0, 1, \ldots, k.$$

We try to find such a solution using the principles set forth in §6.7. Since the nonhomogeneous term in (14-34) is a constant, the equation clearly has the particular solution $w_n = -C$, where

(14-36) $$C = \frac{|c_{k+1}|h^{k+1}Y_{k+2}}{LB_k},$$

(14-37) $$B_k = |b_{k0}| + |b_{k1}| + \cdots + |b_{kk}|.$$

In order to obtain a solution satisfying (14-35), we add to this solution a suitable solution of the homogeneous equation

$$w_{n+1} = w_n + hL\{|b_{k0}|w_n + |b_{k1}|w_{n-1} + \cdots + |b_{kk}|w_{n-k}\}.$$

The characteristic polynomial of the equation is

$$p(z) = z^{k+1} - z^k - hL\{|b_{k0}|z^k + \cdots + |b_{kk}|\}.$$

Evidently, $p(1) \leq 0$. On the other hand, if $z = 1 + hLB_k$, where B_k is given by (14-37), then

$$z^{-k}p(z) = 1 + hLB_k - 1 - hL\{|b_{k0}| + \cdots + |b_{kk}|z^{-k}\}$$
$$\geq hLB_k - hL\{|b_{k0}| + \cdots + |b_{kk}|\}$$
$$= 0.$$

It follows that for some z^* such that $1 \leq z^* \leq 1 + hLB_k$

$$p(z^*) = 0.$$

Thus a solution of (14-34) satisfying (14-35) is given by

$$w_n = (\delta + C)z^{*n} - C.$$

In view of

$$z^{*n} \leq (1 + hLB_k)^n \leq e^{(x_n - a)LB_k},$$

we finally find, remembering the definition of C,

$$(14\text{-}38) \quad |z_n| \leq \delta e^{(x_n - a)LB_k} + |c_{k+1}|h^{k+1}Y_{k+2}\frac{e^{(x_n - a)LB_k} - 1}{LB_k},$$

$$a \leq x_n \leq b.$$

In summary, we have

> **Theorem 14.6** If the starting values of the Adams-Bashforth method are in error by at most δ, then the errors $z_n = y_n - y(x_n)$ are bounded by (14-38), where Y_{k+2} and B_k are defined by (14-33) and (14-37).

As on earlier occasions, the estimate (14-38) should be looked at as a qualitative rather than a quantitative statement. The important thing is that, if the starting errors are sufficiently small (namely at most $0(h^{k+1})$), the error of the Adams-Bashforth algorithm tends to zero like h^{k+1} as $h \to 0$. Interpolating f in (14-30) by an interpolating polynomial of degree k has the same effect as expanding the solution in a Taylor polynomial of degree $k + 1$. However, the coefficient of Y_{k+2} in (14-38) is markedly greater than the corresponding coefficient of Y_{p+1} in (14-15), on account of the fact that both

$$|c_{k+1}| > \frac{1}{(k + 2)!} \quad \text{and} \quad B_k > 1$$

for $k > 0$.

Problems

14*. Assuming that the starting values are exact, show that the errors z_n of the Adams-Bashforth method satisfy

$$z_n = z(x_n)h^{k+1} + 0(h^{k+2}),$$

where z denotes the solution of the initial value problem

$$z' = f_y(x, y(x))z - c_{k+1}y^{(k+2)}(x),$$
$$z(a) = 0.$$

[Analog of theorem 14.4.]

15. Solve problem 10 by the Adams-Bashforth method with $k = 2$, determining the starting values y_{-1} and y_1 by Runge-Kutta or by a series expansion around $x = 0$. Use the steps $h = 0.2$ and $h = 0.1$ and extrapolate to the limit at $x = 0.8$.

14.7   Methods Based on Numerical Integration: The Adams-Moulton Method

A considerable refinement and improvement of the Adams-Bashforth method can be obtained as follows: Suppose a tentative value of y_{n+1} has been obtained, for instance by the Adams-Bashforth method. We shall call this value $y_{n+1}^{(0)}$. Setting $f_{n+1}^{(0)} = f(x_{n+1}, y_{n+1}^{(0)})$, we can construct the polynomial which at the points $x_{n-k}, x_{n-k+1}, \ldots, x_n, x_{n+1}$ takes on the values $f_{n-k}, f_{n-k+1}, \ldots, f_n, f_{n+1}$. According to the Newton backward formula, this polynomial is given by

$$P_{k+1}(x) = \sum_{m=0}^{k+1} (-1)^m \binom{1-s}{m} \nabla^m f_{n+1}^{(0)},$$

where again $s = (x - x_n)/h$, and where the differences are formed with the values $f_{n+1}^{(0)}, f_n, \ldots, f_{n-k}$. Integrating between x_n and x_{n+1} we obtain a new value of y_{n+1}, to be called $y_{n+1}^{(1)}$, according to the formula

$$(14\text{-}39) \quad y_{n+1}^{(1)} = y_n + h\{c_0^* f_{n+1}^{(0)} + c_1^* \nabla f_{n+1}^{(0)} + c_2^* \nabla^2 f_{n+1}^{(0)} + \cdots$$
$$+ c_{k+1}^* \nabla^{k+1} f_{n+1}^{(0)}\}.$$

Here the coefficients c_m^* are defined by (13-8) and are tabulated in table 13.4a.

In general, the value $y_{n+1}^{(1)}$ calculated by (14-39) will disagree with $y_{n+1}^{(0)}$. We therefore compute $f_{n+1}^{(1)} = f(x_{n+1}, y_{n+1}^{(1)})$, correct the difference table of the values of f, and reevaluate the term on the right in (14-39). A new value $y_{n+1}^{(2)}$ will result. The general step of this iteration procedure is described by the formula

$$(14\text{-}40) \quad y_{n+1}^{(i+1)} = y_n + h\{c_0^* f_{n+1}^{(i)} + c_1^* \nabla f_{n+1}^{(i)} + \cdots + c_{k+1}^* \nabla^{k+1} f_{n+1}^{(i)}\},$$

the differences on the right being formed with $f_{n+1}^{(i)} = f(x_{n+1}, y_{n+1}^{(i)})$, $f_n, \ldots, f_{n-k}$. Theoretically, the iteration with respect to i should be continued indefinitely, and the limit as $i \to \infty$ of the sequence $y_{n+1}^{(i)}$ would be accepted as the final value of y_{n+1}. In practice, of course, the iteration continues only to the point where the sequence $y_{n+1}^{(i)}$ has converged numerically in the sense that an inequality

$$|y_{n+1}^{(i+1)} - y_{n+1}^{(i)}| < \varepsilon$$

holds, where ε is some preassigned number. More frequently even, the iteration is performed only a fixed number of times, say for $i = 0, 1, 2, \ldots$, I, where I is a preassigned integer such as 2 or 3. We state the last version of the procedure as a formalized algorithm.

Algorithm 14.7   (The Adams-Moulton method.)   For a given function $f = f(x, y)$, a given constant $h > 0$, given integers $k > 0$ and

$I > 0$, and given values $y_0, y_1, \ldots, y_k$, let $x_m = x_0 + mh$, $f_m = f(x_m, y_m)$ $(m = 0, 1, \ldots, k)$, and generate the sequence y_n by the following set of formulas: Form $\nabla f_k, \nabla^2 f_k, \ldots, \nabla^k f_k$, and calculate for $n = k, k + 1, \ldots$

(14-31)
$$\begin{cases} x_{n+1} = x_n + h, \\ y_{n+1}^{(0)} = y_n + h[c_0 f_n + c_1 \nabla f_n + \cdots + c_k \nabla^k f_n], \end{cases}$$

and

$$y_{n+1} = y_{n+1}^{(I)},$$

where for $i = 0, 1, \ldots, I - 1$

$$f_{n+1}^{(i)} = f(x_{n+1}, y_{n+1}^{(i)}),$$
$$\nabla^m f_{n+1}^{(i)} = \nabla^{m-1} f_{n+1}^{(i)} - \nabla^{m-1} f_n \qquad (m = 1, 2, \ldots, k + 1)$$

and

(14-40) $y_{n+1}^{(i+1)} = y_n + h[c_0^* f_{n+1}^{(i)} + c_1^* \nabla f_{n+1}^{(i)} + \cdots + c_{k+1}^* \nabla^{k+1} f_{n+1}^{(i)}].$

The reader will observe that this algorithm involves three iterations: The "outer" iteration for calculating the sequence $\{y_n\}$, an inner iteration for calculating the sequence of values $\{y_{n+1}^{(i)}\}$ for each n, and a further inner iteration for calculating the differences $\nabla^m f_{n+1}^{(i)}$ for each i. (Problem 19 shows that this last iteration can be avoided, however.) Since the tentative values $y_{n+1}^{(i)}$ are "corrected" at each application of formula (14-40), this formula is called a *corrector formula*. The formula which produces a first approximation $y_{n+1}^{(0)}$ is called a *predictor formula*. The Adams-Moulton method thus belongs to the class of so-called predictor-corrector procedures. Like the Runge-Kutta method, it is one of the most reliable and most widely used procedures for the numerical integration of ordinary differential equations.

Let us convince ourselves that the sequence $\{y_{n+1}^{(i)}\}$ generated by repeated application of the corrector formula would converge for $i \to \infty$. For the purpose of analysis, we express the differences in (14-40) in terms of ordinates and observe that only the foremost values $f_{n+1}^{(i)}$ involve $y_{n+1}^{(i)}$. Setting

$$y_{n+1}^{(i)} = w_i, \qquad i = 0, 1, 2, \ldots,$$

equation (14-40) is of the form

(14-41)
$$w_{i+1} = F(w_i),$$

where

$$F(w) = h(c_0^* + c_1^* + \cdots + c_{k+1}^*) f(x_{n+1}, w) + C,$$

C being a constant independent of w. According to theorem 4.2 the sequence $\{w_i\}$ defined by (14-41) converges—for any choice of the starting value w_0—if there exists a constant $K < 1$ such that, for any two real numbers y and z,

(14-42) $$|F(z) - F(y)| \leqq K|z - y|.$$

Problem 8 of chapter 13 shows that

(14-43) $$c_0^* + c_1^* + \cdots + c_{k+1}^* = c_{k+1},$$

where c_{k+1} is defined by (13-5). Hence if f satisfies a Lipschitz condition with Lipschitz constant L,

$$|F(z) - F(y)| = h|c_{k+1}|\,|f(x_{n+1}, z) - f(x_{n+1}, y)|$$
$$\leqq h|c_{k+1}|L|z - y|.$$

Thus (14-42) is shown to hold with

$$K = h|c_{k+1}|L,$$

and this is less than one whenever

(14-44) $$h < \frac{1}{|c_{k+1}|L}.$$

We thus have obtained:

Theorem 14.7 The inner iteration necessary to obtain each new value y_{n+1} in the Adams-Moulton method converges for any choice of the predicted value $y_{n+1}^{(0)}$ whenever the step h satisfies (14-44).

By theorem 4.2, the quantity

$$y_{n+1} = \lim_{i \to \infty} y_{n+1}^{(i)}$$

is a solution of the equation

(14-45) $$y_{n+1} = y_n + h[c_0^* f_{n+1} + c_1^* \nabla f_{n+1} + \cdots + c_{k+1}^* \nabla^{k+1} f_{n+1}].$$

Theoretically, we could thus define the Adams-Moulton values y_{n+1} as the solution of the non-linear difference equation (14-45). The error $z_n = y_n - y(x_n)$ of these values can be investigated by a method entirely analogous to that used in the proof of theorem 14.6. Expressing differences in terms of ordinates, we write (14-45) in the form

$$y_{n+1} = y_n + h\{b_{k+1,0}^* f_{n+1} + b_{k+1,1}^* f_n + \cdots + b_{k+1,k+1}^* f_{n-k}\}$$

Setting

(14-46) $$B_{k+1}^* = |b_{k+1,0}^*| + |b_{k+1,1}^*| + \cdots + |b_{k+1,k+1}^*|,$$

and assuming that the starting values satisfy (14-32), we can show that

$$(14\text{-}47) \quad |z_n| \leqq \frac{1}{1 - h|c_{k+1}|L} \left\{ \delta e^{(x_n - a)B_{k+1}^* L} \right.$$

$$\left. + |c_{k+2}^*| h^{k+2} Y_{k+3} \frac{e^{(x_n - a)B_{k+1}^* L} - 1}{B_{k+1}^* L} \right\}$$

for $a \leqq x_n \leqq b$. Here L again denotes a Lipschitz constant for the function f. If the starting values are sufficiently accurate, then the error bound (14-47) is considerably better than the corresponding bound (14-38) for the Adams-Bashforth method using equally many points, because h^{k+1} is now replaced by h^{k+2}, and $|c_{k+2}^*| < |c_{k+1}|$, $B_{k+1}^* < B_k$.

Similar bounds can also be obtained for the more practical case where the inner iteration is only performed a finite number I of times, as indicated in algorithm 14.7. It can be shown that the above qualitative conclusions still hold when $I \geqq 2$.

Problems

16. Determine the constants $b_{k,m}$ and $b_{k+1,m}^*$ for $k = 0, 1, 2, 3$ and calculate the values of the constants B_k and B_{k+1}^*.

17. Repeat problem 15, using the Adams-Moulton procedure with $k = 2$.

18. Give a proof of (14-47) along the lines of the proof of theorem 14.6.

19. Show that, by virtue of (14-43), the values $y_{n+1}^{(i)}$ defined in algorithm 14.7 can be calculated more simply from the relation

$$y_{n+1}^{(i+1)} = y_{n+1}^{(i)} + hc_{k+1}[f(x_{n+1}, y_{n+1}^{(i)}) - f(x_{n+1}, y_{n+1}^{(i-1)})], \quad (i \geqq 1).$$

14.8   Numerical Stability

In order to bring the discussions of the two preceding sections to a more concrete level, we now shall determine explicitly the numerical solutions of the initial value problem

$$(14\text{-}48) \qquad\qquad \begin{cases} y(0) = 1, \\ \quad y' = Ay \end{cases}$$

(exact solution: $y(x) = e^{Ax}$) by several methods based on numerical integration.

We begin with the Adams-Bashforth method (algorithm 14.6) for $k = 1$. Since $c_0 = 1$, $c_1 = \frac{1}{2}$, (14-31) yields the recurrence relation

$$y_{n+1} = y_n + h(f_n + \tfrac{1}{2}\nabla f_n)$$
$$= y_n + h(\tfrac{3}{2}f_n - \tfrac{1}{2}f_{n-1})$$

or, since in the present case $f_n = f(x_n, y_n) = Ay_n$,

$$(14\text{-}49) \qquad\qquad y_{n+1} = y_n + Ah(\tfrac{3}{2}y_n - \tfrac{1}{2}y_{n-1}).$$

This is a linear difference equation of order two with constant coefficients. Its characteristic polynomial

$$p(z) = z^2 - (1 + \tfrac{3}{2}Ah)z + \tfrac{1}{2}Ah$$

has the zeros

$$z_1 = \tfrac{1}{2} + \tfrac{3}{4}Ah + \sqrt{\tfrac{1}{4} + \tfrac{1}{4}Ah + \tfrac{9}{16}(Ah)^2}$$

and, by Vieta

$$z_2 = \frac{Ah}{2z_1}.$$

These zeros are certainly distinct for sufficiently small values of h. By §6.3 the general solution is thus given by

$$y_n = c_1 z_1^n + c_2 z_2^n,$$

where c_1 and c_2 are two arbitrary constants. Determining these constants in such a manner that the solution satisfies $y_0 = 1$, we obtain

(14-50) $$y_n = (1 - c_2)z_1^n + c_2 z_2^n,$$

where c_2 is a function of the as yet unspecified value y_1,

(14-51) $$c_2 = \frac{y_1 - z_1}{z_2 - z_1}.$$

As in §14.3, we now shall study the behavior of y_n as $h \to 0$ and $n \to \infty$ while $nh = x$ remains fixed. We begin by expressing z_1^n in such a way that this behavior becomes evident. Expanding the square root in powers of Ah (see Taylor [1959], §15.10) we find

$$z_1 = 1 + Ah + \tfrac{1}{2}(Ah)^2 - \tfrac{1}{4}(Ah)^3 + O(h^4).$$

By virtue of

$$e^{Ah} = 1 + Ah + \tfrac{1}{2}(Ah)^2 + \tfrac{1}{6}(Ah)^3 + O(h^4)$$

this may be written

(14-52) $$z_1 = e^{Ah} - \tfrac{5}{12}(Ah)^3 + O(h^4)$$

or, since $e^{Ah} = 1 + O(h)$,

$$z_1 = e^{Ah}[1 - \tfrac{5}{12}(Ah)^3 + O(h^4)].$$

By virtue of $nh = x$ it follows that

$$z_1^n = e^{Ax}[1 - \tfrac{5}{12}xA^3h^2 + O(h^3)].$$

Since $z_2 = O(h)$, $z_2^n = O(h^n)$ is very small compared with z_1^n. We thus obtain from (14-50), neglecting less important terms,

(14-53) $$y_n \sim e^{Ax} - \tfrac{5}{12}A^3 x e^{Ax} h^2 - c_1 e^{Ax}.$$

Here the first term is clearly the desired exact solution.   The second term arises as a result of approximating the integral in (13-6) by a discrete formula.   In the general case it would be

$$-c_{k+1}A^{k+2}xh^{k+1}e^{Ax}.$$

The third term is a function of the choice of our second starting value y_1. It would be zero if we succeeded in choosing $y_1 = z_1$.   It is small but different from zero, if we take $y_1 = e^{Ah}$, the value of the exact solution at x_1.   The significant fact is that both errors, the one due to discretization as well as the starting error, contain the factor e^{Ax}.   This is particularly important when A is negative, e.g., when $A = -1$.   In this case, the errors do not grow indefinitely; the discretization error, for instance, grows up to $x = 1$ and then tends to zero, like the solution itself.

Let us now examine the integration of the same problem by a different integration formula.   As the error of the interpolating polynomial is smallest near the median interpolating point, one is always tempted to use symmetric central difference formulas for numerical integration.   To consider the simplest case, let us integrate the identity

$$y'(x) = f(x, y(x))$$

between the limits x_{n-1} and x_{n+1}, using the midpoint rule (see problem 4, chapter 13) to approximate the integral.   There results the formula

$$(14\text{-}54) \qquad\qquad y_{n+1} = y_{n-1} + 2hf_n.$$

This formula could be used, in principle, much like the Adams-Bashforth formula.   For the special initial value problem (14-48) we obtain the difference equation

$$y_{n+1} = 2Ahy_n + y_{n-1}.$$

Its characteristic polynomial

$$p(z) = z^2 - 2Ahz - 1$$

has the two distinct zeros

$$z_1 = Ah + \sqrt{1 + (Ah)^2}$$

and

$$z_2 = -\frac{1}{z_1}.$$

All solutions satisfying $y_0 = 1$ can thus again be represented in the form (14-50).

For the purpose of exploring the asymptotic behavior of the solution y_n as $h \to 0$ while $nh = x$ remains fixed, we again expand z_1 in powers of h. Using the binomial formula (see Taylor [1959], p. 479) we obtain

$$z_1 = 1 + Ah + \tfrac{1}{2}(Ah)^2 + 0(h^4)$$

or, by comparison with the exponential series,

$$z_1 = e^{Ah} - \tfrac{1}{6}(Ah)^3 + 0(h^4).$$

Proceeding as above, we thus find

$$z_1^n = e^{Ax}[1 - \tfrac{1}{6}xA^3h^2 + 0(h^3)]$$

and

$$z_2^n = (-z_1)^{-n} = (-1)^n e^{-Ax}[1 + \tfrac{1}{6}xA^3h^2 + 0(h^3)].$$

It is seen that z_1^n and z_2^n have the same order of magnitude, and that the term $c_2 z_2^n$ can now no longer be neglected. We thus find from (14-50), up to less significant terms,

(14-55) $y_n \sim e^{Ax} - \tfrac{1}{6}A^3 x e^{Ax} h^2 + c_2 e^{Ax} + c_2(-1)^n e^{-Ax}.$

In the leading term we again recognize the desired solution of the continuous problem. The second term represents the genuine discretization error, due to the approximation of an integral by a discrete formula. This term is smaller than the corresponding term in (14-53), emphasizing the greater accuracy of central difference formulas. The last two terms are "starting" errors; they owe their presence to the fact that $y_1 \neq z_1$ in general. In particular both terms are present if we take $y_1 = e^{Ah}$, the exact value. These starting errors would not be of much concern if it were not for the fact that the second term $(-1)^n c_2 e^{-Ax}$, shows a behavior whose character is exactly opposite to that of the exact solution. If A is negative, this term will grow at an exponential rate and, no matter how small initially, overshadow all other terms if x is sufficiently large.

EXAMPLE

8. Table 14.8 shows some values of the errors $z_n = y_n - e^{-x_n}$ of the numerical approximations to the solution of $y' = -y$, $y(0) = 1$ obtained by the formulas (14-49) and (14-54) with the step $h = 0.1$.

While for small values of x the Adams-Bashforth method has the larger errors (due to the fact that the discretization error is larger) we have at $x = 5$ reached a point where the error of the midpoint formula is larger by a factor 100, due to the exponentially growing oscillatory component in (14-55).

The presence of oscillatory error terms that grow relative to the exact solution represents a special kind of numerical instability that is typical

Table 14.8

x_n	$10^5 z_n$ Adams	$10^5 z_n$ Midpoint
0.0	0	0
0.1	0	0
0.2	38	30
0.3	70	21
0.4	90	51
$\vdots$		
5.0	15	1129
5.1	14	-1385
5.2	12	1406
5.3	11	-1665
5.4	11	1739

for a number of formulas for integrating differential equations. This phenomenon of numerical instability can be traced to the fact that the characteristic polynomial of the linear difference equation arising from integrating $y' = Ay$ has several zeros of approximate modulus one. This, in turn, will always be the case if the identity $y'(x) = f(x, y(x))$ is integrated over an interval whose length exceeds one integration step.

Problems

20. Carry out the above investigation for the integration algorithm based on Simpson's formula

$$y_{n+1} = y_{n-1} + h(\tfrac{1}{3}f_{n+1} + \tfrac{4}{3}f_n + \tfrac{1}{3}f_{n-1}).$$

Show, in particular, that the starting error now has a component of the form $(-1)^n c_2 e^{-Ax/3}$.

21. Integrating Bessel's interpolating polynomial between the limits x_{-1} and x_2, obtain the following formula for integration:

$$y_{n+2} = y_{n-1} + \tfrac{3}{8}h(f_{n+2} + 3f_{n+1} + 3f_n + f_{n-1}).$$

22*. Investigate the stability of the formula obtained in the preceding problem.

23. In constructing table 14.8, y_1 was chosen as the exact value e^{-h}. Show experimentally that, due to rounding errors, the phenomenon of numerical instability occurs also for $y_1 = z_1 = -h + \sqrt{1 + h^2}$, where theoretically $c_2 = 0$.

Recommended Reading

A large variety of methods for integrating differential equations is discussed in Milne [1953]. Collatz [1960] contains a wealth of valuable information.   The phenomenon of numerical instability of algorithms for the solution of initial value problems was first discussed by Rutis-hauser [1952] in a brief but classical paper.   A comprehensive treatment of errors and numerical stability will be found in Henrici [1962, 1963].   For the numerical solution of boundary value problems and partial differential equations, which had to be ignored here for reasons of space, we refer to the excellent treatises by Fox [1957] and Forsythe and Wasow [1960].

Research Problem

Study the effectiveness of *repeated* extrapolation to the limit in the numerical solution of differential equations.   In particular, compare the accuracy obtained with repeated extrapolation to the limit in the Euler method with that given by the Runge-Kutta method and the Adams-Moulton method, using the same number of evaluations of f.

PART THREE

COMPUTATION

chapter 15 number systems

All numbers that have occurred so far in the theoretical discussions in this book were real (or complex) numbers in the strict mathematical sense. That is, they were to be conceived as infinite decimal fractions, or as Dedekind cuts. For the purposes of computation such numbers have to be approximated by real numbers of a rather special type, such as terminating decimal fractions, or other rational numbers. The present chapter is devoted to a study of the number systems that can be used for the purposes of computation.

15.1 Representation of Integers

Let us take a look at our conventional number system. What do we mean by a symbol such as 247? Evidently

$$247 = 2 \cdot 100 + 4 \cdot 10 + 7$$
$$= 2 \cdot 10^2 + 4 \cdot 10^1 + 7 \cdot 10^0.$$

The number 247 is represented as a polynomial in the base 10, with integral coefficients between 0 and $9 = 10 - 1$.

There is no intrinsic reason why 10 should be used as a base; the number of fingers may have to do with it. There is evidence that in cultures different from ours other number systems have been used. The French word *quatre-vingts* for the number 80 indicates a system with base 20. (Maybe the French counted with their toes as well as with their fingers.) In New Zealand, words for 11^2 and 11^3 have been found. The Babylonian astronomers used a sexagesimal system, i.e., a system with the base 60. A trace of this can be found in our dividing the circumference of the circle into 360 degrees. Also mixed systems, although mathematically much less satisfying, are in use, such as the Anglo-Saxon system for measuring length, and the English monetary system.

In electronic computation, the digits of an integer are represented by various states of a physical quantity, such as an electric current. The technically simplest situation arises when there are only two states to be represented, such as the state "no current" and the state "a unit current." For this reason, modern electronic computers work internally almost exclusively with the base 2. The resulting number system is called the *binary number system*. In this system, only the digits 0 and 1 occur. In order to distinguish them from decimal digits, we shall underline them. Thus, if a given nonnegative integer N is in the binary system represented in the form

$$(15\text{-}1) \qquad N = a_n 2^n + a_{n-1} 2^{n-1} + \cdots + a_1 2 + a_0$$

where the a_1 are either zero or one, it will be written in the form

$$\underline{a_n a_{n-1} \ldots a_1 a_0}.$$

EXAMPLE

1. $1 = \underline{1}, 2 = \underline{10}, 3 = \underline{11}, \underline{101} = 5, 8 = \underline{1000}, \underline{1010} = 10.$

If we wish to communicate with a computer working in the binary system, we (or the computer) must be able to convert a number from the decimal to the binary system and conversely. To convert from binary to decimal, we regard the number N given by (15-1) as the value of the polynomial

$$P(x) = a_n x^n + a_{n-1} x^{n-1} + \cdots + a_1 x + a_0$$

for $x = 2$. To evaluate $P(2)$, we may use algorithm 3.4. (Note that the coefficients of the polynomial are numbered differently now.) It follows that if we calculate the numbers b_k recursively by

$$(15\text{-}2) \qquad b_0 = a_n, \qquad b_k = a_{n-k} + 2b_{k-1} \qquad (k = 1, 2, \ldots, n),$$

then $b_n = P(2) = N$.

EXAMPLE

2. To express the number $N = \underline{11111001111}$ in decimal. The Horner scheme yields

k	0	1	2	3	4	5	6	7	8	9	10
a_{n-k}	1	1	1	1	1	0	0	1	1	1	1
b_k	1	3	7	15	31	62	124	249	499	999	1999

It follows that $N = 1999$.

To convert a given integer from decimal to binary, we make use of the fact that the *last* binary digit a_0 of an integer N is zero if and only if N is even.   The second binary digit a_1 is zero if and only if $(n - a_0)/2$ is even, and so on.   This leads to the following scheme:

Algorithm 15.1   To find the binary representation (15-1) of a given positive integer N let

$$(15\text{-}3) \qquad \begin{cases} N_0 = N, \\ N_{k+1} = \dfrac{N_k - a_k}{2}, & k = 0, 1, 2, \ldots \end{cases}$$

where

$$(15\text{-}4) \qquad a_k = \begin{cases} 1, & \text{if } N_k \text{ is odd}, \\ 0, & \text{if } N_k \text{ is even}. \end{cases}$$

Continue until $N_k = 0$.

EXAMPLE
3. To express $N = 1999$ in binary form.   Algorithm 15.1 yields the scheme

k	0	1	2	3	4	5	6	7	8	9	10
N_k	1999	999	499	249	124	62	31	15	7	3	1
a_k	1	1	1	1	0	0	1	1	1	1	1

It follows that $1999 = 11111001111$.   (Note that the *least* significant digits are obtained first.)   The scheme is an exact reversal of the scheme of example 2.

Problems

1. Express the numbers 1685, 1770, 1882 in the binary system.
2. What are the decimal values of the binary numbers $1'000$, $1'000'000$, $1'000'000'000$?
3. What is, for $N \to \infty$, the ratio of the number of digits in the binary and the decimal representation of an integer N?
4. State the rules for the conversion of a decimal integer N into the ternary system, and vice versa.   Express 10^n in the ternary system ($n = 1, 2, 3$).
5. What is the representation of the positive integer N in the number system to the base N?

15.2   Binary Fractions

A *binary fraction* is a series of the form

$$(15\text{-}5) \qquad z = \sum_{k=1}^{\infty} a_{-k} 2^{-k},$$

where the coefficients $a_{-1}, a_{-2}, \ldots$ are either zero or one. The series
(15-5) always converges, because it is majorized by the geometric series

$$\sum_{k=1}^{\infty} 2^{-k} = \frac{1}{2} \frac{1}{1 - \frac{1}{2}} = 1.$$

The sum z of (15-5) will also be denoted by

$$z = 0.a_{-1}a_{-2}a_{-3}\ldots.$$

The binary fraction (15-5) is said to *terminate* if, for some integer n,
$a_k = 0, k > n$.

The following theorem is fundamental, but will not be proved:

Theorem 15.2a   Any real number z such that $0 < z \leq 1$ can be
represented in a unique manner by a nonterminating binary fraction.

If we drop the condition that the binary fraction shall not terminate, then
the representation may not be unique; for instance, the binary fractions

$$0.1 \quad \text{and} \quad 0.01111\ldots$$

both represent the number 0.5.

A terminating binary fraction $z = 0.a_{-1}a_{-2}\ldots a_{-n}$ can be regarded as
the value of the polynomial

$$P(x) = a_{-1}x + a_{-2}x^2 + \cdots + a_{-n}x^n$$

at $x = \frac{1}{2}$ and thus can be evaluated by algorithm 3.4 (Horner's scheme) as
follows: Let

$$b_0 = a_{-n}, \qquad b_k = a_{-n+k} + \tfrac{1}{2}b_{k-1}, \qquad k = 1, 2, \ldots, n,$$

where $a_0 = 0$.   Then $b_n = z$.

EXAMPLE

4.   To express $z = 0.00110011$ in decimal.   Horner's scheme yields

k	a_{n-k}	b_k
0	1	1
1	1	1.5
2	0	0.75
3	0	0.375
4	1	1.1875
5	1	1.59375
6	0	0.796875
7	0	0.3984375
8	0	0.19921875

It follows that $z = 0.19921875$.

Another method for converting a terminating binary fraction consists in converting the integer

$$2^n z = a_{-1}2^{n-1} + a_{-2}2^{n-2} + \cdots + a_{-n}$$

and dividing the result by 2^n.

Except in special circumstances, non-terminating binary fractions cannot be converted into terminating decimal fractions. To get an approximate decimal representation, we truncate an infinite binary fraction after the nth digit and convert the resulting terminating fraction. The error in this approximation will be less than 2^{-n}.

The inverse problem of converting a given (decimal) fraction into a binary fraction is solved by the following algorithm:

Algorithm 15.2   For a real number z such that $0 \leq z \leq 1$, calculate the sequences $\{z_k\}$ and $\{a_{-k}\}$ recursively by the relations

$$z_1 = z;$$

(15-6)
$$a_{-k} = \begin{cases} 1, & \text{if } 2z_k > 1, \\ 0, & \text{if } 2z_k \leq 1, \end{cases}$$

$$z_{k+1} = 2z_k - a_{-k}, \qquad k = 1, 2, \ldots.$$

Theorem 15.2b   For the sequence $\{a_{-k}\}$ defined by (15-6),

$$z = 0.a_{-1}a_{-2}a_{-3}\cdots.$$

Proof.   According to theorem 15.2a, z has a nonterminating binary representation of the form

$$z = 0.b_{-1}b_{-2}b_{-3}\cdots;$$

we have to show that

(15-7)
$$b_{-k} = a_{-k}, \qquad k = 1, 2, \ldots,$$

where a_{-k} is defined by (15-6).   Incidentally we shall also show that

(15-8)
$$z_k = 0.b_{-k}b_{-k-1}b_{-k-2}\cdots, \qquad k = 1, 2, \ldots.$$

The simultaneous proof of the two assertions is by induction.   Clearly, (15-8) is true for $k = 1$.   Let us assume that (15-7) and (15-8) are true for the integers $k - 1$ and k, where $k \geq 1$.   We then have

$$2z_k = b_{-k} + 0.b_{-k-1}b_{-k-2}\cdots.$$

Since the binary fraction on the right is positive and bounded by one, it follows that $2z_k > 1$, and hence $a_k = 1$, if and only if $b_k = 1$.   This establishes (15-7), and, by the second formula (15-6), (15-8) with k increased by one.

EXAMPLE

5. To express $z = \frac{1}{5}$ as a binary fraction. We have

k	z_k	$2z_k$	a_{-k}
1	0.2	0.4	0
2	0.4	0.8	0
3	0.8	1.6	1
4	0.6	1.2	1
5	0.2		

Since $z_k = 0.2$ has occurred before, the periodic† binary fraction

$$0.2 = 0.001100110011\ldots = 0.\overline{0011}$$

results. (The period is indicated by a cross-bar.)

It is now easy to represent any nonnegative number in the binary system. If z is such a number, let $[z]$ be the greatest integer not exceeding z. According to §15.1 we can write

$$[z] = \sum_{k=0}^{n} a_k 2^k,$$

where the a_k are either zero or one. For the fractional part we have, by (15-5)

$$z - [z] = \sum_{k=1}^{\infty} a_{-k} 2^{-k}$$

with the same limitation on the a_k. Thus

$$z = [z] + (z - [z]) = \sum_{k=-n}^{\infty} a_{-k} 2^{-k}$$

$$= a_n a_{n-1} \ldots a_0 . a_{-1} a_{-2} \ldots .$$

To convert a binary number into a decimal or vice versa, we must convert the integral and the fractional part by themselves.

Problems

6. Give the binary value of π, correct to 12 binary digits after the binary point.

7. Express the binary numbers (a) 0.10101, (b) $0.\overline{10101}$ as ratios of two integers.

8. Determine the representation of $\frac{1}{3}$ in the number systems to the base (a) 2, (b) 5, (c) 7.

---

† It is shown in number theory that every rational fraction can be represented by an infinite binary fraction that is ultimately periodic.

9. Devise an algorithm that yields directly the binary representation of the square root of a given positive binary number. Use your algorithm to determine the binary representations of $\sqrt{2}$, $\sqrt{\pi}$, $\sqrt{5}$, and check your result by converting the known decimal representations of these numbers.

15.3 Fixed Point Arithmetic

As mentioned earlier, in most digital computing machines today numbers are internally represented in the binary system. Each element of a computer can assume only a finite number of (recognizable and repro-ducible) states, and each computer has only finitely many elements. It is thus clear that only finitely many different numbers can be represented in a computer. A number that can be represented exactly in a given computer is called a *machine number*. All other numbers can be represented in the computer only, at best, approximately. There are two principal ways in which a given number z can be represented. They are known as the *fixed point* and as the *floating point* representation of z.

In fixed point operation, the machine numbers are terminating binary fractions of the form

$$(15\text{-}9) \qquad\qquad \pm \sum_{k=1}^{t} a_{-k} 2^{-k}$$

where t is an integer that is either fixed for a given machine ($t = 35$ for the IBM 7090, $t = 48$ for the CDC 1604) or, in some cases, can be selected by the user (e.g., on the IBM 1620). The numbers used by the machine can also be described as the set of numbers $n \cdot 2^{-t}$, where n is an integer, $|n| < 2^t$. The density of machine numbers is uniform in the interval $(-1, 1)$ and zero outside this interval.

If z is any real number, we shall denote by z^* one of the (at most two) terminating binary fractions of the form (15-9) for which $|z - z^*|$ is a minimum. If the interval

$$R = [-1 + 2^{-t-1}, 1 - 2^{-t-1}]$$

is called the *range* of the machine, then

$$(15\text{-}10) \qquad\qquad |z - z^*| \leqq 2^{-t-1} \quad \text{for all } z \in R,$$

that is, any real number within the range of the machine can be approxi-mately represented by a machine number z^* with an error of at most 2^{-t-1}. Any of the (at most two) representations z^* of a number z will be called a *correctly rounded fixed point representation of z*.

If a and b are two numbers within the range of the machine, the numbers $a \pm b$ do not necessarily belong to the range. If they do not belong to the range we say that *overflow* occurs. The ever-present possibility of overflow is one of the serious disadvantages of fixed point arithmetic.

The above remark shows very clearly that even if the data of a computational problem are within range, we can usually not be sure that all intermediate results fall within the range. It is frequently possible, however, to obtain a (possibly very crude) a priori estimate of the size of the intermediate results. One then may try to reformulate the problem (for instance by introducing new units of measurement) in such a manner that the data as well as the intermediate and final results are within range. This reformulation is known as *scaling*. The frequent necessity of scaling is a further disadvantage of fixed point arithmetic.

Let us now consider the accuracy of the arithmetical operations in fixed point arithmetic. If a and b are two machine numbers, and if $a + b$ is within range, then it is clear that $a + b$ is also a machine number. That is, *addition* can be performed without error in a fixed point machine, i.e., we have

$$(15\text{-}11) \qquad (a + b)^* = a + b, \qquad \text{if } a + b \in R.$$

The same is not true of *multiplication*, however. If a and b are two machine numbers, then ab is, in general, not a machine number. However $ab \in R$, hence there exists a correctly rounded machine value of ab, and we have

$$(15\text{-}12) \qquad |(ab)^* - ab| \leq 2^{-t-1}.$$

The *ratio a/b* of two machine numbers is, in general, not in the range of the machine. (The probability for this to be the case is only 50 per cent—a further disadvantage of fixed point arithmetic.) If it is, we again have

$$|(a/b)^* - a/b| \leq 2^{-t-1},$$

independently of the size of $|a/b|$.

Many machines have built-in subroutines for calculating values of elementary functions such as $\sin x$ or $\sqrt{x}$. It is clear that even if the argument x is a machine number, the value of the function is, in general, not. In the case of the sine function, the best we can hope for in general is a correctly rounded value of $\sin x$, but even this ideal is seldom attained. The author knows of a machine where $\sqrt{0}$ was 2^{-t}.

Problems

10. Devise an algorithm for forming the sum of two positive machine numbers of the form (15-9).
11. Reformulate the algorithm obtained in problem 9 in such a manner that it yields, for every nonnegative machine number x, a correctly rounded value of $\sqrt{x}$.

12. Devise an algorithm that yields the correctly rounded product of two machine numbers a and b.

15.4 Floating Point Arithmetic

In floating point operation, the set of machine numbers consists of 0 and of the set of all numbers of the form

(15-13)
$$z = \pm m2^p$$

where m is a terminating binary fraction,

$$m = \sum_{k=1}^{t} a_{-k}2^{-k}, \qquad \tfrac{1}{2} \leq m < 1$$

normalized by the condition $a_{-1} = 1$, and where p is an integer ranging between $-P$ and P, say. The integers t and P are normally fixed for a given machine. (On the IBM 7090 computer, $t = 27$, $P = 128$.) The binary fraction m is called the *mantissa* and the integer p the *exponent* of the floating number (15-13). The machine numbers now cover a much wider range than in fixed point arithmetic. We again denote, for any real z and for a given machine, by $z^\oplus$ any of the at most two numbers of the form (15-13) for which $|z - z^\oplus|$ is minimized. We now define the range of the machine to be the interval

$$R = [-2^P(1 - 2^{-t-1}), \ 2^P(1 - 2^{-t-1})].$$

If $z \in R$, $|z| \geq 2^{-P-1}$, we can define $z^\oplus$ by

(15-14)
$$z^\oplus = \operatorname{sign} z \cdot m \cdot 2^p$$

where

$$p = [\log_2 |z|] + 1, \qquad m = (2^{-p}|z|)^*.$$

Here $\log_2$ denotes the logarithm to the base 2, $[x]$ denotes the greatest integer not exceeding x, and the * refers to correct rounding in fixed point arithmetic. Any of the (at most two) values $z^\oplus$ will be called a correctly rounded floating representation of z. If $|z| < 2^{-P-1}$, we set $z^\oplus = 0$.

If $|z| \geq 2^{-P-1}$, we evidently have

$$
\begin{aligned}
|z - z^\oplus| &\leq 2^p 2^{-t-1} \\
&= 2^{p-1} 2^{-t} \\
&= 2^{[\log_2|z|]} 2^{-t} \\
&\leq 2^{\log_2|z|} 2^{-t}
\end{aligned}
$$

and hence

(15-15)
$$|z - z^\oplus| \leq 2^{-t}|z|.$$

This relation shows:

Theorem 15.4   If, for a given floating point machine, $z \in R$, $|z| \geq 2^{-P-1}$, then the relative error of the correctly rounded floating representation $z^{\oplus}$ of z is at most 2^{-t}.

For $z \to 0$ the relative error tends to ∞, as in fixed point arithmetic.

Since machines with floating arithmetic can handle very large numbers with a small relative error, scaling is, in general, not necessary, and the possibility of overflow is minimal.   For these reasons, floating arithmetic is almost universally used today.   The FORTRAN system, for instance, employs floating arithmetic.

Let us examine briefly the accuracy of the basic arithmetic operations in floating point arithmetic.   If a and b are two floating machine numbers, and if $a + b$ is in the range of the machine, then, unlike the fixed point case, $a + b$ is not necessarily a machine number.   Thus we must also expect rounding errors in addition.   If

$$a = \pm m2^{p}, \qquad b = \pm n2^{q},$$

and if $p \geq q$, we have

$$a + b = \pm m2^{p} \pm n2^{q}$$
$$= (\pm m \pm n2^{q-p})2^{p}.$$

Here $|\pm m \pm n2^{q-p}| < 2$; thus we have, at worst,

$$(a + b)^{\oplus} = \left(\frac{\pm m \pm n2^{q-p}}{2}\right)^{*}2^{p+1}$$

and hence

$$|(a + b)^{\oplus} - (a + b)| \leq 2^{-t-1}2^{p+1}$$
$$= 2^{-t+1}2^{\log_{2}|a|}$$

and consequently

(15-16) $$|(a + b)^{\oplus} - (a + b)| \leq 2^{-t+1}|a|.$$

Thus the error of the machine value of a sum (or difference) is at most 2^{-t+1} times the absolute value of the larger summand.

The *product ab* of two machine numbers a and b is now (contrary to fixed-point arithmetic) not always a number in the range of the machine. If the absolute value of the product lies in the interval $[2^{-P}, 2^{P}]$, we find by considerations similar to those just given that

(15-17) $$|(ab)^{\oplus} - ab| \leq 2^{-t}|ab|.$$

Similarly we find for division, if a/b is in the range of the machine,

(15-18) $$\left|\left(\frac{a}{b}\right)^{\oplus} - \frac{a}{b}\right| \leq 2^{-t}\left|\frac{a}{b}\right|.$$

Concerning built-in subroutines for elementary functions, the remarks made above apply also to machines with floating point arithmetic.

Multiple precision arithmetic. As was noted above, both fixed point numbers and the mantissae of floating numbers are represented in the machine in the form $N \times u$, where the number $u = 2^{-t}$ may be called the *basic unit* of the machine, and where N is an integer, $|N| < 2^t$. For some machines and/or some problems this approximation is not sufficiently accurate. In such cases one can increase the accuracy by working instead with numbers of the form

(15-19) $N_1 u + N_2 u^2 + \cdots + N_q u^q,$

where the N_k are integers satisfying $0 \leq N_k < u^{-1}$, $k = 1, 2, \ldots, q$. This enlargement of the set of available machine numbers is called working with *q-fold precision.* Double precision ($q = 2$) is fairly common; on some machines, its use is encouraged by built-in circuitry to facilitate arithmetical operations with multiple precision numbers of the form (15-19). Multiple precision with $q > 2$ is normally restricted to special experimental programs. From a mathematical point of view, multiple precision is equivalent to single precision with t replaced by qt.

Problems

13. Prove the inequalities (15-17) and (15-18).

14. Assume that in the approximate formula of numerical differentiation

$$D_h f(x_0) = \frac{f_1 - f_{-1}}{2h},$$

f_1 and f_{-1} are known only with errors $\leq 2^{-16}$, whereas h is an exact binary number.

(a) If floating operations are used, how big is (at most) the resulting error in $D_h f(x_0)$?

(b) If the function to be differentiated is $f(x) = J_0(x)$, for which value of h is the maximum error in the numerical value of $J_0'(x)$ due to rounding equal to the maximum error due to numerical differentiation?

(c) For which value is the sum of the two errors a minimum?

Suggested Reading

Wilkinson [1960] gives a careful account of rounding errors in the elementary arithmetic operations. For a wealth of detail from the point of view of computer technology see Speiser [1961].

Research Problem

Assume that in a fixed point machine it is possible to identify both the more and the less significant half of the exact value of the product of two machine numbers. Formulate an algorithm for obtaining the correctly rounded value of the product of any two complex numbers whose real and imaginary parts are machine numbers.

chapter 16 propagation of rounding error

16.1 Introduction and Definitions

The process of replacing a real number z by a machine number z^* or $z^\oplus$ is called *rounding*. The difference $z^* - z$ or $z^\oplus - z$ is called *rounding error*. Due to rounding, the final result of a numerical computation usually differs from the theoretically correct result. The difference of the numerical result and the theoretically correct result is called the *accumulated rounding error*. In order to distinguish them from the accumulated rounding error, the rounding errors committed at each step of a computation will also be called the *local rounding errors*. Except in simple cases, the accumulated rounding error is not merely the sum of the local errors. Each local rounding error is propagated through the remaining part of the computation, during which process its influence on the accumulated rounding error may either be amplified or diminished.

Different numerical procedures for solving the same theoretical problem (such as integrating a differential equation) may show a different behavior with respect to the propagation of local rounding error. This varying sensitivity with regard to rounding off operations is frequently referred to as the *numerical stability* of the process. In chapter 8 we noted that the numerical stability even of one and the same numerical procedure (the QD algorithm) can depend strongly on the way the arithmetical operations are performed.

In the present chapter we shall study the propagation of local rounding error in some simple but typical cases.

16.2 Finite Differences

As a first example we consider a case where the accumulated rounding error is due exclusively to errors in the data of the computation; all intermediate arithmetic operations are performed exactly. Let $x_n =$

302

$a + nh$, $n = 0, 1, 2, \ldots$, let f be a given function, and let $f_n = f(x_n)$. We consider the effect of replacing the exact values f_n by machine values f_n^* on the differences of the sequence of values f_n.

In fixed point arithmetic, we clearly have

(16-1) $$f_n^* = f_n + \varepsilon_n, \qquad n = 0, 1, 2, \ldots$$

where the local rounding errors ε_n satisfy

(16-2) $$|\varepsilon_n| \leq \varepsilon = \tfrac{1}{2}u, \qquad n = 0, 1, 2, \ldots$$

u being the basic unit of the machine (or the table of f). According to (6-27)

$$\Delta^k f_n = \sum_{m=0}^{k} (-1)^{k-m} \binom{k}{m} f_{n+m},$$

$$\Delta^k f_n^* = \sum_{m=0}^{k} (-1)^{k-m} \binom{k}{m} f_{n+m}^*.$$

Hence, in view of (16-1), if we denote by $r_n^{(k)}$ the accumulated rounding error in the kth difference,

$$r_n^{(k)} = \Delta^k f_n^* - \Delta^k f_n = \Delta^k(f_n^* - f_n)$$
$$= \sum_{m=0}^{k} (-1)^{k-m} \binom{k}{m} \varepsilon_{n+m}.$$

It follows by (16-2) that

$$|r_n^{(k)}| \leq \sum_{m=0}^{k} \binom{k}{m} |\varepsilon_{n+m}|$$
$$\leq \varepsilon \sum_{m=0}^{k} \binom{k}{m}.$$

In view of

$$\sum_{m=0}^{k} \binom{k}{m} = (1 + 1)^k = 2^k,$$

we obtain for $r_n^{(k)}$ the bound

(16-3) $$|r_n^{(k)}| \leq 2^k \varepsilon.$$

This bound cannot be improved in general; equality holds whenever

$$\varepsilon_{n+m} = (-1)^m \varepsilon.$$

EXAMPLE

1. We consider an excerpt of a table of sines, tabulated with a step $h = 0.01$ to six decimal places. In view of

$$\Delta^k y_n = h^k y^{(k)}(\xi), \qquad x_n < \xi < x_{n+k}$$

(see problem 11, chapter 11), the exact differences satisfy

$$|\Delta^k y_n| \leq 10^{-2k}, \qquad k = 0, 1, 2, \ldots$$

and are zero from Δ^4 on to the number of digits given. The following table shows the numerical differences $\Delta^k y_n^*$:

Table 16.1

x_n	$y_n^* = (\sin x_n)^*$	Δy_n^*	$\Delta^2 y_n^*$	$\Delta^3 y_n^*$	$\Delta^4 y_n^*$	$\Delta^5 y_n^*$
0.25	0.247 404					
0.26	0.257 081	9677	−27			
0.27	0.266 731	9650	−25	2	−6	
0.28	0.276 356	9625	−29	−4	5	11
0.29	0.285 952	9596	−28	1	−2	−7
0.30	0.295 520	9568	−29	−1	−1	1
0.31	0.305 059	9539	−31	−2	1	2
0.32	0.314 567	9508	−32	−1	1	0
0.33	0.324 043	9476	−32	0	1	−2
0.34	0.333 487	9444	−33	−1	−1	
0.35	0.342 898	9411				

The example shows that the local rounding errors ε_n, although hardly noticeable in the function values f_n themselves, show up very strongly in the higher differences. The propagation of rounding error is real.

Problems

1. A tablemaker wishes to add, in a table of $f(x) = J_0(x)$ giving eight decimal digits after the decimal point, values of $\delta^2 f_n$ and $\delta^4 f_n$ that differ from the exact values by at most $0.6 \cdot 10^{-8}$. To what accuracy does $J_0(x)$ have to be calculated? Does the result depend on the step h or on the function f?
2. *The impossibility of making a correctly rounded table.* A recurrent nightmare of tablemakers is the following situation. Assume a table of a function f is to be prepared, giving six digits after the decimal point. For a certain value of x_n, a very accurate computation yields

$$f(x_n) = 0.123456499996,$$

with an uncertainty of $7 \cdot 10^{-12}$. Should this value be rounded up or down? A classical result in number theory states that if α is an irrational number (such as $\sqrt{2}$ or e), the numbers

$$\alpha n - [\alpha n], \qquad n = 0, 1, 2, \ldots$$

come arbitrarily close to every number in the interval $(0, 1)$. Show that this result implies the following: Given any two positive integers N and M, and any finite procedure for calculating square roots, there always exists an interval such that it is impossible to construct a correctly rounded table of the function $f(x) = \sqrt{2}\,x$ with the step 10^{-M} in that interval, giving N digits after the decimal point.

16.3 Statistical Approach

Example 1 above shows that the propagated rounding error in a table of differences can be considerable. It also shows that the errors in the differences probably rarely have the maximum values given by (16-3) (8 and 16 for the last two difference columns). This is due to the fact that consecutive local rounding errors ε_n only rarely have the maximum value permitted by (16-2) and, in addition, occur with strictly alternating signs such as $\varepsilon, -\varepsilon, \varepsilon, -\varepsilon, \ldots$. Such an occurrence would contradict the intuitive notion that the local rounding errors are, somehow, distributed in a random fashion.

It is plain that, on a given machine and for a given problem, the local rounding errors are not, in fact, random variables. If the same problem is run on the same machine a number of times, there will result always the same local rounding errors, and therefore also the same accumulated error. We may, however, adopt a stochastic *model* of the propagation of rounding error, where the local errors are treated *as if they were* random variables. This stochastic model has been applied in the literature to a number of different numerical problems and has produced results that are in complete agreement with experimentally observed results in several important cases.

The most natural assumption concerning the local rounding error in the stochastic model seems to be the following: We assume that the local rounding errors are uniformly distributed between their extreme values $-\varepsilon$ and ε. The probability density $p(x)$—defined as $(\Delta x)^{-1}$ times the probability that the error lies between x and $x + \Delta x$—of this random distribution is given by

$$(16\text{-}4) \qquad p(x) = \begin{cases} 0, & x < -\varepsilon, \\ c, & -\varepsilon \leq x \leq \varepsilon, \\ 0, & x > \varepsilon. \end{cases}$$

The constant c is determined by the condition that the sum of all probabilities, that is, the integral

$$\int_{-\infty}^{\infty} p(x)\,dx$$

equals 1. This yields the value

$$c = \frac{1}{2\varepsilon}$$

in (16-4) (see Fig. 16.3a).

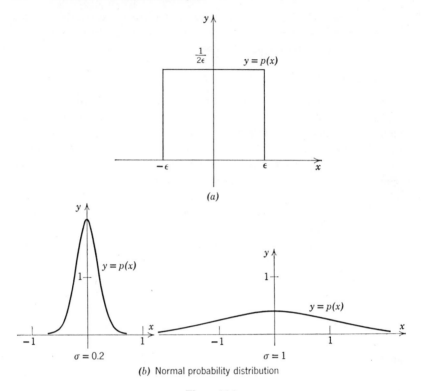

(a)

(b) Normal probability distribution

Figure 16.3

For theoretical purposes it is more convenient to assume that the local rounding errors are *normally distributed*. The normal distribution is defined by the probability density

$$(16\text{-}5) \qquad p(x) = \frac{1}{\sqrt{2\pi}\,\sigma}\, e^{-(x^2/2\sigma^2)}.$$

Here σ is a parameter, called the *standard deviation* of the distribution, that measures the spread of the distribution. For small values of σ the distribution is narrowly concentrated around $x = 0$, with a sharp peak at $x = 0$. For large values of σ the distribution is flattened out (see Fig. 16.3b).

The absolute value of a normally distributed random variable exceeds σ in only 31.7 per cent of all cases and 2.576σ in only 1 per cent of all cases.

Both distributions considered above have mean value 0. For a random variable ξ with arbitrary probability density $p(x)$, the *mean or expected value* is defined by

$$(16\text{-}6) \qquad E(\xi) = \int_{-\infty}^{\infty} xp(x)\, dx.$$

The *standard deviation* is defined as the square root of the *variance* of ξ, which in turn is defined as the mean of the square of the deviation of ξ from its mean.   Thus, if $\mu = E(\xi)$,

$$(16\text{-}7) \qquad \text{var}(\xi) = \int_{-\infty}^{\infty} (x - \mu)^2 p(x)\,dx; \qquad \text{s.d.}(\xi) = \sqrt{\text{var}(\xi)}.$$

EXAMPLES

2.   The variance of the distribution defined by (16-4) is

$$\int_{-\varepsilon}^{\varepsilon} x^2 \frac{1}{2\varepsilon}\,dx = \frac{\varepsilon^2}{3}.$$

3.   The variance of the normal distribution is

$$\frac{1}{\sqrt{2\pi}\,\sigma} \int_{-\infty}^{\infty} x^2 e^{-x^2/2\sigma^2}\,dx = \frac{2\sigma^2}{\sqrt{\pi}} \int_{0}^{\infty} t^{1/2} e^{-t}\,dt = \sigma^2.$$

Thus, if we wish the distributions (16-4) and (16-5) to have the same standard deviation, we must choose

$$\sigma = \frac{\varepsilon}{\sqrt{3}}.$$

If $\xi_1, \xi_2, \ldots, \xi_m$ are random variables with means $\mu_1, \mu_2, \ldots, \mu_m$, and if $a_1, a_2, \ldots, a_m$ are arbitrary constants, the quantity

$$(16\text{-}8) \qquad \xi = a_1\xi_1 + a_2\xi_2 + \cdots + a_m\xi_m$$

is again a random variable, and its mean is given by

$$(16\text{-}9) \qquad E(\xi) = a_1\mu_1 + a_2\mu_2 + \cdots + a_m\mu_m.$$

The importance of the normal probability distribution (16-5) is based on the following fact: If the variables $\xi_1, \xi_2, \ldots, \xi_m$ are *independent* (i.e., if the value of any ξ_i has no bearing on the value of any other ξ_j), and if they are normally distributed with standard deviations $\sigma_1, \sigma_2, \ldots, \sigma_m$, then the variable defined by (16-8) is also normally distributed, and its standard deviation is given by

$$(16\text{-}10) \qquad \text{s.d.}(\xi) = \sqrt{a_1^2\sigma_1^2 + a_2^2\sigma_2^2 + \cdots + a_m^2\sigma_m^2}.$$

In a limiting sense, the above statement is also true if the random variables ξ_j themselves are not normally distributed.

Let us now apply the above results to the problem of forming differences considered in §16.2. Assuming that the local rounding errors ε_n are independent, normally distributed random variables with mean zero and standard deviation σ, we find that $r_n^{(k)}$ is a normally distributed variable with variance

$$\text{var}\,(r_n^{(k)}) = \sigma^2 \sum_{m=0}^{k} \binom{k}{m}^2.$$

Comparing coefficients of x^k in the identity

$$\left[1 + \binom{k}{1}x + \cdots + \binom{k}{k}x^k \right]\left[\binom{k}{k} + \binom{k}{k-1}x + \cdots + \binom{k}{0}x^k \right]$$
$$= 1 + \binom{2k}{1}x + \binom{2k}{2}x^2 + \cdots + \binom{2k}{2k}x^{2k},$$

we find that

$$\sum_{m=0}^{k} \binom{k}{m}^2 = \binom{2k}{k} = \frac{(2k)!}{(k!)^2}.$$

We thus obtain

$$\text{var}\,(r_n^{(k)}) = \frac{(2k)!}{(k!)^2}\,\sigma^2,$$

$$\text{s.d.}\,(r_n^{(k)}) = \frac{\sqrt{(2k)!}}{k!}\,\sigma.$$

In order to obtain an intuitive notion of the size of the coefficient appearing here, we approximate the factorials by Stirling's formula (see §9.3):

$$\frac{(2k)!}{(k!)^2} \sim \frac{\sqrt{4\pi k}\left(\dfrac{2k}{e}\right)^{2k}}{2\pi k\left(\dfrac{k}{e}\right)^{2k}} = \frac{2^{2k}}{\sqrt{\pi k}}.$$

This yields

(16-11) $$\text{s.d.}\,(r_n^{(k)}) \sim \frac{2^k\sigma}{\sqrt[4]{\pi k}} = \frac{2^k\varepsilon}{\sqrt{3}\sqrt[4]{\pi k}}$$

in the case of the rectangular distribution (16-4). This relation shows that the ratio of the standard deviation and the theoretical maximum $2^k\varepsilon$ is not as small as might be expected.

EXAMPLE

4. Twenty-four consecutive values of the exponential function given in Comrie's table (Comrie [1961]) gave the following experimental values of the standard deviations of the differences $\Delta^k y_n$ (in units of the least significant digit):

Table 16.3

k	3	4	5	6
experimental s.d.	1.31	2.28	4.22	7.64
theoretical s.d. (16-11)	1.29	2.42	4.57	8.78

Problem

3. For n given machine numbers $a_1, a_2, \ldots, a_n$ the sums

$$s_k = a_1^2 + a_2^2 + \cdots + a_k^2,$$
$$s_k' = s_1 + s_2 + \cdots + s_k \qquad (k = 1, 2, \ldots, n)$$

are formed by means of the recurrence relations

$$s_0 = 0, \quad s_k = s_{k-1} + a_k^2; \qquad s_0' = 0, \quad s_k' = s_{k-1}' + s_k$$

$(k = 1, 2, \ldots, n)$. What is the resulting standard deviation in s_n and s_n', if the rounding errors at each step are considered independent random variables with the distribution (16-4)? What are the theoretically possible maximum errors?

16.4   A Scheme for the Study of Propagated Error

In the example considered above, the accumulated error was due exclusively to the rounding of the initial data of the computation. All intermediate computations were performed exactly. We shall now discuss a scheme of fairly wide applicability which permits us to take into account rounding in the intermediate results.

Many numerical algorithms consist in generating a sequence of numbers $q_0, q_1, q_2, \ldots$ that are defined by recurrence relations of the form

$$(16\text{-}12) \qquad q_n = Q_n(q_0, q_1, q_2, \ldots, q_{n-1}), \qquad n = 1, 2, \ldots.$$

In actual numerical computation not the numbers q_n are generated, but certain machine numbers $\tilde{q}_n$ (not necessarily the correctly rounded values q_n^*!), which satisfy the recurrence relations

$$(16\text{-}13) \qquad \tilde{q}_n = \tilde{Q}_n(\tilde{q}_0, \tilde{q}_1, \ldots, \tilde{q}_{n-1}), \qquad n = 1, 2, \ldots.$$

Here the symbols $\tilde{Q}_n$ denote rounded and otherwise approximated values of the functions Q_n. Instead of working with relation (16-13), which is difficult mathematically, we write (16-13) in the form

$$(16\text{-}14) \qquad \tilde{q}_n = Q_n(\tilde{q}_0, \tilde{q}_1, \ldots, \tilde{q}_{n-1}) + \varepsilon_n.$$

We consider this relation as the *definition* of the local rounding error ε_n. Thus,

$$\varepsilon_n = \tilde{Q}_n(\tilde{q}_0, \tilde{q}_1, \ldots, \tilde{q}_{n-1}) - Q_n(\tilde{q}_0, \tilde{q}_1, \ldots, \tilde{q}_{n-1}).$$

By analyzing the computational process used to evaluate Q_n, some statement can usually be made about size and distribution of the ε_n.

Let now the accumulated rounding errors r_n be defined by

$$(16\text{-}15) \qquad r_n = \tilde{q}_n - q_n, \qquad n = 0, 1, 2, \ldots.$$

By subtracting (16-12) from (16-14) we find

$$r_n = Q_n(\tilde{q}_0, \tilde{q}_1, \ldots, \tilde{q}_{n-1}) - Q_n(q_0, q_1, \ldots, q_{n-1}) + \varepsilon_n$$

or, using the mean value theorem,

$$(16\text{-}16) \qquad r_n = Q_n^{(0)} r_0 + Q_n^{(1)} r_1 + \cdots + Q_n^{(n-1)} r_{n-1} + \varepsilon_n,$$

where $Q_n^{(k)}$ denotes a partial derivative,

$$Q_n^{(k)} = \frac{\partial Q_n}{\partial q_k}, \qquad k = 0, 1, \ldots, n - 1,$$

taken at some point between $(q_0, q_1, \ldots, q_{n-1})$ and $(\tilde{q}_0, \tilde{q}_1, \ldots, \tilde{q}_{n-1})$.

If the functions Q_n are linear in the q_k, the partial derivatives $Q_n^{(k)}$ do not depend on the previous rounding errors $r_0, r_1, \ldots, r_{n-1}$, and (16-16) then represents a *linear* difference equation for the quantities r_n. If the $Q_n^{(k)}$ are constants, and if $Q_n^{(k)} = 0$ for $k < n - m$, then (16-16) is a difference equation of order m with constant coefficients and can be solved by the method by §6.7 and §6.8.   There results a solution of the form

$$(16\text{-}17) \qquad\qquad r_n = \sum_{m=0}^{n} d_{nm} \varepsilon_m,$$

where the coefficients d_{nm} themselves satisfy a certain difference equation.

A solution of the form (16-17) also must exist in the general case of a recurrence relation of the form (16-12) with linear functions Q_n, since all ε_m enter into (16-16) only linearly.   In order to determine the coefficients d_{nm} we substitute (16-17) into (16-16).   There results

$$\begin{aligned}
r_n = {} & Q_n^{(0)} d_{00} \varepsilon_0 \\
& + Q_n^{(1)} (d_{10} \varepsilon_0 + d_{11} \varepsilon_1) \\
& + Q_n^{(2)} (d_{20} \varepsilon_0 + d_{21} \varepsilon_1 + d_{22} \varepsilon_2) \\
& + \cdots \\
& + Q_n^{(n-1)} (d_{n-1,0} \varepsilon_0 + d_{n-1,1} \varepsilon_1 + \cdots + d_{n-1,n-1} \varepsilon_{n-1}) \\
& + \varepsilon_n.
\end{aligned}$$

The expression on the right must be identical with (16-17) for all possible values of $\varepsilon_0, \varepsilon_1, \ldots, \varepsilon_n$.   It follows that the coefficients of corresponding ε_m must be equal.   Comparing the coefficients of ε_n we immediately find

$$(16\text{-}18) \qquad\qquad d_{n,n} = 1, \qquad n = 0, 1, 2, \ldots,$$

and hence, comparing the coefficients of $\varepsilon_{n-1}, \varepsilon_{n-2}, \ldots, \varepsilon_0$,

$$(16\text{-}19) \qquad
\begin{cases}
d_{n,n-1} = Q_n^{(n-1)} \\
d_{n,n-2} = Q_n^{(n-1)} d_{n-1,n-2} + Q_n^{(n-2)} \\
\quad \cdot \quad \cdot \quad \cdot \quad \cdot \quad \cdot \quad \cdot \quad \cdot \quad \cdot \\
d_{n,0} = Q_n^{(n-1)} d_{n-1,0} + Q_n^{(n-2)} d_{n-2,0} + Q_n^{(1)} d_{1,0} + Q_n^{(0)}
\end{cases}$$

These relations have the form of difference equations for the d_{nm} with respect to the first subscript and can be solved in special cases.

We summarize the above result as follows:

Theorem 16.4 If (16-12) is a linear algorithm, i.e., if

$$(16\text{-}20)\qquad q_n = Q_n^{(0)}q_0 + Q_n^{(1)}q_1 + \cdots + Q_n^{(n-1)}q_{n-1} + p_n$$

for certain constants $Q_n^{(k)}$ and p_n ($n = 0, 1, 2, \ldots$; $k = 0, 1, \ldots$, $n - 1$), then the dependence of the accumulated rounding errors on the local rounding errors is given by (16-17), where the coefficients d_{nm} are determined by (16-19).

The coefficients d_{nm} may be called *influence coefficients*, because they indicate the influence of the local errors on the accumulated error.

Once the influence coefficients have been calculated, the relation (16-17) may be used to make either a deterministic or a probabilistic statement about the accumulated rounding error. Assuming

$$(16\text{-}21)\qquad |\varepsilon_m| \leqq \varepsilon, \qquad m = 0, 1, 2, \ldots$$

we find

$$(16\text{-}22)\qquad |r_n| \leqq \varepsilon \sum_{m=0}^{n} |d_{nm}|, \qquad n = 0, 1, 2, \ldots.$$

The bound on the right is attained for $\varepsilon_m = \varepsilon \operatorname{sign} d_{nm}$ and thus cannot be improved. Assuming that the local errors are independent random variables with

$$(16\text{-}23)\qquad E(\varepsilon_m) = \mu, \qquad \operatorname{var}(\varepsilon_m) = \sigma^2, \qquad m = 0, 1, 2, \ldots$$

we find for the random variables r_n, using (16-9) and (16-10)

$$(16\text{-}24)\qquad \begin{cases} E(r_n) = \mu \sum_{m=0}^{n} d_{nm}, \\[2mm] \operatorname{var}(r_n) = \sigma^2 \sum_{m=0}^{n} d_{nm}^2, \qquad n = 0, 1, 2, \ldots. \end{cases}$$

In the case where the expected values and variances of the local errors depend on m (which is the appropriate assumption for floating point arithmetic) the assumptions (16-21) are to be replaced by

$$(16\text{-}25)\qquad E(\varepsilon_m) = \mu_m, \qquad \operatorname{var}(\varepsilon_m) = \sigma_m^2, \qquad m = 0, 1, 2, \ldots,$$

yielding

$$E(r_n) = \sum_{m=0}^{n} \mu_m d_{nm},$$

$$\operatorname{var}(r_n) = \sum_{m=0}^{n} \sigma_m^2 d_{nm}^2.$$

In a qualitative sense, the above results remain even true when the functions Q_n are not linear. The derivatives $Q_n^{(k)}$ in (16-16) are then evaluated at $(q_0, q_1, \ldots, q_{n-1})$, and terms of higher order in Taylor's expansion are neglected. This can frequently be justified if an a priori bound for the accumulated errors r_n is known.

16.5 Applications

We now shall apply the results of §16.4 to a number of special algorithms. Most of these have already been discussed in preceding chapters. For the sake of simplicity we shall assume that fixed point arithmetic is used. The hypotheses (16-21) and (16-23) are then appropriate. In many cases the results remain qualitatively true for floating point arithmetic.

(*i*) *Evaluation of a sum.* We begin by considering the extremely simple example of evaluating numerically the sum

$$S = \sum_{n=1}^{N} a_n,$$

where $a_1, a_2, \ldots, a_N$ are given real numbers. This can be put into algorithmic form by setting

(16-27)
$$\begin{cases} q_0 = 0, \\ q_n = q_{n-1} + a_n, & n = 1, 2, \ldots, N. \end{cases}$$

Clearly, $q_N = S$. We thus have

$$Q_n(q_0, q_1, \ldots, q_{n-1}) = q_{n-1} + a_n.$$

In actual computation, in place of the exact numbers q_n numerical values $\tilde{q}_n$ are generated according to the scheme

(16-28)
$$\begin{cases} \tilde{q}_0 = 0, \\ \tilde{q}_n = \tilde{q}_{n-1} + a_n^*. \end{cases}$$

(In floating arithmetic, the second of the relations (16-28) would have to be replaced by $q_n = (q_{n-1} + a^\oplus)^\oplus$.) According to our scheme, we replace the second relation (16-28) by

$$\tilde{q}_n = \tilde{q}_{n-1} + a_n + \varepsilon_n,$$

showing that the local error ε_n is given by

$$\varepsilon_n = a_n^* - a_n$$

and thus, in fixed point binary arithmetic, satisfies (16-21) with $\varepsilon = 2^{-t-1}$, or (16-23) with

(16-29)
$$\mu = 0, \qquad \sigma^2 = \tfrac{1}{12} 2^{-2t}.$$

In view of

$$Q_n^{(k)} = 0, \qquad k = 0, 1, 2, \ldots, n - 2,$$
$$Q_n^{(n-1)} = 1,$$

the relations (16-19) reduce to

$$d_{n,\,n-1} = 1, \qquad d_{n,\,k} = d_{n-1,\,k}, \qquad n > k$$

and thus yield $d_{n,\,k} = 1$, $n \geq k$.   Relation (16-17) thus shows that

$$(16\text{-}30) \qquad\qquad r_n = \sum_{m=1}^{n} \varepsilon_m.$$

(In the present simple example this result could have been obtained directly, see problem 3.)   We thus find from (16-22) and (16-24)

$$(16\text{-}31) \qquad\qquad |r_n| \leq n\varepsilon, \qquad \text{s.d. } (r_n) = \sqrt{n}\,\sigma.$$

The result shows that while the theoretically greatest possible error in a sum of n terms grows like n, the standard deviation grows only like $\sqrt{n}$.

(*ii*) *Iteration.*   We next consider the algorithm of solving the equation

$$x = f(x)$$

by determining the limit of the sequence $\{x_n\}$ defined by algorithm 4.1:

$$x_n = f(x_{n-1}).$$

Here the functions Q_n in (16-12) are given by

$$Q_n(q_0, q_1, \ldots, q_{n-1}) = f(q_{n-1}).$$

To simplify matters we shall consider only the idealized situation where

$$f(x) = ax + b$$

and where a and b are constant machine numbers.   For convergence we must assume (see theorem 4.2) that $|a| < 1$.   On the machine, the theoretical recurrence relation

$$q_n = aq_{n-1} + b$$

is replaced by

$$\tilde{q}_n = (a\tilde{q}_{n-1})^* + b.$$

Writing the latter relation in the form

$$\tilde{q}_n = a\tilde{q}_{n-1} + b + \varepsilon_n,$$

we see that

$$\varepsilon_n = (a\tilde{q}_{n-1})^* - a\tilde{q}_{n-1}.$$

Thus, in fixed point arithmetic, (16-21) again holds with $\varepsilon = 2^{-t-1}$, and (16-23) with the values (16-29).   In view of

$$Q_n^{(k)} = 0, \qquad k = 0, 1, \ldots, n - 2,$$
$$Q_n^{(n-1)} = a,$$

the recurrence relations (16-19) simplify to

$$d_{nn} = 1, \qquad d_{n, m} = ad_{n-1, m} \qquad (n > m \geqq 0)$$

and permit the immediate solution

$$d_{nm} = a^{n-m}, \qquad n = m, m + 1, \ldots.$$

It follows that

$$r_n = \sum_{m=0}^{n} a^{n-m}\varepsilon_m.$$

In view of $|a| < 1$ we thus find

(16-32) $$|r_n| \leqq \varepsilon \frac{1 - |a|^{n+1}}{1 - |a|} \leqq \frac{1}{1 - |a|}\varepsilon$$

and for the stochastic model

(16-33) $$\operatorname{var}(r_n) = \sigma^2 \frac{1 - a^{2n+2}}{1 - a^2} \leqq \sigma^2 \frac{1}{1 - a^2}.$$

The bounds (16-32) and (16-33) are remarkable for the fact that they are independent of n, the number of iteration steps.   An algorithm with this property must be called *stable* under any reasonable definition of this term.   Newton's method in particular, which corresponds to the case $a = 0$ of the above simplified model, enjoys an extreme degree of stability.

(iii) *Generating* $\cos n\varphi$ *and* $\sin n\varphi$ *by recurrence relations.*   As our next example, we consider the algorithm for generating the sequences $\{\cos n\varphi\}$ and $\{\sin n\varphi\}$ by the algorithm suggested by example 3, §6.3.   (These sequences are required, e.g., for computing the sums of Fourier series.) The difference equation

(16-34) $$t_n - 2\cos\varphi t_{n-1} + t_{n-2} = 0$$

has the solutions $t_n = \cos n\varphi$ and $t_n = \sin n\varphi$ and can therefore be used to generate these functions recursively if the values of $\cos\varphi$ and $\sin\varphi$ are known.   Here the theory of §16.4 applies with

$$Q(t_0, t_1, \ldots, t_{n-1}) = 2\cos\varphi t_{n-1} - t_{n-2}.$$

In numerical computation, (16-34) is replaced by

(16-35) $$\tilde{t}_n = (2\widetilde{\cos\varphi\tilde{t}_{n-1}})^* - \tilde{t}_{n-2},$$

where $\widehat{\cos \varphi}$ denotes a machine value of $\cos \varphi$. Writing (16-35) in the form

$$\tilde{i}_n = 2 \cos \varphi \tilde{i}_{n-1} - \tilde{i}_{n-2} + \varepsilon_n,$$

we see that

$$\varepsilon_n = (2 \widehat{\cos \varphi \tilde{i}_{n-1}})^* - 2 \cos \varphi \tilde{i}_{n-1}.$$

Adding and subtracting $2 \widehat{\cos \varphi \tilde{i}_{n-1}}$, we also have

$$\varepsilon_n = (2 \widehat{\cos \varphi \tilde{i}_{n-1}})^* - 2 \widehat{\cos \varphi \tilde{i}_{n-1}} + 2(\widehat{\cos \varphi} - \cos \varphi)\tilde{i}_{n-1}.$$

Here we see that ε_n is due to two sources: (a) rounding the product $2 \widehat{\cos \varphi \tilde{i}_{n-1}}$; (b) replacing $\cos \varphi$ by $\widehat{\cos \varphi}$. Both errors can be estimated and permit us to make assumptions such as (16-21) and (16-23).

To determine the d_{nm}, we observe that

$$\begin{aligned} Q_n^{(k)} &= 0, \qquad k = 0, 1, \ldots, n-3, \\ Q_n^{(n-2)} &= -1, \\ Q_n^{(n-1)} &= 2 \cos \varphi. \end{aligned}$$

The relations (16-19) thus yield

$$(16\text{-}36) \quad \begin{cases} d_{mm} = 1, \\ d_{m+1,m} = 2 \cos \varphi, \\ d_{mm} = 2 \cos \varphi d_{n-1,m} - d_{n-2,m}, \qquad n \geqq m + 2. \end{cases}$$

These equations show that, for a fixed value of m, the quantities d_{nm} themselves are a solution of the difference equation (16-34), with the initial values given at $n = m$ and $n = m + 1$. We thus must have

$$d_{nm} = A_m \cos n\varphi + B_m \sin n\varphi, \qquad n = m, m + 1, \ldots,$$

where A_m and B_m are to be determined from the first two relations (16-36). Some algebra yields (if $0 < \varphi < \pi$)

$$A_m = -\frac{\sin (m-1)\varphi}{\sin \varphi}, \qquad B_m = \frac{\cos (m-1)\varphi}{\sin \varphi},$$

and we thus get, after some further simplification,

$$d_{nm} = \frac{\sin (n-m+1)\varphi}{\sin \varphi}, \qquad n \geqq m.$$

Since $\varepsilon_0 = 0$, the general theory thus yields

$$r_n = \sum_{m=1}^{n} \frac{\sin (n-m+1)\varphi}{\sin \varphi} \varepsilon_m.$$

Using the crude estimate $|\sin (n - m + 1)\varphi| \leqq 1$, we thus have

$$|r_n| \leqq \frac{n\varepsilon}{\sin \varphi}, \qquad n = 1, 2, \ldots.$$

Using the known formula

$$\sum_{m=1}^{n} (\sin \varphi)^2 = \frac{n}{2} + \frac{1}{4} - \frac{\sin (2n + 1)\varphi}{4 \sin \varphi},$$

we get for the variance the expression

$$\text{var} (r_n) = \sigma^2 \sum_{m=1}^{n} \left(\frac{\sin (n - m + 1)\varphi}{\sin \varphi}\right)^2$$

$$= \frac{\sigma^2}{(\sin \varphi)^2} \left[\frac{n}{2} + \frac{1}{4} - \frac{\sin (2n + 1)\varphi}{4 \sin \varphi}\right].$$

As in the algorithm for finding the sum of n numbers, the standard deviation of the accumulated error grows only like $\sqrt{n}$, while the rigorous estimate grows like n. This relatively high stability of the recursive method for generating $\cos n\varphi$ and $\sin n\varphi$ (or, what amounts to the same, the Chebyshev polynomials) has been observed experimentally (Lanczos [1955]).

(iv) *Horner's scheme.*   We next consider the propagation of error in the evaluation of a polynomial by Horner's scheme (algorithm 3.4), assuming that the coefficients $a_0, a_1, \ldots, a_N$ of the polynomial and the value of the variable x are machine numbers. The scheme then consists in generating a finite sequence of numbers $q_0, q_1, \ldots, q_N$ by the formulas

(16-37)
$$\begin{cases} q_0 = a_0, \\ q_n = a_n + xq_{n-1}, \qquad n = 1, 2, \ldots, N; \end{cases}$$

the desired value is q_N. The functions Q_n are evidently given by

$$Q_n(q_0, q_1, \ldots, q_{n-1}) = a_n + xq_{n-1}.$$

In numerical work, the second relation (16-37) is replaced by $\tilde{q}_n = a_n + (x\tilde{q}_{n-1})^*$. If the last term is written as $a_n + x\tilde{q}_{n-1} + \varepsilon_n$, we see that, as in (ii) above, the local rounding error equals the rounding error in the fixed point multiplication $x\tilde{q}_{n-1}$. Thus the assumptions (16-21) and (16-23) are again justified with $\varepsilon = 2^{-t-1}$, $\mu = 0$, $\sigma^2 = \frac{1}{12}2^{-2t}$. We now have

$$Q_n^{(k)} = 0, \qquad k = 0, 1, \ldots, n - 2,$$
$$Q_n^{(n-1)} = x,$$

and thus find, as above,

$$d_{nm} = x^{n-m}, \qquad n \geqq m \geqq 0.$$

Since $\varepsilon_0 = 0$ we thus have

$$r_n = \sum_{m=1}^{n} x^{n-m}\varepsilon_m$$

and find in the usual way

$$|r_n| \leq \begin{cases} \dfrac{|x|^n - 1}{|x| - 1}\,\varepsilon, & x \neq 1, \\ n\varepsilon, & x = 1, \end{cases}$$

and, under the statistical hypothesis,

$$\text{s.d. } (r_n) = \begin{cases} \sqrt{\dfrac{x^{2n} - 1}{x^2 - 1}}\,\sigma, & x \neq 1, \\ \sqrt{n}\,\sigma, & x = 1. \end{cases}$$

Thus, for the evaluation of high degree polynomials in fixed point arithmetic, Horner's scheme is unstable when $|x| > 1$ and stable when $|x| < 1$.

(v) *Numerical integration of differential equations.* We finally consider the numerical integration of the simple but typical initial value problem

$$y' = Ay, \qquad y(0) = 1,$$

where A is a real constant, $A \neq 0$, (a) by Euler's method; (b) by the integration scheme based on the midpoint formula discussed in §14.8. The approximate values generated by Euler's method satisfy

$$y_0 = 1,$$
$$y_n = y_{n-1} + Ahy_{n-1}, \qquad n = 1, 2, \ldots.$$

This is an algorithm of the form (16-12) where $Q_n^{(k)} = 0$ ($k = 0, 1, \ldots, n - 2$), $Q_n^{(n-1)} = 1 + Ah$. If the local errors are defined by

$$\tilde{y}_n = \tilde{y}_{n-1} + Ah\tilde{y}_{n-1} + \varepsilon_n,$$

we find, by computations similar to the ones carried out earlier, that the accumulated error $r_n = \tilde{y}_n - y_n$ can be expressed in the form

$$r_n = \sum_{m=1}^{n} (1 + Ah)^{n-m}\varepsilon_m.$$

We thus obtain

$$|r_n| \leq \frac{(1 + Ah)^n - 1}{hA}\,\varepsilon,$$

$$\text{var } (r_n) = \frac{(1 + Ah)^{2n} - 1}{2Ah + h^2A^2}\,\sigma^2.$$

Some simplification is possible by observing that for $h \to 0$, $nh = x$ fixed and positive, $(1 + Ah)^n = e^{Ax} + 0(h)$. We thus find

(16-38)
$$|r_n| \le \frac{\varepsilon}{h}\left[\frac{e^{Ax} - 1}{A} + 0(h)\right],$$

$$\text{var}\,(r_n) = \frac{\sigma^2}{h}\left[\frac{e^{2Ax} - 1}{2A} + 0(h)\right].$$

These relations show that Euler's method is numerically *stable* if $A \le 0$, provided that h is sufficiently small. The method is *unstable* if $A > 0$, but this instability is not serious, since then the solution itself grows exponentially. Analogous statements hold for all methods based on Taylor's expansion, and also for all Adams' methods.

If the midpoint rule is used, the values y_n satisfy

$$y_0 = 1, \qquad y_1 = e^{Ah} \text{ (ideally)}$$
$$y_n - y_{n-2} = 2hAy_{n-1}, \qquad n \ge 2.$$

This is of the form (16-12) where

$$Q_n(y_0, y_1, \ldots, y_{n-1}) = 2hAy_{n-1} + y_{n-2}.$$

A consideration similar to that under (*iii*) above shows that the coefficients d_{nm} satisfy, for a fixed value of m, the difference equation

(16-39) $$d_{nm} - 2hAd_{n-1,m} - d_{n-2,m} = 0 \qquad (n \ge m + 2)$$

and the starting conditions

(16-40) $$d_{mm} = 1, \qquad d_{m+1,m} = 2Ah.$$

The characteristic polynomial p of (16-39) is given by

$$p(z) = z^2 - 2hAz - 1;$$

its two zeros are

$$z_1 = Ah + \sqrt{1 + (Ah)^2} = e^{Ah} + 0(h^2),$$
$$z_2 = -z_1^{-1}.$$

The solution d_{nm} of (16-39) satisfying (16-40) is found to be

$$d_{nm} = \frac{z_1^{n-m+1} - z_2^{n-m+1}}{z_1 - z_2}.$$

A somewhat elaborate but elementary computation using the approximations (see §14.8)

$$z_1^n = e^{Ax} + 0(h), \qquad z_2^n = (-1)^n e^{-Ax} + 0(h),$$

valid for $h \to 0$, $nh = x$ fixed and positive, shows that

$$(16\text{-}41) \qquad |r_n| \leqq \varepsilon \sum_{m=0}^{n} |d_{nm}| = \frac{\varepsilon}{h} \left[\frac{\cosh Ah - 1}{2A} + O(h) \right],$$

and, if var $(\varepsilon_m) = \sigma^2$,

$$(16\text{-}42) \qquad \text{var } (r_n) = \frac{\sigma^2}{h} \left[\frac{\cosh 2Ax - 1}{4A} + O(h) \right].$$

These relations confirm the result of §14.8 that the midpoint formula is always unstable, also when $A < 0$. It produces exponential growth of the rounding error also when the exact solution is exponentially decreasing. It is clear that such a method cannot be used for the numerical integration of a differential equation such as $y' = -y$. In a qualitative sense, this negative result is true for all methods based on numerical integration where the numerical integration is performed over an interval comprising several steps. Simpson's rule, in particular, if applied to the numerical solution of differential equations, also suffers from this kind of instability, see problem 11.

Problems

4. In floating arithmetic it is frequently permissible to replace the statistical hypotheses (16-23) by

$$(16\text{-}43) \qquad E(\varepsilon_m) = \mu q_m, \qquad \text{var } (\varepsilon_m) = \sigma^2 q_m^2,$$

where μ and σ are again constants. Discuss the propagation of error in the computation of the sum

$$(16\text{-}44) \qquad q_n = a_1 + a_2 + \cdots + a_n,$$

where $a_i > 0$, $i = 1, 2, \ldots$, by the algorithm described under (*i*) above, if floating arithmetic is used. Show that, contrary to the fixed point arithmetic case, the propagated error depends on the order of the terms in (16-44), and that both the maximum error and the standard deviation are minimized if the terms are summed in increasing order.

5. Discuss the propagation of rounding error if the geometric series $\sum_{n=0}^{\infty} a^n$, where $|a| < 1$, is summed in floating point arithmetic, generating a^n by the formula $a^n = a(a^{n-1})$. Is the resulting process numerically stable?

6. The Fibonacci numbers (see example 7, §6.3) are generated from their recurrence relation

$$x_0 = x_1 = 1, \qquad x_n = x_{n-1} + x_{n-2}$$

in floating arithmetic. Assuming that the relations (16-43) hold with

$q_n = x_n$, what is the standard deviation of the numerical value $\tilde{x}_n$ for large values of n? If the $\tilde{x}_n$ are used to compute the number

$$\frac{1 + \sqrt{5}}{2} = \lim_{n \to \infty} \frac{\tilde{x}_{n+1}}{\tilde{x}_n},$$

is this a stable process?

7*. (Generalization of problem 6.) Discuss the propagation of rounding error in Bernoulli's method for determining a single dominant zero of a polynomial whose coefficients are exact machine numbers. Discuss both floating and fixed point arithmetic.

8. Discuss the propagation of rounding error in algorithm 5.6 for extracting a quadratic factor, assuming fixed point arithmetic.

9. Let, for non-integral $s > 0$, the binomial coefficients $q_n = \binom{s}{n}$ be generated by the recurrence relation

$$q_0 = 1, \qquad q_n = \frac{s - n + 1}{n} q_{n-1}, \qquad n = 1, 2, \ldots.$$

Assuming fixed point arithmetic, show that

$$\operatorname{var}(r_n) = \sigma^2 S_n(s),$$

where

$$S_n(s) = \left[\binom{s}{n}\right]^2 \sum_{m=1}^{n} \left[\binom{s}{m}\right]^{-2}.$$

(It can be shown that $S_n(s) = n/(3 + 2s) + 0(1)$ as $n \to \infty$.)

10. Study the propagation of rounding error in algorithm 12.4 (repeated extrapolation to the limit), if the same error distribution (16-23) is assumed in the zeroth column $A_{m,0}$ and in the relations defining the elements of the $(m + 1)$st column in terms of those of the mth column.

11. Show that if the problem discussed under (v) above is solved by Simpson's formula

$$y_{n+1} - y_{n-1} = \frac{h}{3}(f_{n+1} + 4f_n + f_{n-1})$$

in fixed point arithmetic, the standard deviation of the accumulated rounding error behaves for x large like $\cosh(\tfrac{1}{3}Ax)$.

Recommended Reading

The detailed study of the propagation of rounding error in numerical computation is of recent origin. The classical paper is by Rademacher [1948]. Very detailed accounts of the propagation of error in a large number of methods for solving ordinary differential equations are given in the author's books [1962, 1963].

Research Problems

1. Make a study, both experimental and (as far as possible) theoretical, of the propagation of error in the Quotient-Difference algorithm.

2. Discuss the stability of Horner's scheme in floating point arithmetic.

3. The hypotheses underlying the statistical theory of propagation of rounding error have been criticized as unreliable (Huskey [1949]). Carry out some statistical experiments on rounding error propagation and compare the results with the theoretical results given above. (See the author's books quoted above for some ideas on how to perform such experiments.)

bibliography

This list of books and papers contains the titles quoted in the text; it is not intended to be complete in any sense of the word. More complete bibliographies on many topics of numerical analysis will be found in some of the references given below; see in particular Hildebrand [1956] and Todd [1962].

Aitken, A. C. [1926]: On Bernoulli's numerical solution of algebraic equations, *Proc. Roy. Soc. Edinburgh*, **46**, 289–305.

Bareiss, E. H. [1960]: Resultant procedure and the mechanization of the Graeffe process, *J. Assoc. Comp. Mach.*, **7**, 346–386.

Bauer, F. L., H. Rutishauser, and E. Stiefel [1963]: New aspects in numerical quadrature, *Proc. of Symp. in Appl. Math.*, vol. 15: *High speed computing and experimental arithmetic*, American Mathematical Society, Providence, R.I.

Birkhoff, G., and S. MacLane [1953]: *A survey of modern algebra*, rev. ed., Macmillan, New York.

Brown, K. M., and P. Henrici [1962]: Sign wave analysis in matrix eigenvalue problems, *Math. of Comput.*, **16**, 291–300.

Buck, R. C. [1956]: *Advanced Calculus*, McGraw-Hill, New York, Toronto, London.

Coddington, E. A. [1961]: *An introduction to ordinary differential equations*, Prentice-Hall, Englewood Cliffs, N.J.

Comrie, L. J. [1961]: *Chambers's shorter six-figure mathematical tables*, W. R. Chambers Ltd., Edinburgh and London.

Erdelyi, A. (ed.) [1953]: *Higher transcendental functions*, vol. 1, McGraw-Hill, New York, Toronto, London.

Forsythe, G. E. [1958]: Singularity and near singularity in numerical analysis, *Amer. Math. Monthly*, **65**, 229–240.

—— and W. Wasow [1960]: *Finite difference methods for partial differential equations*, Wiley, New York.

Fox, L. [1957]: *The numerical solution of two-point boundary problems in ordinary differential equations*, Clarendon, Oxford.

Goldberg, S. [1958]: *Introduction to difference equations*, Wiley, New York.

Henrici, P. [1956]: A subroutine for computations with rational numbers, *J. Assoc. Comput. Mach.*, **3**, 10–15.

—— [1958]: The quotient-difference algorithm, *Nat. Bur. Standards Appl. Math. Series*, **49**, 23–46.

322

Henrici, P. [1962]: *Discrete variable methods in ordinary differential equations,* Wiley, New York.

—— [1963]: *Error propagation for difference methods,* Wiley, New York.

—— [1963a]: Some applications of the quotient-difference algorithm, *Proc. Symp. Appl. Math.,* vol. 15, *High speed computing and experimental arithmetic,* American Mathematical Society, Providence, R.I.

Hildebrand, F. B. [1956]: *Introduction to numerical analysis,* McGraw-Hill, New York, Toronto, London.

Householder, A. S. [1953]: *Principles of numerical analysis,* McGraw-Hill, New York and London.

Huskey, H. D. [1949]: On the precision of a certain procedure of numerical integration, with an appendix by Douglas R. Hartree, *J. Res. Nat. Bur. Stand.,* **42,** 57–62.

Jahnke, E., and F. Emde [1945]: *Tables of functions with formulae and curves,* Dover, New York.

Kantorovich, L. V. [1948]: Functional analysis and applied mathematics, *Uspekhi Mat. Nauk.,* **3,** 89–185. Translated by C. D. Benster and edited by G. E. Forsythe as *Nat. Bur. Stand. Rept.* No. 1509.

Kaplan, W. [1953]: *Advanced Calculus,* Addison-Wesley, Cambridge.

Lanczos, C. [1955]: Spectroscopic eigenvalue analysis, *J. Wash. Acad. Sci.,* **45,** 315–323.

Liusternik, L. A., and W. I. Sobolev [1960]: *Elemente der Funktionalanalysis* (translated from the Russian). Akademieverlag, Berlin.

McCracken, D. D. [1961]: *A guide to FORTRAN programming,* Wiley, New York.

—— [1962]: *A guide to ALGOL programming,* Wiley, New York.

Milne, W. E. [1949]: *Numerical Calculus,* Princeton University Press, Princeton, N.J.

—— [1953]: *Numerical solution of differential equations,* Wiley, New York.

Milne-Thomson, L. M. [1933]: *The calculus of finite differences,* Macmillan, London.

Muller, D. E. [1956]: A method for solving algebraic equations using an automatic computer, *Math. Tables Aids Comput.,* **10,** 208–215.

National Physical Laboratory [1961]: *Modern Computing Methods* (Notes on Applied Science No. 16), 2nd ed., H.M. Stationery Office, London.

Naur, P. et al. [1960]: Report on the algorithmic language ALGOL 60, *Comm. Assoc. Comp. Mach.,* **3,** 299–314.

Nautical Almanac Office [1956]: *Interpolation and allied tables,* H.M. Stationery Office, London.

Ostrowski, A. [1940]: Recherches sur la méthode de Graeffe et les zéros des polynomes et les séries de Laurent, *Acta Math.,* **72,** 99–257.

—— [1960]: *Solution of equations and systems of equations,* Academic Press, New York.

Rademacher, H. [1948]: On the accumulation of errors in processes of integration on high-speed calculating machines. Proceedings of a symposium on large scale digital calculating machinery, *Annals Comput. Labor. Harvard Univ.,* **16,** 176–187.

Richardson, L. F. [1927]: The deferred approach to the limit, I—Single lattice, *Trans. Roy. Soc. London,* **226,** 299–349.

Rutishauser, H. [1952]: Ueber die Instabilität von Methoden zur Integration gewöhnlicher Differentialgleichungen, *Z. angew. Math. Physik,* **3,** 65–74.

—— [1956]: *Der Quotienten-Differenzen-Algorithmus.* Mitteilungen aus dem Institut für angew. Math. No. 7. Birkhäuser, Basel and Stuttgart.

Rutishauser, H. [1963]: Ausdehnung des Rombergschen Prinzips, *Numer. Math.*, **5**, 48–53.

Schwarz, H. R. [1962]: An Introduction to ALGOL, *Comm. Assoc. Comp. Mach.*, **5**, 82–95.

Sokolnikoff, I. S., and R. M. Redheffer [1958]: *Mathematics of Physics and modern Engineering*, McGraw-Hill, New York.

Speiser, A. P. [1961]: *Digitale Rechenanlagen*, Springer, Berlin.

Steffensen, J. F. [1933]: Remarks on iteration, *Skand. Aktuar. Tidskr.*, **16**, 64–72.

Taylor, A. E. [1959]: *Calculus with analytic geometry*, Prentice-Hall, Englewood Cliffs, N.J.

Todd, J. (ed.) [1962]: *Survey of numerical analysis*, McGraw-Hill, New York.

—— [1963]: *Introduction to the constructive theory of functions*, Birkhäuser, Basel and Stuttgart.

Watkins, B. O. [1964]: Roots of a polynomial using the QD method. To be published.

Watson, G. N. [1944]: *A treatise on the theory of Bessel functions*, 2nd. ed., University Press, Cambridge.

Whittaker, E. T., and G. Robinson [1924]: *The calculus of observations*, Blackie, London.

Wilkinson, J. H. [1959]: The evaluation of the zeros of ill-conditioned polynomials, *Numer. Math.*, **1**, 150–180.

—— [1960]: Error analysis of floating-point computation, *Numer. Math.*, **2**, 319–340.

—— [1963]: *Rounding errors in algebraic processes*, National Physical Laboratory Notes on Applied Science No. 32, H.M. Stationery Office, London.

INDEX

325

ANSWERS FOR PROBLEMS

Chapter 1

1) If $n > 1$ is the given integer, divide n by all integers m from 2 to the greatest integer not exceeding $\sqrt{n}$. n is prime if no such m divides n without remainder.

2) Algorithm infinite because $\sqrt{2}$ irrational, hence decimal fraction neither terminating nor periodic.

3) Existence of $\sqrt{a}$ for real a > 0; proof of the fact that every bounded monotone sequence has a limit; existence of supremum and infimum of a bounded set of real numbers.

Chapter 2

1) a) $-\frac{1}{2} + i\frac{\sqrt{3}}{2}$ b) -4 c) $\cos 2\varphi + i \sin 2\varphi$
d) $\cos\varphi + i \sin\varphi$ e) 1 .

2) Not defined for $z = -1$, $z = 0$.

3)
$$s = \frac{1 - z^{n+1}}{1 - z}$$

4) a) 2, $\varphi = -\frac{\pi}{2} + 2k\pi$ b) 10, $\varphi = 126°45' + k\,360°$
c) $2^{-1/2}$, $\varphi = \frac{\pi}{4} + 2k\pi$.

5) $\arg\omega = \frac{2\pi}{3} + 2k\pi$, $|\omega| = \sqrt[3]{1} = 1$

8) If k = 1, then locus is imaginary axis. If $k \neq 1$, let
$$q = \frac{1 + k^2}{1 - k^2} .$$

The equation is
$$z\,\bar{z} - q(z + \bar{z}) + 1 = 0$$
or in real form
$$x^2 + y^2 - 2qx + 1 = 0$$
(circle with center at (q, 0) and radius $q^2 - 1$).

9) $i^n = (-1)^m$ if $n = 2m$
              $i^n = i$ if $n = 4m + 1$
              $i^n = -i$ if $n = 4m + 3$

10) $z = e^{ik\varphi}$, k = 1, 2, ... , n, where
$$\varphi = \frac{2\pi}{n + 1} .$$

11) $A_n = \dfrac{1 - r\cos\varphi - r^{n+1}\cos(n+1)\varphi + r^{n+2}\cos n\varphi}{1 - 2r\cos\varphi + r^2}$

$B_n = \dfrac{r\sin\varphi - r^{n+1}\sin(n+1)\varphi + r^{n+2}\sin n\varphi}{1 - 2r\cos\varphi + r^2}$

12) a) $\pm 1 \pm i$

b) $\exp i(\dfrac{\pi}{12} + n\,\dfrac{\pi}{3})$, $n = 0, 1, \ldots , 5$

c) 3 , $\dfrac{-\sqrt{3} + i3}{2}$, $\dfrac{-\sqrt{3} - i3}{2}$.

13) $\cos 4\varphi = 8(\cos\varphi)^4 - 8(\cos\varphi)^2 + 1$

14) $\sqrt{x + iy} =$

$\pm\left[\sqrt{\dfrac{x + \sqrt{x^2 + y^2}}{2}} + (\operatorname{sign} y)i\sqrt{\dfrac{-x + \sqrt{x^2 + y^2}}{2}}\right]$

15) $\dfrac{1}{64}(\cos 7\varphi - \cos 5\varphi - 3\cos 3\varphi + 3\cos\varphi)$

18) a) ellipse with semiaxes $|a + b|$ and $|a - b|$

b) parabolic arc $x = 2y^2 - 1$, $-1 \leqq y \leqq 1$.

22) a) $\pm 1 \pm 2i$ b) $\pm 2 \pm i$ c) $\pm 1 \pm 2i$, $\pm 2 \pm i$

23) $a_1 = a_2 = a_3 = 0$, $a_4 = 4$.

24) three.

25) two.

Chapter 3

1) $x_n = \dfrac{1 - a^{n+1}}{1 - a}$, $\lim\limits_{n\to\infty} x_n = \dfrac{1}{1 - a}$

2) $x_n = \dfrac{n!}{z^n}(1 + z + \dfrac{z^2}{2!} + \ldots + \dfrac{z^n}{n!})$

3) $x_n = b_0 + b_1 + \ldots + b_n = $ n-th partial sum of series.

4) $x_n = \prod\limits_{k=1}^{n}(1 - \dfrac{z^2}{k^2})$

8) $p(x) = 2.392 - 3.92(x + 0.3) - 3.6(x + 0.3)^2 + 4(x + 0.3)$

9) $P(1.5) = -22.625$, $P'(1.5) = -37.625$, $P''(1.5) = -39$

Chapter 4

1) a) $|f'(x)| = |\dfrac{3}{4}\sin x| \leqq \dfrac{3}{4}$, thus $L = \dfrac{3}{4}$

b) $|f'(x)| = \dfrac{1}{2}$, thus $L = \dfrac{1}{2}$

c) $|f'(x)| = \dfrac{1}{x^2}$, thus $|f'(x)| \leqq \dfrac{1}{4}$ for $x \geqq 2$,

thus $L = \dfrac{1}{4}$

2) $f(x) = m - \varepsilon \sin x$ maps $[m - \pi, m + \pi]$ into itself and satisfies Lipschitz condition with $L = |\varepsilon| < 1$.

3) $f(x) = \sqrt{x}$, $0 \leq x \leq 1$. f' is unbounded!

5) $x = 0.96433388$

7) $x_{32} = 1.839287$

9) Any M such that $-\frac{2}{5} < M < 0$ will do.

10) $s = 2$. For $x \geq 0$, $y \geq 0$, $|f(x) - f(y)| \leq \frac{1}{2\sqrt{2}}|x - y|$.

11) $|x_{10} - s| \leq 1.8 \times 10^{-8}$.

13) Prob. 7: $x'_{15} = 1.839286$, Prob. 5: $x'_3 = 0.964318$

16) 2.766204

19) $M = -\frac{1}{g'(s)}$

20) $\lim_{n \to \infty} \frac{d_{n+1}}{d_n^3} = \frac{f'''(s)}{6}$

21) $f(x) = x\frac{x^2 + 3z}{3x^2 + z}$; $x_1 = 3.16216216$, $x_2 = 3.162277660$

22) $x = 0.5671433$

23) $\sqrt{\pi} = 1.772453851$

24) $x_0 = 0.3$, $x_1 = 0.33$, $x_2 = 0.3333$, $x_3 = 0.33333333$, ... (x_n has 2^n 3's).

27) $h(x) = -\frac{1}{2}\frac{F''(x)}{F'(x)}$

28) $x = 1.8392867$

29) $x = 0.9024738$

32) $x_n = 2^{-n}$; $x_n = 0$, $n = 1$

34) 1.782191

35) 3.16227766 (sequence identical with that obtained by Newton's method, if starting values are the same).

39) $s = 1.4655712$

Chapter 5

7) $x = 0.7718445$, $y = 0.4196434$

8) $x = 4.9322865$, $y = 5.0673188$

9) $x = 1.0309038$, $y = 1.0033246$

10) $3.0053220 \pm i\ 3.9963692$, $0.9946779 \pm i\ 1.0248789$.

11) Quadratic factors $x^2 + 1.5415682\,x + 1.4873158$, $x^2 - 1.1272386\,x + 0.8316443$, real zeros -1.8620342, -0.7236362.

Chapter 6

1) a) $x_n = c_1 + c_2(-1)^n$

   b) $x_n = c_1 \cos\frac{n\pi}{4} + c_2 \sin\frac{n\pi}{4}$

   c) $x_n = (-1)^n(c_1 + c_2 n)$

2) $x_n = c_1 \left[-i(1 - \sqrt{2})\right]^n + c_2 \left[-i(1 + \sqrt{2})\right]^n$

3) $x_n = r^n(c_1 \cos n\varphi + c_2 \sin n\varphi)$, where

   $$r = \sqrt{b}, \quad \varphi = \text{arctg}\,\frac{\sqrt{4b - a^2}}{a}$$

7) a) $x_n = \dfrac{1 + \sqrt{5}}{2\sqrt{5}} \left(\dfrac{3 + \sqrt{5}}{2}\right)^n + \dfrac{\sqrt{5} - 1}{2\sqrt{5}} \left(\dfrac{3 - \sqrt{5}}{2}\right)^n$

   b) $x_n = 2^{n/2} \cos\frac{n\pi}{4}$ is solution of $x_n - 2x_{n-1} + 2x_{n-2} = 0$

12) a) $x_n = c_1 3^n + c_2 2^n + 1$

    b) $x_n = c_1 2^{-n} + c_2 4^{-n} + \frac{4}{21} 2^n$

    c) $x_n = c_1 2^n + c_2 (-1)^n - \frac{1}{2}n^2 - \frac{5}{2}n - 4$

13) $x_n = -2\times3^n + 2\times2^n + 1$

14) $x_n = c_1\left(\dfrac{3 + \sqrt{17}}{4}\right)^n + c_2\left(\dfrac{3 - \sqrt{17}}{4}\right)^n + 1$;

    $x_1 - 1 = (x_0 - 1)\dfrac{3 - \sqrt{17}}{4}$.

15) $y_n = 2\left[1 + 2^{-n'/2} \sin\frac{n'\pi}{4}\right]$, $n' = n - 1946$;

every 8 years; if a and b are non-negative, the condition

is $a \leqq 1$, $ab \leqq 1$.

16) $x_n = \dfrac{na^{n-1}}{a - b}$, if $b \neq a$; $x_n = \frac{1}{2} n^2 a^n$, if $b = a$.

20) a) $x_n - 6x_{n-1} + 11x_{n-2} - 6x_{n-3} = 0$

    b) $x_n - 4x_{n-1} + 6x_{n-2} - 4x_{n-3} + x_{n-4} = 0$

    c) $x_n - 9x_{n-1} + 32x_{n-2} - 58x_{n-3} + 57x_{n-4}$
       $- 29x_{n-5} + 6x_{n-6} = 0$

22) $(\nabla^2 T(t))_{n+1} = 2(t - 1)T_n(t)$,

    $(\nabla^4 T(t))_{n+2} = 4(t - 1)^2 T_n(t)$.

23) $\nabla^k f(x) = (1 - e^{-h})^k e^x$.

25) $x_{n-k} = \displaystyle\sum_{m=0}^{k} (-1)^m \binom{k}{m} \nabla^m x_n$

Chapter 7

1) $x = 0.933$

2) $x = 0.933012$

5) $z = 4.189791$

6) $x_9' = 0.933013$, $x_{10}' = 0.933011$

7) $|z_2| = |z_3|$ $(= \sqrt{2})$ (compare condition $|z_2| > |z_3|$ at top of p. 150)

10) Dominant zero has multiplicity 2 .

16) Dominant zeros are $z = \pm 2$.

17) In the left half-plane.

18) All three zeros are dominant.

Chapter 8

1) Scheme cannot be generated in a stable manner. If constructed according to §8.5, scheme is stable and yields the zeros $z = 2 \pm i$, $z = 1$.

2) $e_n^{(k)} = 1$, $n \geq 0$, $k > 0$; $q_n^{(k)} = n + k$, $n \geq 0$, $k > 0$.

3) $$q_n^{(1)} = \frac{1 + q^{n+1}}{1 + q^n} \quad , \quad e_n^{(1)} = \frac{(1 - q)^2 q^n}{(1 + q^n)(1 + q^{n+1})} \quad ,$$

$$q_n^{(2)} = \frac{q(1 + q^n)}{1 + q^{n+1}} \quad ,$$

9) $z = 1 \pm i \frac{1}{3}$

11) $z_1 = 0.933013$, $z_2 = 0.500000$, $z_3 = 0.066987$.

12) 0.9305682, 0.6699905, 0.3300095, 0.0694318 .

13) Real zeros: 4.70967, −4.41484, 1.00608, quadratic factor has zeros $1.85562 \pm i0.534194$.

16) a) $3.0053221 \pm i\, 3.9963933$, $0.9446779 \pm i\, 1.0248789$

b) 5.7074300, $1.0725870 \pm i\, 0.4775376$, $3.5736981 \pm i\, 0.1.5751498$

c) 0.1700599, 0.9024738, $0.7351475 \pm i\, 0.9018407$, $-0.6464144 \pm i\, 0.8862372$.

334

Chapter 9

2) No; if $f(x) = c$, the polynomial $P(x) = c$ interpolates at arbitrarily many points.

7) From $x = 1^\circ 42'$ onward.

8) From $x = 16$ onward.

9) a) $h \lesssim 0.003$ b) $h = 0.08$ suffices.

10) $|f(x) - P(x)| \lesssim 0.063\, h^4 M_4$, $0 \lesssim x \lesssim h$.

11) The error is at least 1; Error $= (e - 1)^{n+1}$.

13) Choose $x_0 = \dfrac{2 - \sqrt{2}}{4}$, $x_1 = \dfrac{2 + \sqrt{2}}{4}$, then
$$\max_{0 \lesssim x \lesssim 1} |L(x)| = \tfrac{1}{8} \cdot$$

15) For $h < 1$.

18) $P(x) = Q(x) - (\tfrac{b-a}{2})^n a_0 \cos(n \arccos\tfrac{2x-b-a}{b - a})\, 2^{-n+1}$;
$$\left| P(x) - Q(x) \right| \lesssim (\tfrac{b - a}{2})^n |a_0|\, 2^{-n+1} \quad.$$

19) $P(x) = 10x^4 - 35x^3 + 50x^2 - 25x + 2$

20) $P(x) = \tfrac{3}{2}x^2 - \tfrac{9}{16}x + \tfrac{1}{32}$ ($|error| \lesssim \tfrac{1}{32}$)

21) $P^*(x) = \tfrac{15}{16} x - \tfrac{5}{32}$ ($|error| \lesssim \tfrac{3}{16} + \tfrac{1}{32} = \tfrac{7}{32} = \tfrac{1}{4.571}$)

22) $P(x), = x - \tfrac{1}{3\sqrt{3}}$ ($|error| \lesssim \tfrac{1}{3\sqrt{3}} = \tfrac{1}{5.196}$)

Chapter 10

1) See example 8-6.

2) See problem 8-16a.

3) (as stated) $J_0(2.4068) \sim -0.00101$
   (corrected) $J_0(2.4048) \sim 0.00002$

10) $f(\tfrac{1}{2}) \sim 0.6718$

11) See problem 10-3

13) $\sqrt{2} \sim \tfrac{1449}{1024} = 1.415\ldots$

14) Ten extrapolated terms give only 1.646109.

15) $x = 5.5200299$

Chapter 11

1) Define f arbitrarily for $0 \leqq x < h$ and set (for $hc \neq 1$) $f(x + nh) = (1 + hc)^n f(x)$, $0 \leqq x < h$, $n = 0, \pm 1, \pm 2, \ldots$. For $hc = -1, f(x) = 0$ is the only solution.

5) $\Delta^n e^x = (e^h - 1)^n e^x$. Use $\dfrac{e^h - 1}{h} \to 1$.

10) Convergent for $h < \log 2$.

Chapter 12

1) $f'(0) - P'(0) = \dfrac{1}{n + 1} f^{(n+1)}(\xi_0)(-h)^n$, $0 \leqq \xi_0 \leqq nh$.

2) By preceding problem, error is at least $\dfrac{1}{n + 1}$.

3) $f'(0) = \dfrac{f(h) - f(-h)}{2h} + \dfrac{h^2}{6} f'''(\xi)$, $-h < \xi < h$.

4) $P_k'(x_0) = \dfrac{1}{h} \left[\Delta f_0 - \dfrac{1}{2} \Delta^2 f_0 + \dfrac{1}{3} \Delta^3 f_0 - \ldots + \dfrac{(-1)^{k-1}}{k} \Delta^k f_0 \right]$.

7) (12-7): $\dfrac{\varepsilon}{h}$, (12-8): $\dfrac{3\varepsilon}{h}$.

8) $h = 0.0145$.

11) $1.667 \dfrac{\varepsilon}{h}$ and $1.889 \dfrac{\varepsilon}{h}$.

17) $n = 22$.

Chapter 13

1) actual error $= 2$; error bound $= \dfrac{\pi^3}{12} = 2.583\ldots$

6) $\sum\limits_{k=1}^{\infty} a_k t^k = \log(1 + t)$

9) 1.85193704

Chapter 14

1) a) yes, $L = 1$ b) yes, $L = 1$ c) no d) no

3) a) $f^{(n)}(x,y) = y + (n + 1)e^x$

b) $f^{(n)}(x,y) = P_n(e^x) y$, where P_n = polynomial of degree n satisfying recurrence relation $P_1(z) = z$, $P_{n+1}(z) = z \left[P_n'(z) + P_n(z) \right]$.

5)　　　　　　h　　　　solution
　　　　　　　0.5　　　　0.81250
　　　　　　　0.25　　　0.71710
　　　　　　　0.125　　0.67314

8) Twice extrapolated value 0.65524.

11)　　　　　　h　　　　y_n
　　　　　　　0.4　　　0.304405
　　　　　　　0.2　　　0.304588
　　　　　　　0.1　　　0.304599

Extrapolated value: 0.304600.

Chapter 15

1) <u>11010010101</u>, <u>11011101010</u>, <u>11100111010</u>.

2) $8 = 2^3$, $64 = 2^6$, $512 = 2^9$.

3) $1/\log 2 = 3.3219\ldots$.

4) $10 = \underline{101}$, $100 = \underline{10201}$, $1000 = \underline{1101001}$.

5) $N = \underline{10}$

6) $\pi = \underline{11.001001000100}$

7) a) $\frac{21}{32}$　b) $\frac{21}{31}$

8) a) $\frac{1}{3} = \underline{0.\overline{01}}$　　b) $\frac{1}{3} = \underline{0.\overline{13}}$　c) $\frac{1}{3} = \underline{0.\overline{2}}$

9) $\sqrt{2} = \underline{1.01101010\ldots}$, $\sqrt{\pi} = \underline{1.1100010\ldots}$,

　　　$\sqrt{5} = \underline{10.001111000\ldots}$

Chapter 16

1) 3.75×10^{-10} ; No.